Recent Trends and The Future of Antimicrobial Agents (Part 2)

Edited by

Tilak Saha

*Immunology and Microbiology Laboratory
Department of Zoology, University of North Bengal
Darjeeling
West Bengal, India*

Manab Deb Adhikari

*Department of Biotechnology
University of North Bengal Darjeeling
West Bengal, India*

&

Bipransh Kumar Tiwary

*Department of Microbiology
North Bengal St. Xavier's College Rajganj, Jalpaiguri,
West Bengal, India*

Recent Trends and the Future of Antimicrobial Agents

(Part 2)

Editors: Tilak Saha, Manab Deb Adhikari and Bipransh Kumar Tiwary

ISBN (Online): 978-981-5123-97-5

ISBN (Print): 978-981-5123-98-2

ISBN (Paperback): 978-981-5123-99-9

First published in 2023.

need for a court order if at any point you breach any terms of this License Agreement. In no event will any delay or failure by Bentham Science Publishers in enforcing your compliance with this License Agreement constitute a waiver of any of its rights.

3. You acknowledge that you have read this License Agreement, and agree to be bound by its terms and conditions. To the extent that any other terms and conditions presented on any website of Bentham Science Publishers conflict with, or are inconsistent with, the terms and conditions set out in this License Agreement, you acknowledge that the terms and conditions set out in this License Agreement shall prevail.

Bentham Science Publishers Pte. Ltd.
80 Robinson Road #02-00
Singapore 068898
Singapore
Email: subscriptions@benthamscience.net

BENTHAM SCIENCE

CONTENTS

FOREWORD

The book "Recent Trends and The Future of Antimicrobial Agents" tries to explore various alternatives of multi drug resistant bacteria which are the major causes of therapeutic failure. The book provides various approaches to the solution and each section describes and analyses the approach towards the problem. Research is going on globally on various alternatives to treatment like Plant based antimicrobials, Photodynamic therapies, enzyme based and antibody based antimicrobial approaches, chemical compounds that act as antimicrobial agents, nano-materials which act as antimicrobial agents, probiotic, prebiotic and peptides compounds or agents. The writers have taken up each scenario to make the readers understand about the macro and micro factors associated with the approach.

The book attempts to throw light on the various aspects of the pathogenic multi drug resistant bacteria and takes a wide horizon on the impact of antibiotics on them. The discovery of penicillin paved the way for the antibiotics to become popular but as the bacteria can accumulate on multiple genes making them resistant to a particular drug, similarly the resistance can also be caused by an increased expression of genes responsible for multi-drug efflux pumps forcing out a lot many drugs. Hence the need to develop an alternative strategy is very critical for therapeutic success. The book describes all these scenarios in two subsequent volumes of the title. Volume-1 includes the naturally derived antimicrobial remedies/strategies. The Volume-2 of the same title incorporates the chemical and advanced nanomaterial based strategies along with sustainable antimicrobial strategies *viz.* use of probiotics and photodynamic therapy. I would like to thank the authors for their dedicated effort and the publishers in converting that effort into a reality. I am sure that the information will be very useful for Clinicians as well as Microbiologists.

Dhrubajyoti Chattopadhyay PhD, F.A.Sc , F.Na.Sc , F.A.Sc.T
Vice Chancellor
Sister Nivedita University
Kolkata
India

PREFACE

Many microbial pathogens have evolved as drug resistant due to indiscriminate and injudicious use of drugs. This has compelled researchers to find novel antimicrobial agents with diverse chemical structures and novel mechanisms over the conventional antimicrobial agents rendering the pathogens with minimum scope to develop resistance. Last few decades have witnessed profound research on different areas for the development of alternative antimicrobial agents. These include novel chemically synthesized molecules, nanomaterials and probiotic/prebiotic mediated immunity boosters, *etc*. **"Recent Trends and the Future of Antimicrobial Agents"**, Part 2 is a continuation of the Part 1 of the same title that dealt with the naturally derived antimicrobial remedies/strategies. The present Part 2 of the same title deals with the chemically synthesized compounds, nanomaterials and probiotics.

The devastating pandemic caused by the severe acute respiratory syndrome-causing corona virus-2 (SARS CoV-2) virus has once again taught us that "Prevention is better than cure". The overburden of xenobiotic drugs can be drastically reduced by boosting our immune system and fighting the disease causing microbes in association with the helpful bacteria and their metabolites. Three chapters of the book uncover the probiotic/prebiotic/antibacterial peptide compounds as novel antimicrobial approaches and disorder-management therapies. All of these "-biotics" are designed to modulate the gut microbiota in a way that improves health and reduces the need to gulp antibiotics indiscriminately and thus indirectly assist in fighting potential bacterial threats. But, prevention may not always be able to protect us from infiltrating microbes. Chemical synthesis enables researchers to develop target based prospective drug molecules to fight against ever-changing microorganisms. The potent synthetic pathways are discussed in a chapter. The plant-based products have traditionally been used as natural healing systems. Although, modern scientific approaches focus on active compounds. Bioactive natural compounds and synthetic drug candidates are promising therapeutic agents for human health and disease management. Their therapeutic efficacy can be enhanced if their bioavailability is raised to the optimum level and/or delivered to the target cells/tissue involving nanocarriers. The membrane targeting bactericidal agents are also emerging as potent antimicrobials since developing resistance against them demands extensive restoration of membrane compounds, which is a conceivably formidable challenge for the bacteria. In this regard, membrane-targeting nanoscale materials, amphiphiles, and antimicrobial peptides bear special merit. Two chapters discuss the potential of cationic amphiphiles as promising antimicrobial entities and amphiphilic nanocarriers as delivery vehicles. Another chapter discusses the design, synthesis and antimicrobial applications of Metal-Organic Frameworks (MOFs). Thus, amphiphiles of this new genre have enough potential to deliver several antibacterial molecules in years to come. The emergence of nanoscience and technology in recent years offers great promise in therapeutics. Nanomaterials are emerging as a novel class of antimicrobial agents to overcome the challenges faced by conventional antimicrobials. Using nanomaterials as bactericidal agents represents a novel approach to antibacterial therapeutics. Three chapters of this book cover the recent development, antimicrobial prospects of biogenic metal or metalloid nanoparticles, bactericidal QDs and MoS_2 based antibacterial nanocomposites. A new-age approach to combat microbes, antimicrobial photodynamic therapy (aPDT), is discussed in a chapter. PDT uses a nontoxic and lightsensitive dye, a photosensitizer (PS), in combination with nontoxic visible light of the appropriate wavelength to excite the PS and oxygen that can selectively control bacterial infections by the generation of highly cytotoxic reactive oxygen species (ROS).

In the process of editing the book we have received needful assistance and inspiration from different spheres of academy. We express our sincere gratitude to Prof. (Dr.) Dhrubajyoti Chattopadhyay, Vice Chancellor, Sister Nivedita University, Kolkata, West Bengal for his motivation throughout the project. We express our gratitude to the Vice Chancellor, University of North Bengal, Darjeeling, for all necessary facilities and support. We are thankful to Fr. (Dr.) Lalit P. Tirkey, Principal, North Bengal St. Xavier's College (NBSXC), Jalpaiguri, for his continuous encouragement. Our sincere thanks go to all authors for their hard work and professionalism in making this book a reality. Their expertise in the contributed chapters is acknowledged and appreciated. Finally, we appreciate Bentham Science Publishers for their assistance and constant support in publishing the book.

Tilak Saha
Immunology and Microbiology Laboratory
Department of Zoology, University of North Bengal
Darjeeling, West Bengal
India

Manab Deb Adhikari
Department of Biotechnology
University of North Bengal
Darjeeling, West Bengal
India

&

Bipransh Kumar Tiwary
Department of Microbiology
North Bengal St. Xavier's College
Rajganj, Jalpaiguri, West Bengal
India

List of Contributors

Amaresh Kumar Sahoo Department of Applied Sciences, Indian Institute of Information Technology Allahabad, India

Amit Jaiswal School of Basic Science, Indian Institute of Technology Mandi, amand, Mandi-175005,Himachal Pradesh, India

Amit Sarder Biotechnology and Genetic Engineering Discipline, Khulna University, Khulna-9208, Bangladesh

Areeba Khayal Department of Chemistry, Aligarh Muslim University, Uttar Pradesh, India

TotiCell Limited, Dhanmondi, Dhaka-1206, Bangladesh

Department of Microbiology, North Bengal St. Xavier's College, Rajganj, Jalpaiguri, West Bengal, India

Chirantan Kar Amity Institute of Applied sciences, Amity University Kolkata, Kolkata 700135, India

Kabirun Ahmed Department of Chemical Sciences, Tezpur University, Tezpur, Assam, India

Kinkar Biswas Department of Chemistr, Raiganj University, Raiganj, Uttar Dinajpur 733134, India
Department of Chemistry, University of North Bengal, Darjeeling 734013, India

Manoj Lama Molecular Immunology Laboratory, Department of Zoology, University of Gour Banga, Malda -732103, India

Md Palashuddin Sk Department of Chemistry, Aligarh Muslim University, Uttar Pradesh, India

Mintu Thakur Department of Chemistr, Raiganj University, Raiganj, Uttar Dinajpur 733134, India

Moushree Pal Roy Department of Microbiology, Ananda Chandra College, Jalpaiguri, West Bengal, India

Nazeer Abdul Azeez Department of Biotechnology, Bannari Amman Institute of Technology, athyamangalam, Erode,Tamil Nadu-638 401, India

Palas Samanta Department of Environmental Science,Sukanta Mahavidyalay, University of North Bengal, Dhupguri, West Bengal, India

Praveen Kumar School of Basic Science, Indian Institute of Technology Mandi, amand, Mandi-175005, Himachal Pradesh, India

Rejuan Islam Immunology and Microbiology Laboratory, Department of Zoology, University of North Bengal, Siliguri, West Bengal, India

Sapna Pahil Department of Microbiology and Immunology, University of Michigan Medical School, Ann Arbor, Michigan, USA

Sovik Dey Sarkar Amity Institute of Applied sciences, Amity University Kolkata, Kolkata 700135, India

Sudarshana Deepa Vijaykumar	Department of Biotechnology, National Institute of Technology, Adepalligudem, Andhra Pradesh 534101, India
Sukhendu Dey	Department of Environmental Science, University of Burdwan, West Bengal, India
Surendra H. Mahadevegowda	Department of Chemistry, School of Sciences, National Institute of Technology, Tadepalligudem, Andhra Pradesh – 534 10, India
Sushobhon Sen	Department of Biotechnology, University of North Bengal, Darjeeling, West Bengal, India
Tilak Saha	Immunology and Microbiology Laboratory, Department of Zoology, University of North Bengal, Siliguri, West Bengal, India

CHAPTER 1

Probiotics as Potential Remedy for Restoration of Gut Microbiome and Mitigation of Polycystic Ovarian Syndrome

Rejuan Islam[1] and **Tilak Saha**[1,*]

[1] *Immunology and Microbiology Laboratory, Department of Zoology, University of North Bengal, Siliguri, West Bengal, India*

Abstract: Polycystic ovarian syndrome (PCOS) is the most frequent endocrine disorder currently plaguing women. There are many factors associated with high androgenicity in the female body. Dysbiosis of gut microbiota may be one of the primary reasons that initiate PCOS. Emerging evidence suggests that some plastics, pesticides, synthetic fertilizers, electronic waste, food additives, and artificial hormones that release endocrine-disrupting chemicals (EDCs) cause microbial Dysbiosis. It is reported that the permeability of the gut is increased due to an increase of some Gram-negative bacteria. It helps to promote the lipopolysaccharides (LPS) from the gut lumen to enter the systemic circulation resulting in inflammation. Due to inflammation, insulin receptors' impaired activity may result in insulin resistance (IR), which could be a possible pathogenic factor in PCOS development. Good bacteria produce short-chain fatty acids (SCFAs), and these SCFAs have been reported to increase the development of Mucin-2 (MUC-2) mucin in colonic mucosal cells and prevent the passage of bacteria. Probiotic supplementation for PCOS patients enhances many biochemical pathways with beneficial effects on changing the colonic bacterial balance. This way of applying probiotics in the modulation of the gut microbiome could be a potential therapy for PCOS.

Keywords: Endocrine-disrupting chemicals, Gut microbiome, Insulin resistance, Mucin-2, PCOS, Probiotics, SCFAs, *Bifidobacteria*, *Lactobacillus*, Gut bacteria dysbiosis, hypertension, central obesity, dyslipidemia, progesterone, estrogen, luteinizing hormone, Infertility, cardiovascular disease, type 2 diabetes mellitus, visceral obesity, and endothelial dysfunction, Hyper-insulinemia, Androgens, lipopolysaccharides, reproductive abnormalities.

* **Corresponding author Tilak Saha:** Immunology and Microbiology Laboratory, Department of Zoology, University of North Bengal, Siliguri, West Bengal, India; E-mail- tilaksaha@nbu.ac.in

Tilak Saha, Manab Deb Adhikari and Bipransh Kumar Tiwary (Eds.)

INTRODUCTION

Polycystic ovarian syndrome (PCOS) is a condition of hormonal imbalance that causes female reproductive abnormalities, especially in reproductive age common disorder in women, with a wide range of prevalence rate from 6 to 20% [1, 2]. The main characteristics of PCOS are polycystic ovaries, hyperandrogenism, anovulation, abnormal menstruation [3], hypertension, central obesity, and dyslipidemia [4]. Though the pathologic process of PCOS is complex and mostly unknown, the symptoms are often associated with internal secretory problems, such as decreased progesterone, estrogen, and sex hormone binding globulin (SHBG) and elevated testosterone, luteinizing hormone (LH), among other things [5]. Progesterone is one of the key hormones linked to PCOS and whose primary role is to aid in the maintenance of pregnancy [6]. PCOS patients are unable to produce a corpus luteum due to low progesterone levels and irregularities in the fertilization process [7]. Infertility, cardiovascular disease, insulin resistance (IR), type 2 diabetes mellitus (T2DM), visceral obesity, and endothelial dysfunction are all common symptoms of PCOS. As a result, this syndrome is classified as a metabolic disease that affects the quality of women's lives as well as a fertility concern [8].

While PCOS is known to cause genetic, neuroendocrine, metabolic, environmental, and lifestyle factors, the etiology of PCOS remains unclear. According to new data, the gut microbiome may have a role in the development of PCOS [9]. It is suggested that differences in gut microbiota composition are correlated with metabolic and clinical changes in PCOS patients [10, 11]. Imbalances in gut microbiology may result in Dysbiosis of gut microbiota and may cause activation of the host's immune system. The activation of the immune system causes chronic activation of inflammatory response and initiates a state of IR due to improper function of insulin receptors. It is known earlier that IR interferes with the growth of follicles for the excess production of androgen by the ovary's thecal cells [12].

An unhealthy lifestyle, consuming junk food, and various inflammatory mediators increase the risk of PCOS [13, 14]. Emerging evidence suggests interactions between endocrine-disrupting chemicals (EDCs) and the microbiome, affecting host health. EDC exposure has been shown to disrupt the microbiome, which can lead to Dysbiosis and the induction of xenobiotic-related pathways, microbiome-associated genes, enzymes and metabolite production, which can play a key role in EDCs biotransformation. This Dysbiosis of gut microbiota may be associated with the globalization of industry and the manufacture of plastics, synthetic fertilizers, pesticides, electronic trash, and additives in food that release EDCs into the environment and food chain [15]. Gut bacteria dysbiosis helps to promote the Lipopolysaccharides (LPS) from the gut lumen to the systemic circulation. LPS causes chronic stimulation of hepatic and tissue macrophages, and insulin tolerance is increased due to impaired activity of insulin receptors. Hyper-insulinemia then increases the secretion of androgens in the ovaries and prevents normal processes of ovulation [12].

WHAT IS PCOS?

PCOS is a complex condition marked by high testosterone levels, irregular menstruation cycles, and/or small cysts on one or both ovaries [16]. Later it was redefined to establish different diagnostic criteria. It was first redefined by the National Institutes of Health (NIH) in 1990, and according to it, patients with hyperandrogenism and oligo-anovulation are diagnosed with PCOS [17]. It was further redefined by Rotterdam Consensus in 2003 that postulates patients should have at least two among the three classic features which are irregular menstrual cycle, hyperandrogenism and enlarged "polycystic" ovaries in pelvic ultrasonography [18]. In addition to the main hyperandrogenic findings, those with oligo anovulation or polycystic ovarian criteria are considered to have PCOS, according to the Androgen Excess and PCOS Society (AE-PCOS) in 2006 [19]. The three Rotterdam criteria, which are currently accepted according to the international PCOS guidelines [20], can divide the condition down into four phenotypes. These are (1) classic PCOS (chronic anovulation, hyperandrogenism, and polycystic ovaries); (2) non-polycystic ovary PCOS (hyperandrogenism, chronic anovulation, and normal ovaries); (3) ovulatory PCOS (hyperandrogenism, polycystic ovaries, and regular menstrual cycles); and (4) mild/norm androgenic PCOS (chronic anovulation, normal androgens, and polycystic ovaries) [21].

The various components of the diagnostic criteria cause changes in prevalence across the NIH criteria 1990, the Rotterdam 2003 criteria and the AE-PCOS 2006 criteria [22]. A meta-analysis was performed on all published studies which reported PCOS prevalence was 6%, 10%, and 10% according to the diagnostic criteria of NIH, Rotterdam, and AE-PCOS Society, respectively, based on at least one subset of diagnostic criteria [23]. In India, PCOS prevalence was 22.5% by Rotterdam and 10.7% by Androgen Excess Society criteria. Mild PCOS is amongst the most common phenotypes occurring in about 52.6% of women [24].

Etiopathology of PCOS

Though the main reason for PCOS is unknown, it is known to be a multifunctional disorder with genetic, endocrinological, and environmental factors having a role to play [25]. Hyperandrogenism, seen in 90 percent of women with PCOS, plays an important role in the disease etiology [26]. Androgen excess may cause hirsutism, acne, and alopecia in PCOS patients. Not only androgen hormones but also the level of gonadotrophin-releasing hormone, follicular stimulating hormone (FSH), luteinizing hormone (LH) and prolactin are also disturbed in the case of PCOS [27]. It is also linked with many metabolic disorders, like glucose intolerance, T2DM, dyslipidemia, hepatic steatosis, hypertension and increased cardiovascular surrogate markers [28].

Since 1968, studies have shown a significant genetic function that contributes to the etiology of PCOS [29]. Patients with first-degree relatives of PCOS are at greater risk of being influenced by the syndrome relative to the general population. There are so many candidate genes that are responsible for the involvement of various biochemical pathways that may lead to an increase of androgen hormone, leading to an ovary dysfunction. It had shown that there are 100 candidate genes associated with the reproductive axis, IR and chronic inflammation, which have not shown reproducible results [30].

It is reported that obesity is another factor for PCOS because obese women take a much longer time for pregnancy than non-obese women. Obese women face higher infertility risk than non-obese women because of adipokines, a bioactive cytokine that regulate so many functions like IR, inflammation, hypertension, cardiovascular risk, coagulation, and oocyte differentiation and maturation [31]. It was seen that these abnormal levels of these factors are strongly associated with IR and T2DM in PCOS patients. Another factor associated with PCOS is insulin resistance, in which blood glucose levels increase dramatically. It is reported that CC chemokine ligand 18 (CCL18) is a chemokine whose expression was much higher in obese people and decreased after they lost weight [32]. In PCOS patients, there is a strong association between serum CCL 18 levels and IR, and this can serve as a marker for PCOS [33]. Due to IR, high levels of glucose increase in blood; as a result, functional disturbances in ovaries occur, and androgen hormone levels increase in the female body [12]. There are so many chemical toxicants that are used in industry or as fertilizer that act as endocrine disruptor that plays a crucial role in the dysfunction of our endocrine system. In experimental use of BPE and BPS, structurally similar BPA analogs accelerate the onset of puberty, disrupt estrous cyclicity, and impair adult reproductive functions with age which indicates that there is a strong association between the endocrine disruptors and PCOS [34].

A further mechanism has been thought to be a key factor in PCOS development in recent years. This mechanism is intestinal microbiota, which is thought to contribute to the pathogenesis of many disorders [9].

GUT MICROBIOTA

The human gastrointestinal tract (GI) comprises an abundant and complex microbial population that contains over 100 trillion micro-organisms [35]. The gut microbiome encodes more than three million genes that contain thousands of metabolites, while the human genome consists of about 23,000 genes [36]. The microbiota starts to evolve immediately after birth, and its composition is damaged by various factors such as age, birth type, diet, lifestyle, genetic predisposition and antibiotic usage [37]. Every healthy human being has unique gut microbiota. The makeup of a balanced gut microbiome is described by the richness and diversity of gut microbiota produced in early life [38]. *Firmicutes, Bacteroidetes, Proteobacteria* and *Actinobacteria* are the four main dominant phyla that belong to more than 99% of intestinal bacteria [39, 40]. Among healthy adults, two phyla, *Bacteroidetes* and *Firmicutes,* dominate the intestinal microbiota [41, 42]. Most probiotic products contain high levels of *Lactobacillus* or *Bifidobacterium spp.* that support the host's immune system, stimulate the host's defense by increasing anti-microbial defensive production, regulate intestinal permeability, and are often the main metabolite producers, such as vitamins [43].

Importance of Gut Microbiota

The gut microbiota maintains a symbiotic relationship with the gut mucosa and provides significant roles in metabolic, protective, structural, immunomodulation, and neuroendocrine functions in a healthy person [44, 45]. Gut microbiota performs various metabolic functions like vitamins, short-chain free fatty acids (SCFAs) and the processing of conjugated linoleic acids, amino acid synthesis, bile acid biotransformation, nondigestible food fermentation and hydrolysis, ammonia synthesis and detoxification [46]. The SCFA is produced by good bacteria through the fermentation of carbohydrates such as soluble fibers that are delivered to the colon without digestion [47]. Pyruvate is produced primarily from carbohydrates, and gut microbiota helps to further fermentation to generate energy as it is catabolized into succinate, lactate, or acetyl-CoA. However, these catabolized products are not available in large concentrations in typical faecal samples because they can be further metabolized by cross-feeders, producing SCFAs such as acetate, propionate, and butyrate [48]. These three main SCFAs have several important functions in metabolism as they play different roles in anti-inflammatory, anticarcinogenic, and immunomodulatory effects. Consumption of butyrate improves the integrity of intestinal epithelial cells (IECs) by promoting tight junctions, cell proliferation, and increasing mucin production by Goblet cells [49]. Another important function of butyrate is to stimulate both IECs and antigen-presenting cells (APCs) to produce the cytokines

TGF-β, IL-10, and IL-18, and induce the differentiation of naive T cells to T regulatory cells [50]. Butyrate also has an anti-carcinogenic effect as it helps to increase the apoptosis of colon cancer cells. Propionate participates in the process of gluconeogenesis, while acetate plays an important role in cholesterol metabolism and lipogenesis [51].

Good bacteria have the ability to produce some nutrients and regulatory substances that improve the function of the colonic epithelium. The production of mucus is maintained in the human body by the mucin gene. The major intestinal mucus is produced by the MUC2 gene. SCFA produced by good bacteria have been reported to increase the development of MUC-2 mucin in colonic mucosal cells and prevent the passage of bacteria [52, 53]. Good bacteria such as *Bifidobacteria* and *Lactobacillus* help in maintaining the growth of bad bacteria and hold bad bacteria numbers in control. *Bifidobacteria* and *Lactobacillus* reduce the pH of the colonic lumen by producing SCFA and lactic acid and create a condition that is unfavorable for bad bacteria. In this way, the bad bacterial number is maintained and minimized the production of colonic luminal endotoxin (LPS) [54].

An imbalance in the composition and metabolic capacity of our microbiota is termed Dysbiosis. It is a process that results from a reduction in the ratio of beneficial/ harmful bacteria, and as a result, it increases the risk of developing some chronic diseases such as allergies, inflammatory bowel disease, cancer, lupus, asthma, multiple sclerosis, Parkinson's disease, celiac disease, obesity, IR, T2DM, and cardiovascular diseases [46, 55].

Relationship between PCOS and Gut Microbiota

Early it was very confusing whether gut microbiota dysbiosis causes PCOS or PCOS leads to gut microbiota dysbiosis, but recently, it has been confirmed that microbiota dysbiosis may play a role in PCOS pathogenesis [11]. PCOS women have a lowered α diversity compared with healthy women. It has been shown that hyperandrogenism, total testosterone, and hirsutism have a negative correlation with α diversity of gut microbiota, and there is also a correlation of β diversity with hyperandrogenism [56].

There are some possible mechanisms, which explain the role of gut microbiota in PCOS pathogenesis. One of them is gut microbiota dysbiosis which activates the host's immune system. IR and chronic inflammation are the two main biochemical features known to be present in PCOS patients in the vast majority. It is known that there are 10^{14} bacteria present in the human gut [57], and Dysbiosis of these bacteria may cause the activation of the immune system of the host. As a result, the activation of immune system causes chronic activation of inflammatory response and initiates a state of IR due to improper function of insulin receptors. Earlier it is well-known that IR interferes with the growth of follicles for the production of excess androgen by the ovary's thecal cells. This novel microbiological paradigm for PCOS is called the DOGMA theory-Dysbiosis of Gut Microbiota [9].

Gut Microbiota and SCFA

The consumed carbohydrate is degraded into simple sugar and further fermented into hydrogen, carbon-di-oxide, methane, and SCFA to provide energy to the host [58]. A vicious circle is formed in obese individuals because they consume more energy from food. Due to this phenomenon, the composition of gut microbiota is altered, and this alteration leads to the disease. Gut microbiota decomposes organic materials to produce SCFA, which can stimulate the release of peptide YY(PYY) in the ileum and colon. PYY has three major functions; it inhibits intestinal peristalsis, decreases the secretion of the pancreas and promotes the absorption of energy in the intestinal tract [59]. According to certain studies, the distribution of different gut microbes may have different abilities to absorb energy. An experiment observed that obese mice absorb more energy as they have more Firmicutes and fewer Bacteroides than lean mice after treating the same diet. In addition, wild aseptic mice were given a gut microbiota transplant from obese mice, and they turned became obese with high energy intake capacity. More than half of the PCOS patients show characteristics of the overweight or obese condition [10, 60].

Gut Microbiota and Cytokines

PCOS is associated with chronic activation of the immune system by some proinflammatory cytokines such as TNF-α and interleukin-6 (IL-6). TNF-α induces the NF-κB signaling pathway, which affects the barrier function of pancreatic duct epithelial cells by altering tight junction-related proteins [61]. Increasing LPS in the systemic circulation causes low-grade chronic inflammation and may result in metabolic endotoxemia [62]. The introduction of some endocrine-disrupting chemicals in the human intestine may cause Dysbiosis of the gut microbiome, and there is an increase of excessive Gram-negative bacteria introduced that causes chronic endotoxemia by increasing the amount of circulating bacterial LPS. The enterocytes that bear toll-like receptors on their surface can recognize bacterial LPS that help to induce inflammation through the activation of nuclear factor Kappa B (NF-κB). Another factor that causes metabolic endotoxemia is the decrease of tight junction protein due to changes in the intestinal flora. There is a biomarker for intestinal permeability called zonulin that was found higher in 78 women with PCOS compared to the healthy controls.

A positive correlation was also found between serum zonulin levels and IR that may lead to the severity of menstrual problems [63]. Studies suggest that there is a significant alteration of T-lymphocyte subsets and different profiles of the leukocyte population that clearly show a connection between excess androgen, chronic inflammation, and immune-mediated diseases in patients with polycystic ovary syndrome [64].

IR is the most common feature of PCOS that affects 70% of obese and lean women. A substantial combination of low-grade chronic inflammation with sympathetic dysfunction and hyperandrogenism indicates the role of chronic inflammation in mediating the effect of sympathetic dysfunction on PCOS hyperandrogenism and IR [65]. IR is caused in obese PCOS patients by increased cytokine levels such as TNF-α, and IL-6 and by the movement of LPS throughout systemic circulation that results from increased bowel permeability. If levels of both fasting blood glucose and insulin rise, then LPS is directly applied to the circulation of mice and humans [66, 67]. Inflammation-induced IR raises blood testosterone levels in two ways. The first cause of hyperinsulinemia results in the overproduction of androgen synthesis by ovaries [68]. Second, hyperinsulinemia, by decreasing SHBG, raises the free (bioactive) testosterone levels [69]. Higher levels of androgens caused by hyperinsulinemia may cause acne as well as hirsutism [70]. In addition, the synthesis of androgen by singleton cells is stimulated by hyperinsulinemia as it raises the level of free insulin-like growth factor 1 (IGF-1) in the blood, which suppresses the production of the insulin-like growth factor 1 binding protein (IGFBP-1) [71]. High insulin levels, as well as IGF-1 activity, hamper the natural process of follicle growth from primary follicles and trigger ovarian polycystic structures and also menstrual irregularities [72].

Gut Microbiota and the Gut-brain Axis

Metabolites of gut microbiota stimulate the secretion of gut-brain peptides and regulate inflammatory pathway activation that may cause IR as well as hyperinsulinemia. The pathway activation occurs through a complex process by the brain-gut axis, the central nervous system, and the gastrointestinal system. Some gut-brain axis mediators have been identified, including serotonin, ghrelin, and peptide YY (PYY), which play various roles in folliculogenesis. In primordial follicle oocytes, serotonin is found to a greater degree and decreases at the later stages of folliculogenesis. Serotonin is uptaken by the specific receptor serotonin transporter (SERT) and has a significant impact on the folliculogenesis process [73]. Gut-brain mediators also play an important role in the psychological well-being of women with PCOS by appetite regulation. Another function of these mediators is energy homeostasis and LH secretion [10, 74, 75]. Some spore-forming bacteria, such as *Clostridial* bacteria, help in serotonin biosynthesis but species belonging to *Bacteroides* are not associated with this function. Some intestinal bacteria which produce SCFA also regulate PYY secretion [76]. An investigation reported that the ghrelin, serotonin, and PYY levels were lower in PCOS patients than in the control group. These mediators also negatively correlate with PCOS-related parameters such as waist circumference and testosterone [10]. Patients with PCOS show lower levels of Clostridial species and higher levels of *Bacteroides* species. Researchers found that the level of ghrelin is negatively correlated with the abundance of *Bacteroides*, *Blautia*, *Escherichia/Shigella* [10]. There are few studies that support the possible mechanism of PCOS pathogenesis in relation to the gut microbiome. Therefore, more thorough and in-depth studies are needed to strongly prove the link between gut bacteria, mediators of the brain-intestinal axis, and PCOS phenotype.

Gut Microbiota and Androgen Hormone

Another factor that may contribute to the development of PCOS is the androgen hormone, which is influenced by the composition of the gut flora [77 - 79]. This has been demonstrated that changing the composition of the gut bacteria causes an increase in androgen hormone in PCOS patients [78, 79]. However, little evidence is there that supports how gut microbiota is affected by sex steroids. Gut microbiota composition is directly influenced by sex steroids by energy production and changing the activity of beta-glucuronidase [77, 80]. In addition, the gut microbiota is also regulated indirectly by sex steroids activating the steroid receptors in the body of the host [77]. Due to changes in sex steroids, the integrity of the intestinal barrier is also regulated, which can alter the immune response. Intestinal barrier integrity may cause peripheral inflammation due to infiltration of some Gram-negative bacteria in the circulation, which bears LPS in the periphery. The gut microbiome may have a role in PCOS through regulating sex steroids [78]. It is important to remember that the impact of gut bacteria on steroid regulation is not entirely understood. That's why there should be more investigation to know the related mechanism of the relationship between hyperandrogenism and the gut microbiome.

All the mechanisms that are discussed so far are somehow responsible for gut microbiota dysbiosis, which later are responsible for PCOS pathogenesis (Fig. **1**). The current study suggests that Dysbiosis may have occurred for a variety of reasons, such as consumption of junk food (high fat-low fiber diet) and endocrine-disrupting chemicals [81]. So far, there has been a lot of research done about a high-fat diet for Dysbiosis, but the EDCs have not been discussed in such a way. So, now we will discuss the EDCs and their effects in shaping the gut microbiota composition.

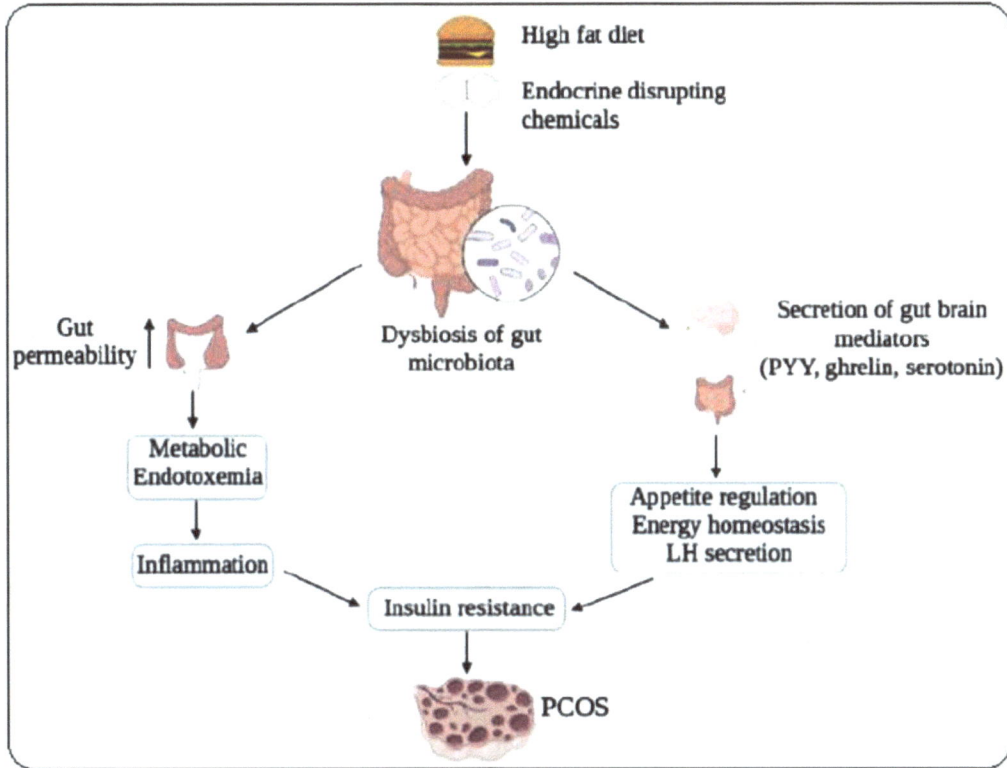

Fig. (1). A possible mechanisms that could explain the function of gut microbiota in polycystic ovary syndrome pathogenesis.

ENDOCRINE DISRUPTING CHEMICALS

The planet has seen huge production and release of toxic chemicals into the environment through manufacturing, chemical-based agriculture and food processing, and electronic waste [82]. Most of the chemicals released from these activities interfere with the hormonal system by affecting the growth, release, transport and action of hormones; these are called endocrine-disrupting chemicals [83]. There are many items that are commonly used that can release EDCs, such as plastics, pesticides, synthetic fertilizers, electronic waste, food additives, and artificial hormones. Recent data suggests that unhealthy lifestyles and environmental pollutants exposure lead to ovulatory dysfunction in PCOS [84]. It is also suggested that some metabolic disturbances, such as dyslipidemia, IR, and obesity, are associated in girls with PCOS [85]. In this respect, the primary cause of human exposure to EDCs is the consumption of food or water. Exposure to EDCs has been shown to disrupt the microbiome, leading to Dysbiosis and activation of xenobiotics-related pathways, microbiome-associated genes, enzymes and metabolite development that can play a significant role in EDCs biotransformation [15]. The products and by-products released after the host will take over the microbial metabolism of EDCs, thereby affecting the development of host health and diseases. The gut microbiome, however, can change the EDCs profile by different possible mechanisms.

There are so many endocrine disruptors consumed in our bodies in different ways, but various pesticides are the most important. We have seen a dramatic increase in the world population to date, and meanwhile, demand for food has also increased significantly. In order to obtain better quality agricultural products and increase crop yields, various types of pesticides are widely used and bring considerable and more economic benefits. That is why pesticides are widely used globally, and minimizing human, and wildlife exposure remain a challenge [86, 87]. Due to excessive use, residues of pesticides or their metabolites have been found in food, drinking water [88] and groundwater, suggesting that pesticides can effectively reach animals or humans through different pathways. In most situations, the gut comes into direct contact with food pollutants, and pesticides are likely to reach the human gastrointestinal tract and intestinal flora. However, we will mainly discuss the mechanisms by which different types of EDCs cause changes in the gut microbiota's composition and function.

EDCs and their Effect on Gut Microbiota

The gut microbiota's composition, diversity, and enzymatic capacities are readily affected by various environmental factors, including the lifestyle of the host, diet, and antibiotic use. Several environmental chemicals have been seen to suppress the growth of gut bacteria populations or induce Dysbiosis. These populations include high abundant population levels of *Bacteroidetes* (27.8% relative abundance) and *Firmicutes* (38.8% relative abundance), or relatively low abundant *Proteobacteria* (2.1% relative abundance; a marker of gut inflammation) [89].

An increase in metabolic diseases occurred through interaction between persistent organic pollutants, and gut microbiome was found to be mediated via aryl hydrocarbon receptor activation [63]. A knock-out study in mice suggests that TLR5 or inflammasomes, which are the innate immune system components, play a critical role in modulating the gut microbiome and contribute to the development of an abnormal metabolic phenotype [90, 91]. It also suggests that both perinatal-period alterations in microbial colonization and early-life exposure to environmental chemicals can promote dysregulated immune response [92 - 94]. Therefore, it seems that exposure to toxic chemicals, toxic substances have the potential to disrupt the natural colonization of bacteria in the gut, which later affects host physiology.

In addition, there is a growing consensus that a healthy microbiome may not be characterized as an idealized assembly of specialized microbe species, but that the microbe community should be capable of performing a set of metabolic functions together with its host, although this set of metabolic functions is still to be established [95]. This is particularly important since some xenobiotics could affect the physiology of gut microbiota without inducing Dysbiosis. After all, when fresh human faecal samples were incubated with antibiotics or with host-targeted drugs, all host-targeted drugs resulted in major changes in the expression of microbial genes, including genes involved in xenobiotic metabolism, and also because of minimal short-term impact on the structure of microbial communities [96]. These interactions can significantly impact the metabolism, toxicity and risk assessment of many environmental compounds that become toxic upon microbiome-mediated metabolism. Several different classes of xenobiotic chemicals have been known to interact with the biochemical and enzymatic activity of gut microbes affecting the composition of the bacterial community and the overall homeostasis of the gut microbiome, with possible harmful effects for the host [97]. It is also suggested that the herbicide glyphosate is specifically known to block the synthesis of three essential aromatic amino acids tyrosine, phenylalanine and tryptophan. Glyphosate inhibits the 5-enolpyruvylshikimate-3-phosphate synthase (EPSPS) in the shikimate pathway of some bacterial species as well as in plants [98, 99].

EDCs could alter the composition of the gut microbiota as well as its metabolites, such as TMA, bile acids, SCFAs or other metabolites, which cause adverse effects on hosts from some known and unknown signaling pathways (Fig. 2). For example, pesticides also target bile acids which are considered very important signals between the liver and the gut axis. The primary bile acids are generally first produced in the liver, and can then be transformed by the gut microbiota into secondary bile acids. Different bile acids can bind to different receptors (such as TGR5 and FXR), which can cause problems in the near future like atherosclerosis, liver lipid metabolism disorders and fat accumulation dysbiosis [100]. In addition, SCFAs can inhibit histone deacetylases and serve as energy substrates by activating the G-coupled receptors directly [48, 101]. The effect of SCFAs in fat cells on GPR41/43 can lead to Dysbiosis of fat accumulation [102]. Aside from the metabolic pathways mentioned above, there are other pathways. For example, bacterial endotoxin called LPS present in Gram-negative bacteria is released into the gut and interact with the Toll-like receptor 4 (TLR-4) on the surface of innate immune cells. Many pro-inflammatory cytokines are released from activated innate immune cells, which cause low-grade inflammation and even neuroinflammation. Consumption of pesticides leads to Dysbiosis of gut microbiota. As a result, there is an increase in intestinal permeability takes place. Due to intestinal permeability, a potential risk for the entry of LPS into the body is increased and, finally, low-grade inflammation [103].

Pesticides can act on gut microbes, affect metabolites, destroy gut mucosa and intestinal cells, and so on. By acting on receptor sites, these alterations produce pathogenic changes in various tissues and organs [104, 105]. We, therefore, emphasize the importance and need to regard the gut microbiota as an unwanted recipient of pesticide pollution. The long-term consequences on microbial diversity at a societal level of chronically consuming pesticides (via dietary exposure) should be addressed and figured out.

More research into the mechanisms of EDC-induced gut microbiota dysbiosis and its long-term implications on host health is needed in the future. We believe the gut microbiota will become a new target for analyzing pesticide toxicity and bring new knowledge of the occurrence and impacts of contamination with EDCs.

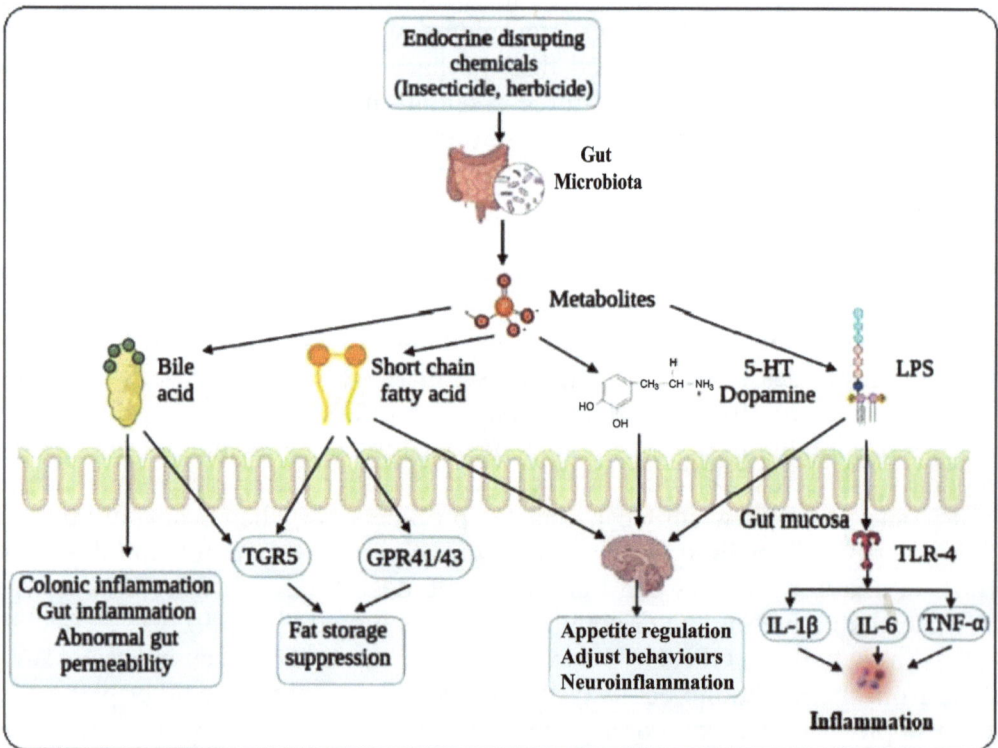

Fig. (2). Effect of endocrine disrupting chemicals on gut microbiota. EDCs could alter gut microbiota composition, gut barrier and metabolites, which further cause inflammation to the host.

INFLAMMATION, INSULIN RESISTANCE, GUT DYSBIOSIS AND PCOS

As far as we have already discussed the gut microbiota dysbiosis by several EDCs, this Dysbiosis may favor the growth of some Gram-negative bacteria while reducing the growth of beneficial "good" bacteria [81]. Dysbiosis of the Gut Microbiome can result in the activation of the host's immune system that, triggers a chronic inflammatory response. An inflammatory response is further involved in impairing the function of insulin receptors and initiates IR. The resulting hyper-insulinemic interferes with follicular development, while driving excess androgen production by the ovary's thecal cells – thus producing all three PCOS classical features [9].

Loss of Gut Integrity Due to Dysbiosis of Gut

The gastrointestinal tract is a wide portion of the body. The surface of the gastrointestinal body contains a mucosal surface. There is a circular layer of muscle, which prevents food from going backward. The mucosal surface provides a large area where the intestinal bacteria are exposed. The colonic mucosa forms an intricate selective barrier, and it stops intestinal bacteria from entering the systemic circulation and saves from the lethal systemic condition of infection. It acts as a selective barrier, as it only facilitates the movement of valuable nutrients and water across the intestinal wall while preventing potentially harmful bacteria from passing through [106].

Within the mucosa, there is a type of cell called the goblet cell that produces a dense mucus barrier that prevents the luminal bacteria from closely contacting the mucosal surface [107]. Only small-sized molecules can pass through this barrier, but this heavily glycosylated mucous layer prevents large molecules. An important barrier function is provided by proteins of cell-to-cell adhesion that allow selective para-cell passage between mucosal epithelial cells of colonic luminal content [106]. This is due to the fact that most hydrophilic solutes pass through the mucosal epithelial cells of the lipid membranes due to the lack of particular "shuttle" proteins. Tight junctions between epithelial cells, produced by the interaction of adhesive proteins such as Claudins and Occludins (zonula Occludens 1 and 2), provide a tight seal between neighboring epithelial cells and thus control the passing of solutes into the circulation from the luminal room via the colon wall. Large molecules such as whole bacteria cannot move through these paracellular pathways if tight junctions operate normally [106].

Mucins are required to maintain an adequate layer of mucus that covers the intestinal epithelium, providing a physical barrier that protects the intestinal epithelium from toxic agents and interrupts the passage of bad bacteria from the gut to systemic circulation. The MUC2 gene is responsible for major intestinal mucus production. The good bacteria *Bifidobacteria* and *Lactobacillus* produce SCFA through the fermentation of carbohydrates, and it has been reported that these SCFA increase the development of MUC-2 mucin in colonic mucosal cells and prevent bacteria from passing through the gut epithelium [52, 53]. Good bacteria help in maintaining the growth of bad bacteria and hold bad bacteria numbers in control. There is another function of good bacteria is to reduce the pH of the colonic lumen by producing SCFA and lactic acid, and as a result, a condition is created which is unfavorable for bad bacteria. In this way, bad bacterial numbers are maintained, and colonic luminal endotoxin (LPS) production is minimized [54].

Several independent studies demonstrated that the number of *Firmicutes* in the PCOS group of rats was significantly fewer than in healthy people where Gram-negative bacteria are extremely higher, specifically those who belong to the *Bacteroidetes* group [108]. On the contrary, other studies suggested that the number of Firmicutes was much greater than *Bacteroides* in PCOS mice [109]. PCOS has been linked to variations in the amount of *Bacteroidetes* and *Firmicutes* bacteria. It is apparent that in PCOS, changes in the abundance of particular *Bacteroidetes* and *Firmicutes* can lead to alterations in short-chain fatty acid synthesis, which can affect metabolism, gut barrier integrity, and immunity. *Firmicutes* help to produce butyrate and among all the SCFA, butyrate is more efficient than acetate and propionate because butyrate is an important source of nutrition for epithelial cells [110, 111] and it promotes mucosal restitution, induces differentiation, and inhibits tumor growth and inflammation [111, 112]. In several experiments, such as cell-line studies, experimental animal models and fresh specimens of human intestinal tissue, it was shown that butyrate acts to express in a dose-dependent manner [113 - 117]. Reports from *in-vitro* studies have indicated that low concentrations of butyrate induce MUC2 mucin expression, resulting in increased in vivo activation of the intestinal epithelial barrier. In comparison, high concentrations of butyrate's decrease the expression of MUC2, which may reduce the role of the intestinal barrier [53]. Butyrate is known to use the transcription factor AP-1 and prevents proliferation in HT-29 cells by inducing cyclin D3, a progression cycle blocker, and p21, a stimulator for cell differentiation [118]. AP-1 is a multiprotein complex made up of proto-oncogenes from c-Jun and c-Fos. In MUC2 promoter there is a putative consensus binding site (ATGAGTCAGA) for AP-1 at -817/-808 and MUC2 gene is regulated by AP1 transcription factor. Butyrate induces the expression of c-Jun and c-Fos to trigger the AP1 reporter construct and induce MUC2 expression through AP1 binding to the promoter of MUC2 gene (Fig. **3**) [53]. As we have discussed earlier, only low butyrate concentration can induce the expression of

MUC2 mRNA and protein. This suggests that MUC2 promoter activation and upregulation of MUC2 RNA and protein levels, at low concentrations of butyrate, was, at least partly, regulated by AP-1. It has also shown that in a dose-dependent manner, butyrate changes gene expression and can also reflect changes in the status of histone modification and chromatin marks. Methylation of histone H3 at Lys-9 (H3-Lys^9) was considered a chromatin sign synonymous with heterochromatin and silencing of the gene for a long time. High concentration butyrate causes H3-Lys^9 methylation of gene and forms heterochromatin; as a result, silencing of the gene for a long time has occurred. On the other hand, low butyrate treatment caused an increase in trimethylation of histone H3 at Lys 27 (H3-Lys^{27}) at the MUC2 promoter [53, 119]. It was reported that higher levels of H3-Lys^9 mono-methylation are found in active promoters surrounding gene transcription sites, indicating that transcription activation may be associated with this alteration [120].

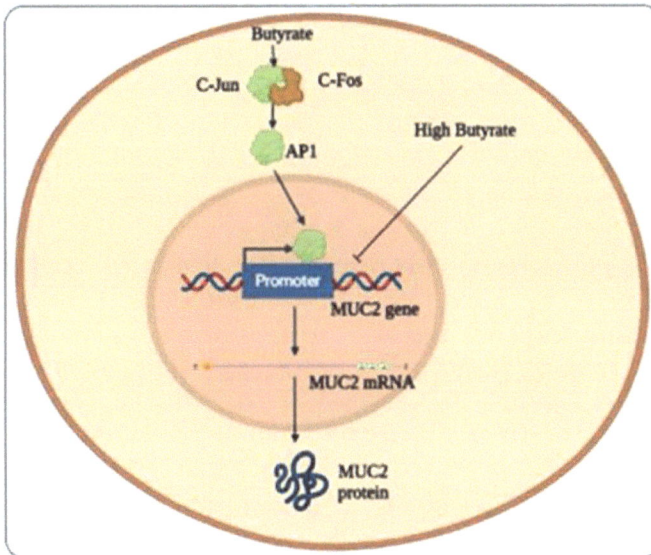

Fig. (3). Regulation of MUC-2 gene expression by butyrate.

Another mechanism reported by which MUC-2 gene expression is regulated by butyrate. Butyrate enhances gene expression of MUC-2 by regulating the expression of prostaglandins (PG). There are two types of PG, *e.g.*, PG1 and PG2, and it is reported that PG1 is more effective in enhancing MUC-2 gene expression than PG2. In the intestine, there are two layers of cells, the outer epithelial layer and beneath the epithelial cells. There is another subepithelial layer composed of a thin layer of myofibroblast cells. The myofibroblast cells that underlie the epithelium control epithelial cell functions such as proliferation, differentiation, secretion, and motility. These two types of cells that are present in different layers are an important source of PG and thus play an important role in the expression of the MUC-2 gene in epithelial cells. An important enzyme Cyclooxygenase (COX)-1, is present in both epithelial cells as well as subepithelial cells that can induce the expression of PG. Epithelial cells express COX-1 in a constitutive way, while myofibroblast regulates its expression by butyrate. SCFA increased the PGE1/PGE2 ratio provided by subepithelial myofibroblasts by enhancing the expression of MUC-2 in epithelial cells, and it was found that PGE1 was superior to PGE2. This indicates that bacterial fermentation products can have beneficial effects on the health of the gut while maintaining the integrity of the mucosal barrier [52].

Metabolic Endotoxemia and Insulin Resistance

It is now evident from the above discussion that Gram-negative bacteria may be responsible for the loss of gut integrity and may cause the level of pro-inflammatory mediators to increase. The majority of the Gram-negative bacteria belonging to the phyla *Bacteroidetes,* whose cell wall consists of LPS, an important pro-inflammatory mediator. This situation of extremely high circulating LPS level is called metabolic endotoxemia [121]. These Gram-negative bacteria may invade the intestinal barrier with a loss of gut integrity, and may come into contact with immune cells such as macrophages and dendritic cells. These cells express some high-affinity immune receptors, such as TLRs, the inflammatory NLRP3 and NLRs [122]. These receptors can bind specifically to the LPS and cause a local immune response that leads to chronic low-grade inflammation.

LPS and its endotoxic moiety were reported as potential TLR4 activators. LPS is composed of oligosaccharides and acylated saturated fatty acids (SFAs). LPS can activate TLR-4/MyD 88/NFκβ pathway, which can trigger some inflammatory responses by releasing various pro-inflammatory molecules such as TNFα, IL-1, IL-6 and inducible synthase nitric oxide (iNOS) [123]. This pro-inflammatory molecule produced by bacterial endotoxins can result in IR.

An inflammatory cascade is initiated after binding LPS to the TLR-4, which then dimerizes and recruits downstream adaptor molecules MyD 88/MyD 88 adaptor-like protein (MAL). The activated MyD 88/MAL further activates IL-1 receptor-associated kinase (IRAK), TNF receptor-associated factor-6 (TRAF-6), Transforming growth factor B associated kinase-1 (TAK-1), and Janus kinase (JNK) and IKK complexes. Initially, IKK was inactivated by iκβ, but after phosphorylation of iκβ, it can no longer inhibit IKK. Now the free IKK activates the transcription factor NFκβ. After activation, NFκβ translocates inside the nucleus to activate an inflammatory response. In the meantime, activation of serine kinases (JNK and IKK) induces insulin receptor substrates (IRS) serine phosphorylation [124]. In addition, high circulating LPS may increase iNOS expression via the TLR-4 pathway [125] (Fig. **4**). The increased iNOS causes protein S-nitrosation/ S-nitrosylation, in which the protein functions are altered as the nitric oxide (NO) interacts with the protein's cysteine residues [126]. Therefore, LPS could be responsible for impaired insulin signaling because the processes of S-nitrosation/S-nitrosylation may disrupt the IR, IRS and Akt in insulin-sensitive tissues [127], and this will generate systemic IR.

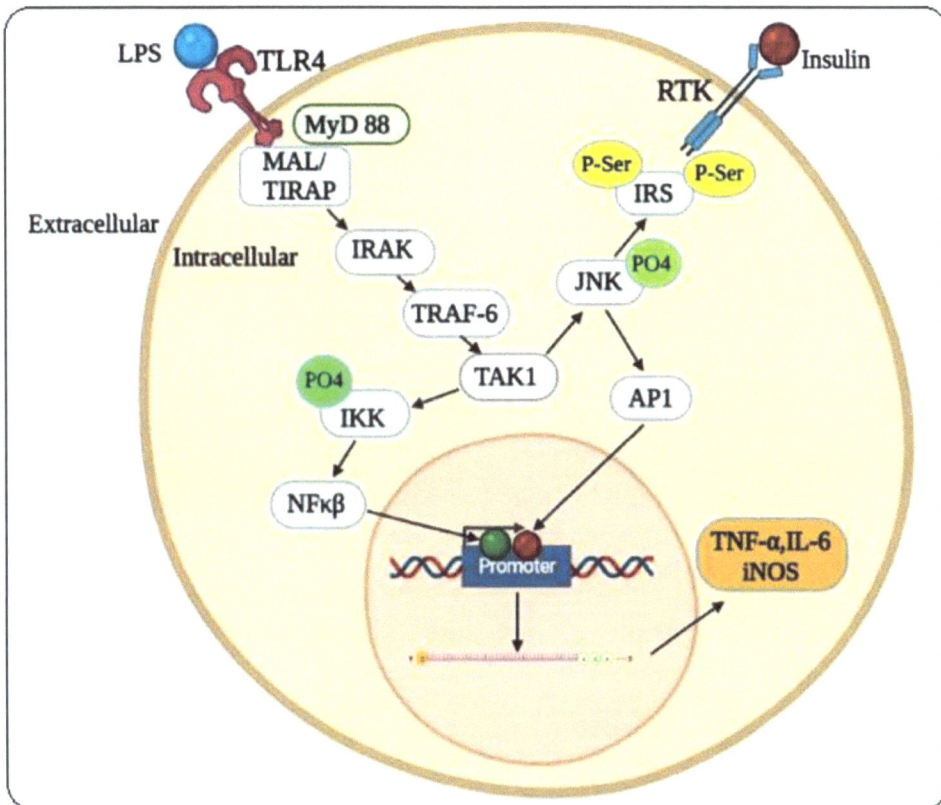

Fig. (4). Mechanism of the induction of systemic insulin resistance by LPS (PO$_4$: phosphate group; P-Ser: Serine phosphorylation).

Hyperinsulinemia Leads to PCOS Pathogenesis

Oligo-anovulation or anovulation is a prevalent symptom of PCOS. In the growth of PCOS, IR and its compensatory hyperinsulinemia are the main factors and play a key role in generating excessive androgen in PCOS [128 - 131].

Hyperinsulinemia significantly increases the synthesis of ovarian androgen. Insulin also causes hyperandrogenemia by directly activating mitogenic pathways in ovarian cells, as well as increasing steroidogenic acute regulatory protein (StAR) transcription and several primary steroidogenic enzymes. Insulin may play a role in developing the usual increased amplitude and frequency of pulse secretion of GnRH and LH, as seen in PCOS. This effect can be mediated by insulin in the hypothalamus GnRH-secreting cells, potentiating transcription of the GnRH gene through the MAPK pathway. This results in increased synthesis and secretion of GnRH, leading to a subsequent elevation of LH levels. This continuous stimulation would result in increased synthesis of ovarian steroid hormones, particularly androgens [132].

On the other hand, by potentiating hypothalamus-hypophysis-adrenal axis (HHAA) activity at several key sites, insulin also strengthens the adrenal glands as an alternate androgen source parallel to ovaries. The hippocampus is an essential mediator of HHAA negative feedback by inhibiting hippocampal activity; insulin indirectly increases hypothalamic Corticotropin-releasing hormone (CRH) secretion [133], although it can also play a direct role in both the hypothalamus [134] and the hypophysis [135]. Although the mechanisms are not clear, insulin tends to increase the responsiveness of the adrenal cortex to ACTH stimulation, with increased androgen secretion [136].

Lower levels of SHBG have been associated with elevated insulin concentrations, leading to increased bioavailability of androgens [137]. On the other hand, insulin has been shown to suppress IGFBP-1 synthesis in both the liver and the ovaries. This allows greater availability of IGF-1 and boosts insulin production not only in the liver, but also in the ovaries, which in turn stimulate thecal androgen production [138].

A local ovarian environment with high androgen, insulin and IGF-1 activity can impair normal follicle development from small antral follicles (2–10 mm) to maturation. This would, of course, provide the mechanism for developing the last two cardinal features of PCOS-multiple small ovarian subcapsular cysts and impaired ovulation/menstrual irregularity.

Probiotics as a Potential Treatment for PCOS

With a better understanding of the function of intestinal microbiota in the pathophysiology of PCOS, the use of microbiota-targeted medicines in the treatment of PCOS has recently been addressed. For this aim, probiotic therapy may be a new treatment choice for PCOS.

Probiotics can be defined as "live micro-organisms which give the host a benefit for health when given in adequate amounts" [139]. These micro-organisms are found naturally in foods that are made possible through fermentation, such as kefir, yogurt, ayran, tarhana, pickles, (sundried tomato, flour, and yogurt), turnip, soy products, wine, olives, beer, bacon, and dried or fermented meat products [140].

Only a few studies are found that assess the effects of probiotic supplementation on women's hormonal profile with PCOS. It was found that supplementation of probiotics (*L. acidophilus*, *L. casei*, and *Bifidobacterium bifidum*) among PCOS women for 12 weeks increased SBHG and plasma total antioxidant capacity, but decreased the level of serum testosterone as compared to the control group [141]. When *Lactobacillius* and faecal microbiota were transplanted together, serum androgen levels were decreased, and estrogen levels were increased and a regulated menstrual cycle was seen [108]. In another study, probiotic administration could increase total testosterone, SHBG by improving insulin sensitivity, decreasing the level of proinflammatory cytokines and raising the faecal pH [142].

High C-reactive protein (CRP) and increased oxidative stress induced by oxygen species (ROS) may contribute to the progress of IR and hyperandrogenism, which are the main features of PCOS [143, 144]. A recent study suggested that PCOS women supplemented with for 12 weeks had beneficial effects on total testosterone, SHBG, modified Ferriman-Gallwey (mFG) scores, serum high-sensitivity C-reactive protein (hs-CRP), plasma total antioxidant capacity (TAC) and plasma malondialdehyde (MDA) levels. A randomized placebo-controlled trial with probiotics, such as *Lactobacillus acidophilus*, *L. casei* and *Bifidobacterium bifidum* supplementation has shown that there is a decrease in serum hs-CRP and plasma MDA and increased plasma TAC [141]. In another study, it was found that Vitamin D and probiotic co-administration for 12 weeks in women with PCOS had beneficial effects on parameters of mental well-being, total serum testosterone, hs-CRP, hirsutism, plasma MDA, TAC, and GSH but did not affect levels of plasma NO, acne, serum SHBG and alopecia [145].

So many studies have reported that supplementation with probiotics has advantageous results on the metabolic profile of women with PCOS. A meta-analysis of seven randomized control trials found significant impacts on metabolic profiles. They found that probiotic supplementation in women with PCOS had positive effects on High-density lipoprotein (HDL), Triglycerides (TG), and fasting insulin, but no significant effect was found on Homeostatic model assessment of insulin resistance (HOMA-IR), CRP, hsCRP, LDL, and total cholesterol as well as anthropometric measurements including weight, BMI, and waist circumference [146]. In another study, probiotics supplementation (*Bifidobacterium bifidum, Lactobacillus acidophilus, L. casei*) for 12 weeks showed that there is a significant reduction in fasting plasma glucose, serum insulin concentration, HOMA-IR, TG, very low-density lipoprotein (VLDL) cholesterol levels in women with PCOS compared with the controls [147]. Plasma glucose and serum insulin levels are important parameters for PCOS patients. Probiotic supplementation (*L. casei, L. acidophilus, L. bulgaricus, L. rhamnosus, Streptococcus thermophiles, B. breve, B. longum*) in PCOS women for 8 weeks was linked with a substantial decrease in fasting plasma glucose and serum insulin levels [142]. Another study demonstrated that probiotic supplementation with *L. fermentum and L. delbruekii* has a meaningful impact on macrophage migration inhibitory factor (MIF), HOMA-IR, and fasting blood glucose. Rashad *et al.* [148] found that supplementation of probiotics for 12 weeks significantly reduced MIF, fasting plasma glucose, and HOMA-IR levels, and it also improved the lipid profile. According to Ghanei *et al.* [149], probiotic supplementation can also modulate inflammation in women with PCOS as they found significant results with *Acidophilus*, *L. plantarum*, *L. fermentum,* and *L. gasseri* supplementation for 12 weeks.

Supplementation of probiotics has also shown positive results on the gut microbiome. Zhang *et al.* [150] reported that 210 mg/kg body weight probiotic administration (*Bifidobacterium* sp., *Lactobacillus acidophilus,* and *Enterococcus faecalis*) showed better results than the treatment of Diane-35 (medication to treat PCOS) and berberine in dihydrotestosterone-induced PCOS rats. The effect of probiotics showed an increased microbial diversity and significant effects on *Firmicutes, Tenericutes, Proteobacteria,* and *Spirochaetae.* A recent study has reported that Through the Gut-Brain Axis, the probiotic *Bifidobacterium lactis* V9 regulates the secretion of sex hormones in Polycystic Ovary Syndrome patients. In PCOS patients, levels of LH, testosterone, and PRL were found to be higher, whereas levels of brain-gut mediators such as ghrelin and PYY were found to be lower, compared to healthy controls. It was reported that consumption of the probiotic *Bifidobacterium lactis* V9 promotes the growth of SCFA-producing microbes, such as *Faecalibacterium prausnitzii, Butyricimonas sp.,* and *Akkermansia sp.* Thus, these microbes produce SCFA together with *Bifidobacterium,* affecting the gut-brain mediators (ghrelin and PYY) secretion. Eventually, variations in the PYY and ghrelin levels result in variability of sex hormone levels secreted by the hypophysis and hypothalamus through the gut-brain axis [151].

CONCLUSION

As a result of the preceding discussion, it is obvious that Dysbiosis of the gut microbiota, inflammation, and IR are three of the essential fundamental components in the development of PCOS. The gut microbiota's composition, diversity, as well as enzymatic capacities, are readily affected by various EDCs.

Studies have shown that due to Dysbiosis of gut microbiota, α-diversity is decreased in the gut; as a result, the abundances of some bacteria species change. PCOS was associated with changes in the abundance of multiple bacteria of *Bacteroidetes* and *Firmicutes* phyla. In PCOS, changes in the number of particular *Bacteroidetes* and *Firmicutes* can lead to changes in short-chain fatty acid synthesis, which can affect metabolism, gut barrier integrity, and immunity. An increase in Gram-negative "bad" bacteria results in increased translocation of immunostimulatory LPS molecules from the intestinal lumen to the systemic circulation. It is also evident that Gram-negative bacterial LPS can cause metabolic endotoxemia, which is a key effector in IR development. This condition of "metabolic endotoxemia" stimulates macrophages of fat, liver and muscle in the body, leading to the release of high levels of TNF-α and the initiation of IR. This condition of hyper-insulinemic interferes with the normal production of follicles in the ovary, causing a halt in follicular growth with multiple small follicles generation and abnormal ovulation with its related menstrual irregularities. High serum insulin also drives excessive androgen production by ovarian thecal cells while depressing SHBG hepatic production, leading to a net increase in the availability of free androgen and the development of acne and hirsutism.

The symbiotic application of some probiotics is likely to improve intestinal barrier function. These improvements would reduce the flow of LPS through the mucosal surface, thus decreasing metabolic endotoxemia. Probiotic supplements help reduce fasting blood glucose and the level of serum insulin concentration and also regulate the sex hormone level through manipulating the intestinal microbiome in PCOS-like patients.

FUTURE PERSPECTIVES AND SCOPE

A number of processes have been presented to support that EDCS may play a role in the establishment of gut microbial dysbiosis. It should be noted that the EDCs consumed via dietary intake require the identification of the compounds and the specific responses of the different species of the gut microbiome. Due to insufficient data obtained from cross-sectional studies regarding the effects of EDCs on the human gut microbiota, the exact mechanism by which EDCs may cause Dysbiosis is unknown. To understand the underlying processes of this association and the involvement of the gut bacteria in PCOS, random control trials are needed.

Studies have demonstrated that gut microbiota dysbiosis exists in PCOS, a reduction in diversity, and changes in abundance of certain types of bacteria linked with metabolic disorders. It is important to remember that the findings were inconsistent with those studies. In some studies, it has shown that certain bacterial (e.g., *Firmicutes, Bacteroides*) numbers are decreasing, while on the contrary, other studies demonstrated that the number of these bacteria is increasing. In fact, it is not clear precisely how to gut microbiota shifts in various ways in PCOS phenotypes; thus, more studies are required.

Probiotic supplementation to PCOS patients has been shown to effectively and beneficially affect many biochemical pathways, but the fundamental process is unknown. Further research is therefore essential to figure out what role these agents play in developing PCOS.

CONSENT FOR PUBLICATION

Not applicable.

CONFLICT OF INTEREST

The authors declare no conflict of interest, financial or otherwise.

ACKNOWLEDGEMENTS

The authors thank the Department of Zoology, University of North Bengal for their support and generous help in completing this research work. We would like to acknowledge the financial support of the University of North Bengal (Ref. No. 2274/R-2021 Dated 24.06.2021) to carry out background work. Fellowship support (for RI) from the University Grants Commission (UGC), Govt. of India is also hereby acknowledged [UGC Ref. No. 711/(CSIR-UGC NET DEC.2018) dated 11.07.2019].

REFERENCES

[1]   Wolf W, Wattick R, Kinkade O, Olfert M. Geographical Prevalence of Polycystic Ovary Syndrome as Determined by Region and Race/Ethnicity. Int J Environ Res Public Health 2018; 15(11): 2589.
[http://dx.doi.org/10.3390/ijerph15112589] [PMID: 30463276]

[2]   Escobar-Morreale HF. Polycystic ovary syndrome: definition, aetiology, diagnosis and treatment. Nat Rev Endocrinol 2018; 14(5): 270-84.
[http://dx.doi.org/10.1038/nrendo.2018.24] [PMID: 29569621]

[3]   Witchel SF, Burghard AC, Tao RH, Oberfield SE. The diagnosis and treatment of PCOS in adolescents. Curr Opin Pediatr 2019; 31(4): 562-9.
[http://dx.doi.org/10.1097/MOP.0000000000000778] [PMID: 31299022]

[4]   Meier RK. Polycystic Ovary Syndrome. Nurs Clin North Am 2018; 53(3): 407-20.
[http://dx.doi.org/10.1016/j.cnur.2018.04.008] [PMID: 30100006]

[5]   Witchel SF, Oberfield SE, Peña AS. Polycystic Ovary Syndrome: Pathophysiology, Presentation, and Treatment With Emphasis on Adolescent Girls. J Endocr Soc 2019; 3(8): 1545-73.
[http://dx.doi.org/10.1210/js.2019-00078] [PMID: 31384717]

[6]   Di Renzo GC, Giardina I, Clerici G, Brillo E, Gerli S. Progesterone in normal and pathological

pregnancy. Horm Mol Biol Clin Investig 2016; 27(1): 35-48.
[http://dx.doi.org/10.1515/hmbci-2016-0038] [PMID: 27662646]

[7]   Ebrahimi-Mamaghani M, Saghafi-Asl M, Pirouzpanah S, *et al.* Association of insulin resistance with lipid profile, metabolic syndrome, and hormonal aberrations in overweight or obese women with polycystic ovary syndrome. J Health Popul Nutr 2015; 33(1): 157-67.
[PMID: 25995732]

[8]   Anagnostis P, Tarlatzis BC, Kauffman RP. Polycystic ovarian syndrome (PCOS): Long-term metabolic consequences. Metabolism 2018; 86: 33-43.
[http://dx.doi.org/10.1016/j.metabol.2017.09.016] [PMID: 29024702]

[9]   Tremellen K, Pearce K. Dysbiosis of Gut Microbiota (DOGMA) – A novel theory for the development of Polycystic Ovarian Syndrome. Med Hypotheses 2012; 79(1): 104-12.
[http://dx.doi.org/10.1016/j.mehy.2012.04.016] [PMID: 22543078]

[10]   Liu R, Zhang C, Shi Y, *et al.* Dysbiosis of gut microbiota associated with clinical parameters in polycystic ovary syndrome. Front Microbiol 2017; 8: 324.
[http://dx.doi.org/10.3389/fmicb.2017.00324] [PMID: 28293234]

[11]   Zeng B, Lai Z, Sun L, *et al.* Structural and functional profiles of the gut microbial community in polycystic ovary syndrome with insulin resistance (IR-PCOS): A pilot study. Res Microbiol 2019; 170(1): 43-52.
[http://dx.doi.org/10.1016/j.resmic.2018.09.002] [PMID: 30292647]

[12]   Baptiste CG, Battista MC, Trottier A, Baillargeon JP. Insulin and hyperandrogenism in women with polycystic ovary syndrome. J Steroid Biochem Mol Biol 2010; 122(1-3): 42-52.
[http://dx.doi.org/10.1016/j.jsbmb.2009.12.010] [PMID: 20036327]

[13]   Hajivandi L, Noroozi M, Mostafavi F, Ekramzadeh M. Food habits in overweight and obese adolescent girls with Polycystic ovary syndrome (PCOS): A qualitative study in Iran. BMC Pediatr 2020; 20(1): 277.
[http://dx.doi.org/10.1186/s12887-020-02173-y] [PMID: 32498675]

[14]   Kshetrimayum C, Sharma A, Mishra VV, Kumar S. Polycystic ovarian syndrome: Environmental/occupational, lifestyle factors; an overview. J Turk Ger Gynecol Assoc 2019; 20(4): 255-63.
[http://dx.doi.org/10.4274/jtgga.galenos.2019.2018.0142] [PMID: 30821135]

[15]   Velmurugan G, Ramprasath T, Gilles M, Swaminathan K, Ramasamy S. Gut Microbiota, Endocrine-Disrupting Chemicals, and the Diabetes Epidemic. Trends Endocrinol Metab 2017; 28(8): 612-25.
[http://dx.doi.org/10.1016/j.tem.2017.05.001] [PMID: 28571659]

[16]   Rosenfield RL, Ehrmann DA. The Pathogenesis of Polycystic Ovary Syndrome (PCOS): The Hypothesis of PCOS as Functional Ovarian Hyperandrogenism Revisited. Endocr Rev 2016; 37(5): 467-520.
[http://dx.doi.org/10.1210/er.2015-1104] [PMID: 27459230]

[17]   Zawadski JK, Dunaif A. Diagnostic criteria for polycystic ovary syndrome: Towards a rational approach. In: Dunaif A, Givens JR, Haseltine F, Eds. Polycystic Ovary Syndrome. Boston: Blackwell Scientific 1992; pp. 377-84.

[18]   Revised 2003 consensus on diagnostic criteria and long-term health risks related to polycystic ovary syndrome. Fertil Steril 2004; 81(1): 19-25.
[http://dx.doi.org/10.1016/j.fertnstert.2003.10.004]

[19]   Azziz R, Carmina E, Dewailly D, *et al.* Positions statement: criteria for defining polycystic ovary syndrome as a predominantly hyperandrogenic syndrome: An Androgen Excess Society guideline. J Clin Endocrinol Metab 2006; 91(11): 4237-45.
[http://dx.doi.org/10.1210/jc.2006-0178] [PMID: 16940456]

[20]   Teede HJ, Misso ML, Costello MF, *et al.* Recommendations from the international evidence-based

guideline for the assessment and management of polycystic ovary syndrome. Fertil Steril 2018; 110(3): 364-79.
[http://dx.doi.org/10.1016/j.fertnstert.2018.05.004] [PMID: 30033227]

[21]   El Hayek S, Bitar L, Hamdar LH, Mirza FG, Daoud G. Poly Cystic Ovarian Syndrome: An Updated Overview. Front Physiol 2016; 7: 124.
[http://dx.doi.org/10.3389/fphys.2016.00124] [PMID: 27092084]

[22]   Azziz R, Carmina E, Dewailly D, *et al*. The Androgen Excess and PCOS Society criteria for the polycystic ovary syndrome: the complete task force report. Fertil Steril 2009; 91(2): 456-88.
[http://dx.doi.org/10.1016/j.fertnstert.2008.06.035] [PMID: 18950759]

[23]   Bozdag G, Mumusoglu S, Zengin D, Karabulut E, Yildiz BO. The prevalence and phenotypic features of polycystic ovary syndrome: A systematic review and meta-analysis. Hum Reprod 2016; 31(12): 2841-55.
[http://dx.doi.org/10.1093/humrep/dew218] [PMID: 27664216]

[24]   Taponen S, Ahonkallio S, Martikainen H, *et al*. Prevalence of polycystic ovaries in women with self-reported symptoms of oligomenorrhoea and/or hirsutism: Northern Finland Birth Cohort 1966 Study. Hum Reprod 2004; 19(5): 1083-8.
[http://dx.doi.org/10.1093/humrep/deh214] [PMID: 15044401]

[25]   De Leo V, Musacchio MC, Cappelli V, Massaro MG, Morgante G, Petraglia F. Genetic, hormonal and metabolic aspects of PCOS: An update. Reprod Biol Endocrinol 2016; 14(1): 38.
[http://dx.doi.org/10.1186/s12958-016-0173-x] [PMID: 27423183]

[26]   Glueck CJ, Goldenberg N. Characteristics of obesity in polycystic ovary syndrome: Etiology, treatment, and genetics. Metabolism 2019; 92: 108-20.
[http://dx.doi.org/10.1016/j.metabol.2018.11.002] [PMID: 30445140]

[27]   Marx TL, Mehta AE. Polycystic ovary syndrome: pathogenesis and treatment over the short and long term. Cleve Clin J Med 2003; 70(1): 31-33, 36-41, 45.
[http://dx.doi.org/10.3949/ccjm.70.1.31] [PMID: 12549723]

[28]   Gunning MN, Fauser BCJM. Are women with polycystic ovary syndrome at increased cardiovascular disease risk later in life? Climacteric 2017; 20(3): 222-7.
[http://dx.doi.org/10.1080/13697137.2017.1316256] [PMID: 28457146]

[29]   Cooper HE, Spellacy WN, Prem KA, Cohen WD. Hereditary factors in the Stein-Leventhal syndrome. Am J Obstet Gynecol 1968; 100(3): 371-87.
[http://dx.doi.org/10.1016/S0002-9378(15)33704-2] [PMID: 15782458]

[30]   Urbanek M, Legro RS, Driscoll DA, *et al*. Thirty-seven candidate genes for polycystic ovary syndrome: Strongest evidence for linkage is with follistatin. Proc Natl Acad Sci USA 1999; 96(15): 8573-8.
[http://dx.doi.org/10.1073/pnas.96.15.8573] [PMID: 10411917]

[31]   Silvestris E, de Pergola G, Rosania R, Loverro G. Obesity as disruptor of the female fertility. Reprod Biol Endocrinol 2018; 16(1): 22.
[http://dx.doi.org/10.1186/s12958-018-0336-z] [PMID: 29523133]

[32]   Eriksson Hogling D, Petrus P, Gao H, *et al*. Adipose and Circulating CCL18 Levels Associate With Metabolic Risk Factors in Women. J Clin Endocrinol Metab 2016; 101(11): 4021-9.
[http://dx.doi.org/10.1210/jc.2016-2390] [PMID: 27459538]

[33]   Suryanarayana KM, Sam JE, Dharmalingam M, Kalra P, Selvan C. Serum CCL 18 levels in women with polycystic ovarian syndrome. Indian J Endocrinol Metab 2020; 24(3): 280-5.
[http://dx.doi.org/10.4103/ijem.IJEM_650_19] [PMID: 33083270]

[34]   Shi M, Whorton AE, Sekulovski N, MacLean JA II, Hayashi K. Prenatal Exposure to Bisphenol A, E, and S Induces Transgenerational Effects on Female Reproductive Functions in Mice. Toxicol Sci 2019; 170(2): 320-9.

[http://dx.doi.org/10.1093/toxsci/kfz124] [PMID: 31132128]

[35]   Thursby E, Juge N. Introduction to the human gut microbiota. Biochem J 2017; 474(11): 1823-36.
       [http://dx.doi.org/10.1042/BCJ20160510] [PMID: 28512250]

[36]   Valdes AM, Walter J, Segal E, Spector TD. Role of the gut microbiota in nutrition and health. BMJ
       2018; 361: k2179.
       [http://dx.doi.org/10.1136/bmj.k2179] [PMID: 29899036]

[37]   Ottman N, Smidt H, de Vos WM, Belzer C. The function of our microbiota: who is out there and what
       do they do? Front Cell Infect Microbiol 2012; 2: 104.
       [http://dx.doi.org/10.3389/fcimb.2012.00104] [PMID: 22919693]

[38]   Rinninella E, Raoul P, Cintoni M, *et al.* What is the Healthy Gut Microbiota Composition? A
       Changing Ecosystem across Age, Environment, Diet, and Diseases. Microorganisms 2019; 7(1): 14.
       [http://dx.doi.org/10.3390/microorganisms7010014] [PMID: 30634578]

[39]   Eckburg PB, Bik EM, Bernstein CN, *et al.* Diversity of the human intestinal microbial flora. Science
       2005; 308(5728): 1635-8.
       [http://dx.doi.org/10.1126/science.1110591] [PMID: 15831718]

[40]   Ley RE, Hamady M, Lozupone C, *et al.* Evolution of mammals and their gut microbes. Science 2008;
       320(5883): 1647-51.
       [http://dx.doi.org/10.1126/science.1155725] [PMID: 18497261]

[41]   Sheehan D, Moran C, Shanahan F. The microbiota in inflammatory bowel disease. J Gastroenterol
       2015; 50(5): 495-507.
       [http://dx.doi.org/10.1007/s00535-015-1064-1] [PMID: 25808229]

[42]   Andoh A. Physiological role of gut microbiota for maintaining human health. Digestion 2016; 93(3):
       176-81.
       [http://dx.doi.org/10.1159/000444066] [PMID: 26859303]

[43]   Rivière A, Selak M, Lantin D, Leroy F, De Vuyst L. *Bifidobacteria* and Butyrate-Producing Colon
       Bacteria: Importance and Strategies for Their Stimulation in the Human Gut. Front Microbiol 2016; 7:
       979.
       [http://dx.doi.org/10.3389/fmicb.2016.00979] [PMID: 27446020]

[44]   Jandhyala SM, Talukdar R, Subramanyam C, Vuyyuru H, Sasikala M, Nageshwar Reddy D. Role of
       the normal gut microbiota. World J Gastroenterol 2015; 21(29): 8787-803.
       [http://dx.doi.org/10.3748/wjg.v21.i29.8787] [PMID: 26269668]

[45]   Kho ZY, Lal SK. The Human Gut Microbiome – A Potential Controller of Wellness and Disease.
       Front Microbiol 2018; 9: 1835.
       [http://dx.doi.org/10.3389/fmicb.2018.01835] [PMID: 30154767]

[46]   Altuntaş Y, Batman A. Mikrobiyota ve metabolik sendrom. Turk Kardiyol Dern Ars 2017; 45(3): 286-
       96. [Microbiota and metabolic syndrome].
       [PMID: 28429701]

[47]   El-Salhy M, Hatlebakk JG, Hausken T. Diet in Irritable Bowel Syndrome (IBS): Interaction with Gut
       Microbiota and Gut Hormones. Nutrients 2019; 11(8): 1824.
       [http://dx.doi.org/10.3390/nu11081824] [PMID: 31394793]

[48]   Koh A, De Vadder F, Kovatcheva-Datchary P, Bäckhed F. From dietary fiber to host physiology:
       short-chain fatty acids as key bacterial metabolites. Cell 2016; 165(6): 1332-45.
       [http://dx.doi.org/10.1016/j.cell.2016.05.041] [PMID: 27259147]

[49]   den Besten G, van Eunen K, Groen AK, Venema K, Reijngoud DJ, Bakker BM. The role of short-
       chain fatty acids in the interplay between diet, gut microbiota, and host energy metabolism. J Lipid
       Res 2013; 54(9): 2325-40.
       [http://dx.doi.org/10.1194/jlr.R036012] [PMID: 23821742]

[50]   Lee WJ, Hase K. Gut microbiota–generated metabolites in animal health and disease. Nat Chem Biol 2014; 10(6): 416-24.
[http://dx.doi.org/10.1038/nchembio.1535] [PMID: 24838170]

[51]   Prakash S, Rodes L, Coussa-Charley M, *et al.* Gut microbiota: next frontier in understanding human health and development of biotherapeutics. Biologics 2011; 5: 71-86.
[http://dx.doi.org/10.2147/BTT.S19099] [PMID: 21847343]

[52]   Willemsen LEM, Koetsier MA, van Deventer SJ, van Tol EA. Short chain fatty acids stimulate epithelial mucin 2 expression through differential effects on prostaglandin E1 and E2 production by intestinal myofibroblasts. Gut 2003; 52(10): 1442-7.
[http://dx.doi.org/10.1136/gut.52.10.1442] [PMID: 12970137]

[53]   Burger-van Paassen N, Vincent A, Puiman PJ, *et al.* The regulation of intestinal mucin MUC2 expression by short-chain fatty acids: implications for epithelial protection. Biochem J 2009; 420(2): 211-9.
[http://dx.doi.org/10.1042/BJ20082222] [PMID: 19228118]

[54]   Gibson GR, McCartney AL, Rastall RA. Prebiotics and resistance to gastrointestinal infections. Br J Nutr 2005; 93(S1) (Suppl. 1): S31-4.
[http://dx.doi.org/10.1079/BJN20041343] [PMID: 15877892]

[55]   Fallucca F, Porrata C, Fallucca S, Pianesi M. Influence of diet on gut microbiota, inflammation and type 2 diabetes mellitus. First experience with macrobiotic Ma-Pi 2 diet. Diabetes Metab Res Rev 2014; 30(S1) (Suppl. 1): 48-54.
[http://dx.doi.org/10.1002/dmrr.2518] [PMID: 24532292]

[56]   Torres PJ, Siakowska M, Banaszewska B, *et al.* Gut Microbial Diversity in Women With Polycystic Ovary Syndrome Correlates With Hyperandrogenism. J Clin Endocrinol Metab 2018; 103(4): 1502-11.
[http://dx.doi.org/10.1210/jc.2017-02153] [PMID: 29370410]

[57]   Neish AS. Microbes in gastrointestinal health and disease. Gastroenterology 2009; 136(1): 65-80.
[http://dx.doi.org/10.1053/j.gastro.2008.10.080] [PMID: 19026645]

[58]   Everard A, Cani PD. Gut microbiota and GLP-1. Rev Endocr Metab Disord 2014; 15(3): 189-96.
[http://dx.doi.org/10.1007/s11154-014-9288-6] [PMID: 24789701]

[59]   Larraufie P, Martin-Gallausiaux C, Lapaque N, *et al.* SCFAs strongly stimulate PYY production in human enteroendocrine cells. Sci Rep 2018; 8(1): 74.
[http://dx.doi.org/10.1038/s41598-017-18259-0] [PMID: 29311617]

[60]   Lindheim L, Bashir M, Münzker J, *et al.* Alterations in Gut Microbiome Composition and Barrier Function Are Associated with Reproductive and Metabolic Defects in Women with Polycystic Ovary Syndrome (PCOS): A Pilot Study. PLoS One 2017; 12(1): e0168390.
[http://dx.doi.org/10.1371/journal.pone.0168390] [PMID: 28045919]

[61]   Su Z, Gong Y, Yang H, Deng D, Liang Z. Activation of the Nuclear Factor-kappa B Signaling Pathway Damages the Epithelial Barrier in the Human Pancreatic Ductal Adenocarcinoma Cell Line HPAF-II. Pancreas 2019; 48(10): 1380-5.
[http://dx.doi.org/10.1097/MPA.0000000000001441] [PMID: 31688605]

[62]   Kuzu F. Bağırsak Mikrobiyotasının Obezite, İnsülin Direnci ve Diyabetteki Rolü. BSHR 2017; 1: 68-80.

[63]   Zhang D, Zhang L, Yue F, Zheng Y, Russell R. Serum zonulin is elevated in women with polycystic ovary syndrome and correlates with insulin resistance and severity of anovulation. Eur J Endocrinol 2015; 172(1): 29-36.
[http://dx.doi.org/10.1530/EJE-14-0589] [PMID: 25336505]

[64]   Moulana M. Immunophenotypic profile of leukocytes in hyperandrogenemic female rat an animal model of polycystic ovary syndrome. Life Sci 2019; 220: 44-9.

[http://dx.doi.org/10.1016/j.lfs.2019.01.048] [PMID: 30708097]

[65]   Shorakae S, Ranasinha S, Abell S, *et al.* Inter-related effects of insulin resistance, hyperandrogenism, sympathetic dysfunction and chronic inflammation in PCOS. Clin Endocrinol (Oxf) 2018; 89(5): 628-33.
[http://dx.doi.org/10.1111/cen.13808] [PMID: 29992612]

[66]   Cani PD, Amar J, Iglesias MA, *et al.* Metabolic endotoxemia initiates obesity and insulin resistance. Diabetes 2007; 56(7): 1761-72.
[http://dx.doi.org/10.2337/db06-1491] [PMID: 17456850]

[67]   Andreasen AS, Larsen N, Pedersen-Skovsgaard T, *et al.* Effects of *Lactobacillus acidophilus* NCFM on insulin sensitivity and the systemic inflammatory response in human subjects. Br J Nutr 2010; 104(12): 1831-8.
[http://dx.doi.org/10.1017/S0007114510002874] [PMID: 20815975]

[68]   Cara JF, Rosenfield RL. Insulin-like growth factor I and insulin potentiate luteinizing hormone-induced androgen synthesis by rat ovarian thecal-interstitial cells. Endocrinology 1988; 123(2): 733-9.
[http://dx.doi.org/10.1210/endo-123-2-733] [PMID: 2969325]

[69]   Nestler J, Powers LP, Matt DW, *et al.* A direct effect of hyperinsulinemia on serum sex hormone-binding globulin levels in obese women with the polycystic ovary syndrome. J Clin Endocrinol Metab 1991; 72(1): 83-9.
[http://dx.doi.org/10.1210/jcem-72-1-83] [PMID: 1898744]

[70]   Evliyaoğlu O. Polycystic ovary syndrome and hirsutism. Turkish Pediatrics Arch 2011; 2: 8-13.

[71]   Bergh C, Carlsson B, Olsson JH, Selleskog U, Hillensjö T. Regulation of androgen production in cultured human thecal cells by insulin-like growth factor I and insulin. Fertil Steril 1993; 59(2): 323-31.
[http://dx.doi.org/10.1016/S0015-0282(16)55675-1] [PMID: 8425626]

[72]   Franks S, Hardy K. Aberrant follicle development and anovulation in polycystic ovary syndrome. Ann Endocrinol (Paris) 2010; 71(3): 228-30.
[http://dx.doi.org/10.1016/j.ando.2010.02.007] [PMID: 20362969]

[73]   Nikishin DA, Alyoshina NM, Semenova ML, Shmukler YB. Analysis of Expression and Functional Activity of Aromatic L-Amino Acid Decarboxylase (DDC) and Serotonin Transporter (SERT) as Potential Sources of Serotonin in Mouse Ovary. Int J Mol Sci 2019; 20(12): 3070.
[http://dx.doi.org/10.3390/ijms20123070] [PMID: 31234589]

[74]   Konturek SJ, Konturek JW, Pawlik T, Brzozowski T. Brain-gut axis and its role in the control of food intake. J Physiol Pharmacol 2004; 55(1 Pt 2): 137-54.
[PMID: 15082874]

[75]   Lang UE, Beglinger C, Schweinfurth N, Walter M, Borgwardt S. Nutritional aspects of depression. Cell Physiol Biochem 2015; 37(3): 1029-43.
[http://dx.doi.org/10.1159/000430229] [PMID: 26402520]

[76]   Tremaroli V, Bäckhed F. Functional interactions between the gut microbiota and host metabolism. Nature 2012; 489(7415): 242-9.
[http://dx.doi.org/10.1038/nature11552] [PMID: 22972297]

[77]   Thackray VG. Sex, microbes, and polycystic ovary syndrome. Trends Endocrinol Metab 2019; 30(1): 54-65.
[http://dx.doi.org/10.1016/j.tem.2018.11.001] [PMID: 30503354]

[78]   Insenser M, Murri M, del Campo R, Martínez-García MÁ, Fernández-Durán E, Escobar-Morreale HF. Gut microbiota and the polycystic ovary syndrome: influence of sex, sex hormones, and obesity. J Clin Endocrinol Metab 2018; 103(7): 2552-62.
[http://dx.doi.org/10.1210/jc.2017-02799] [PMID: 29897462]

[79]   Sherman SB, Sarsour N, Salehi M, *et al.* Prenatal androgen exposure causes hypertension and gut

microbiota dysbiosis. Gut Microbes 2018; 9(5): 1-22.
[http://dx.doi.org/10.1080/19490976.2018.1441664] [PMID: 29469650]

[80]   Pellock SJ, Redinbo MR. Glucuronides in the gut: Sugar-driven symbioses between microbe and host. J Biol Chem 2017; 292(21): 8569-76.
[http://dx.doi.org/10.1074/jbc.R116.767434] [PMID: 28389557]

[81]   Rosenfeld CS. Gut Dysbiosis in Animals Due to Environmental Chemical Exposures. Front Cell Infect Microbiol 2017; 7: 396.
[http://dx.doi.org/10.3389/fcimb.2017.00396] [PMID: 28936425]

[82]   Neel BA, Sargis RM. The paradox of progress: environmental disruption of metabolism and the diabetes epidemic. Diabetes 2011; 60(7): 1838-48.
[http://dx.doi.org/10.2337/db11-0153] [PMID: 21709279]

[83]   Gore AC, Chappell VA, Fenton SE, *et al.* EDC-2: The Endocrine Society's Second Scientific Statement on Endocrine-Disrupting Chemicals. Endocr Rev 2015; 36(6): E1-E150.
[http://dx.doi.org/10.1210/er.2015-1010] [PMID: 26544531]

[84]   Zhang B, Zhou W, Shi Y, Zhang J, Cui L, Chen ZJ. Lifestyle and environmental contributions to ovulatory dysfunction in women of polycystic ovary syndrome. BMC Endocr Disord 2020; 20(1): 19.
[http://dx.doi.org/10.1186/s12902-020-0497-6] [PMID: 32000752]

[85]   Akın L, Kendirci M, Narin F, Kurtoğlu S, Hatipoğlu N, Elmalı F. Endocrine Disruptors and Polycystic Ovary Syndrome: Phthalates. J Clin Res Pediatr Endocrinol 2020; 12(4): 393-400.
[http://dx.doi.org/10.4274/jcrpe.galenos.2020.2020.0037] [PMID: 32431137]

[86]   Davis FR. Banned: A History of Pesticides and the Science of Toxicology. Yale University Press. Environ Hist 2016; 21: 401-3.

[87]   Donkor A, Osei-Fosu P, Dubey B, Kingsford-Adaboh R, Ziwu C, Asante I. Pesticide residues in fruits and vegetables in Ghana: A review. Environ Sci Pollut Res Int 2016; 23(19): 18966-87.
[http://dx.doi.org/10.1007/s11356-016-7317-6] [PMID: 27530198]

[88]   Chaza C, Sopheak N, Mariam H, David D, Baghdad O, Moomen B. Assessment of pesticide contamination in Akkar groundwater, northern Lebanon. Environ Sci Pollut Res Int 2018; 25(15): 14302-12.
[http://dx.doi.org/10.1007/s11356-017-8568-6] [PMID: 28265872]

[89]   D'Argenio V, Salvatore F. The role of the gut microbiome in the healthy adult status. Clin Chim Acta 2015; 451(Pt A): 97-102.
[http://dx.doi.org/10.1016/j.cca.2015.01.003] [PMID: 25584460]

[90]   Vijay-Kumar M, Aitken JD, Carvalho FA, *et al.* Metabolic syndrome and altered gut microbiota in mice lacking Toll-like receptor 5. Science 2010; 328(5975): 228-31.
[http://dx.doi.org/10.1126/science.1179721] [PMID: 20203013]

[91]   Henao-Mejia J, Elinav E, Jin C, *et al.* Inflammasome-mediated dysbiosis regulates progression of NAFLD and obesity. Nature 2012; 482(7384): 179-85.
[http://dx.doi.org/10.1038/nature10809] [PMID: 22297845]

[92]   Russell SL, Finlay BB. The impact of gut microbes in allergic diseases. Curr Opin Gastroenterol 2012; 28(6): 563-9.
[http://dx.doi.org/10.1097/MOG.0b013e3283573017] [PMID: 23010680]

[93]   Gascon M, Morales E, Sunyer J, Vrijheid M. Effects of persistent organic pollutants on the developing respiratory and immune systems: A systematic review. Environ Int 2013; 52: 51-65.
[http://dx.doi.org/10.1016/j.envint.2012.11.005] [PMID: 23291098]

[94]   Menard S, Guzylack-Piriou L, Leveque M, *et al.* Food intolerance at adulthood after perinatal exposure to the endocrine disruptor bisphenol A. FASEB J 2014; 28(11): 4893-900.
[http://dx.doi.org/10.1096/fj.14-255380] [PMID: 25085925]

[95]   Bäckhed F, Fraser CM, Ringel Y, *et al.* Defining a healthy human gut microbiome: current concepts, future directions, and clinical applications. Cell Host Microbe 2012; 12(5): 611-22.
[http://dx.doi.org/10.1016/j.chom.2012.10.012] [PMID: 23159051]

[96]   Maurice CF, Haiser HJ, Turnbaugh PJ. Xenobiotics shape the physiology and gene expression of the active human gut microbiome. Cell 2013; 152(1-2): 39-50.
[http://dx.doi.org/10.1016/j.cell.2012.10.052] [PMID: 23332745]

[97]   Claus SP, Guillou H, Ellero-Simatos S. Erratum: The gut microbiota: A major player in the toxicity of environmental pollutants? NPJ Biofilms Microbiomes 2017; 3(1): 17001.
[http://dx.doi.org/10.1038/npjbiofilms.2017.1] [PMID: 28726854]

[98]   Steinrücken HC, Amrhein N. 5-Enolpyruvylshikimate-3-phosphate synthase of Klebsiella pneumoniae. 2. Inhibition by glyphosate [N-(phosphononmethyl)glycine]. Eur J Biochem 1984; 143(2): 351-7.
[http://dx.doi.org/10.1111/j.1432-1033.1984.tb08379.x] [PMID: 6381057]

[99]   Steinrücken HC, Amrhein N. The herbicide glyphosate is a potent inhibitor of 5-enolpyruvylshikimic acid-3-phosphate synthase. Biochem Biophys Res Commun 1980; 94(4): 1207-12.
[http://dx.doi.org/10.1016/0006-291X(80)90547-1] [PMID: 7396959]

[100] Liu Q, Shao W, Zhang C, *et al.* Organochloride pesticides modulated gut microbiota and influenced bile acid metabolism in mice. Environ Pollut 2017; 226: 268-76.
[http://dx.doi.org/10.1016/j.envpol.2017.03.068] [PMID: 28392238]

[101] Frost G, Sleeth ML, Sahuri-Arisoylu M, *et al.* The short-chain fatty acid acetate reduces appetite *via* a central homeostatic mechanism. Nat Commun 2014; 5(1): 3611.
[http://dx.doi.org/10.1038/ncomms4611] [PMID: 24781306]

[102] Adamovsky O, Buerger AN, Wormington AM, *et al.* The gut microbiome and aquatic toxicology: An emerging concept for environmental health. Environ Toxicol Chem 2018; 37(11): 2758-75.
[http://dx.doi.org/10.1002/etc.4249] [PMID: 30094867]

[103] de Faria Ghetti F, Oliveira DG, de Oliveira JM, de Castro Ferreira LEVV, Cesar DE, Moreira APB. Influence of gut microbiota on the development and progression of nonalcoholic steatohepatitis. Eur J Nutr 2018; 57(3): 861-76.
[http://dx.doi.org/10.1007/s00394-017-1524-x] [PMID: 28875318]

[104] Groh KJ, Geueke B, Muncke J. Food contact materials and gut health: Implications for toxicity assessment and relevance of high molecular weight migrants. Food Chem Toxicol 2017; 109(Pt 1): 1-18.
[http://dx.doi.org/10.1016/j.fct.2017.08.023] [PMID: 28830834]

[105] Gillois K, Lévêque M, Théodorou V, Robert H, Mercier-Bonin M. Mucus: An Underestimated Gut Target for Environmental Pollutants and Food Additives. Microorganisms 2018; 6(2): 53.
[http://dx.doi.org/10.3390/microorganisms6020053] [PMID: 29914144]

[106] Turner JR. Intestinal mucosal barrier function in health and disease. Nat Rev Immunol 2009; 9(11): 799-809.
[http://dx.doi.org/10.1038/nri2653] [PMID: 19855405]

[107] Johansson MEV, Phillipson M, Petersson J, Velcich A, Holm L, Hansson GC. The inner of the two Muc2 mucin-dependent mucus layers in colon is devoid of bacteria. Proc Natl Acad Sci USA 2008; 105(39): 15064-9.
[http://dx.doi.org/10.1073/pnas.0803124105] [PMID: 18806221]

[108] Guo Y, Qi Y, Yang X, *et al.* Association between Polycystic Ovary Syndrome and Gut Microbiota. PLoS One 2016; 11(4): e0153196.
[http://dx.doi.org/10.1371/journal.pone.0153196] [PMID: 27093642]

[109] Kelley ST, Skarra DV, Rivera AJ, Thackray VG. The Gut Microbiome Is Altered in a Letrozole-Induced Mouse Model of Polycystic Ovary Syndrome. PLoS One 2016; 11(1): e0146509.

[http://dx.doi.org/10.1371/journal.pone.0146509] [PMID: 26731268]

[110] Cummings JH, Macfarlane GT. The control and consequences of bacterial fermentation in the human colon. J Appl Bacteriol 1991; 70(6): 443-59.
[http://dx.doi.org/10.1111/j.1365-2672.1991.tb02739.x] [PMID: 1938669]

[111] Awad AB, Kamei A, Horvath PJ, Fink CS. Prostaglandin synthesis in human cancer cells: Influence of fatty acids and butyrate. Prostaglandins Leukot Essent Fatty Acids 1995; 53(2): 87-93.
[http://dx.doi.org/10.1016/0952-3278(95)90134-5] [PMID: 7480078]

[112] D'Argenio G, Mazzacca G. Short-chain fatty acid in the human colon. Relation to inflammatory bowel diseases and colon cancer. Adv Exp Med Biol 1999; 472: 149-58.
[http://dx.doi.org/10.1007/978-1-4757-3230-6_13] [PMID: 10736623]

[113] Shimotoyodome A, Meguro S, Hase T, Tokimitsu I, Sakata T. Short chain fatty acids but not lactate or succinate stimulate mucus release in the rat colon. Comp Biochem Physiol A Mol Integr Physiol 2000; 125(4): 525-31.
[http://dx.doi.org/10.1016/S1095-6433(00)00183-5] [PMID: 10840229]

[114] Hatayama H, Iwashita J, Kuwajima A, Abe T. The short chain fatty acid, butyrate, stimulates MUC2 mucin production in the human colon cancer cell line, LS174T. Biochem Biophys Res Commun 2007; 356(3): 599-603.
[http://dx.doi.org/10.1016/j.bbrc.2007.03.025] [PMID: 17374366]

[115] Barcelo A, Claustre J, Moro F, Chayvialle JA, Cuber JC, Plaisancié P. Mucin secretion is modulated by luminal factors in the isolated vascularly perfused rat colon. Gut 2000; 46(2): 218-24.
[http://dx.doi.org/10.1136/gut.46.2.218] [PMID: 10644316]

[116] Finnie IA, Dwarakanath AD, Taylor BA, Rhodes JM. Colonic mucin synthesis is increased by sodium butyrate. Gut 1995; 36(1): 93-9.
[http://dx.doi.org/10.1136/gut.36.1.93] [PMID: 7890244]

[117] Sakata T, Setoyama H. Local stimulatory effect of short-chain fatty acids on the mucus release from the hindgut mucosa of rats (Rattus norvegicus). Comp Biochem Physiol A Physiol 1995; 111: 429-32.

[118] Siavoshian S, Segain JP, Kornprobst M, *et al.* Butyrate and trichostatin A effects on the proliferation/differentiation of human intestinal epithelial cells: induction of cyclin D3 and p21 expression. Gut 2000; 46(4): 507-14.
[http://dx.doi.org/10.1136/gut.46.4.507] [PMID: 10716680]

[119] Nightingale KP, Gendreizig S, White DA, Bradbury C, Hollfelder F, Turner BM. Cross-talk between histone modifications in response to histone deacetylase inhibitors: MLL4 links histone H3 acetylation and histone H3K4 methylation. J Biol Chem 2007; 282(7): 4408-16.
[http://dx.doi.org/10.1074/jbc.M606773200] [PMID: 17166833]

[120] Barski A, Cuddapah S, Cui K, *et al.* High-resolution profiling of histone methylations in the human genome. Cell 2007; 129(4): 823-37.
[http://dx.doi.org/10.1016/j.cell.2007.05.009] [PMID: 17512414]

[121] Sircana A, Framarin L, Leone N, *et al.* Altered Gut Microbiota in Type 2 Diabetes: Just a Coincidence? Curr Diab Rep 2018; 18(10): 98.
[http://dx.doi.org/10.1007/s11892-018-1057-6] [PMID: 30215149]

[122] Stamler JS, Simon DI, Osborne JA, *et al.* S-nitrosylation of proteins with nitric oxide: synthesis and characterization of biologically active compounds. Proc Natl Acad Sci USA 1992; 89(1): 444-8.
[http://dx.doi.org/10.1073/pnas.89.1.444] [PMID: 1346070]

[123] Lawrence T. The nuclear factor NF-kappaB pathway in inflammation. Cold Spring Harb Perspect Biol 2009; 1(6): a001651.
[http://dx.doi.org/10.1101/cshperspect.a001651] [PMID: 20457564]

[124] Sugita H, Kaneki M, Tokunaga E, *et al.* Inducible nitric oxide synthase plays a role in LPS-induced hyperglycemia and insulin resistance. Am J Physiol Endocrinol Metab 2002; 282(2): E386-94.

[http://dx.doi.org/10.1152/ajpendo.00087.2001] [PMID: 11788371]

[125] Suzuki T. Regulation of intestinal epithelial permeability by tight junctions. Cell Mol Life Sci 2013; 70(4): 631-59.
[http://dx.doi.org/10.1007/s00018-012-1070-x] [PMID: 22782113]

[126] Takeuchi O, Akira S. Pattern recognition receptors and inflammation. Cell 2010; 140(6): 805-20.
[http://dx.doi.org/10.1016/j.cell.2010.01.022] [PMID: 20303872]

[127] Tan J, McKenzie C, Potamitis M, Thorburn AN, Mackay CR, Macia L. The role of short-chain fatty acids in health and disease. Adv Immunol 2014; 121: 91-119.
[http://dx.doi.org/10.1016/B978-0-12-800100-4.00003-9] [PMID: 24388214]

[128] Macut D, Bjekić-Macut J, Rahelić D, Doknić M. Insulin and the polycystic ovary syndrome. Diabetes Res Clin Pract 2017; 130: 163-70.
[http://dx.doi.org/10.1016/j.diabres.2017.06.011] [PMID: 28646699]

[129] Patel S. Polycystic ovary syndrome (PCOS), an inflammatory, systemic, lifestyle endocrinopathy. J Steroid Biochem Mol Biol 2018; 182: 27-36.
[http://dx.doi.org/10.1016/j.jsbmb.2018.04.008] [PMID: 29678491]

[130] Barthelmess EK, Naz RK. Polycystic ovary syndrome: current status and future perspective. Front Biosci (Elite Ed) 2014; 6(1): 104-19.
[PMID: 24389146]

[131] Barber TM, Dimitriadis GK, Andreou A, Franks S. Polycystic ovary syndrome: insight into pathogenesis and a common association with insulin resistance. Clin Med (Lond) 2016; 16(3): 262-6.
[http://dx.doi.org/10.7861/clinmedicine.16-3-262] [PMID: 27251917]

[132] Kim HH, DiVall SA, Deneau RM, Wolfe A. Insulin regulation of GnRH gene expression through MAP kinase signaling pathways. Mol Cell Endocrinol 2005; 242(1-2): 42-9.
[http://dx.doi.org/10.1016/j.mce.2005.07.002] [PMID: 16144737]

[133] Jacobson L, Sapolsky R. The role of the hippocampus in feedback regulation of the hypothalamic-pituitary-adrenocortical axis. Endocr Rev 1991; 12(2): 118-34.
[http://dx.doi.org/10.1210/edrv-12-2-118] [PMID: 2070776]

[134] Grunstein HS, James DE, Storlien LH, Smythe GA, Kraegen EW. Hyperinsulinemia suppresses glucose utilization in specific brain regions: *In vivo* studies using the euglycemic clamp in the rat. Endocrinology 1985; 116(2): 604-10.
[http://dx.doi.org/10.1210/endo-116-2-604] [PMID: 3967621]

[135] Unger JW, Lange W. Insulin receptors in the pituitary gland: morphological evidence for influence on opioid peptide-synthesizing cells. Cell Tissue Res 1997; 288(3): 471-83.
[http://dx.doi.org/10.1007/s004410050833] [PMID: 9134860]

[136] Alesci S, Koch CA, Bornstein SR, Pacak K. Adrenal androgens regulation and adrenopause. Endocr Regul 2001; 35(2): 95-100.
[PMID: 11563938]

[137] Wallace IR, McKinley MC, Bell PM, Hunter SJ. Sex hormone binding globulin and insulin resistance. Clin Endocrinol (Oxf) 2013; 78(3): 321-9.
[http://dx.doi.org/10.1111/cen.12086] [PMID: 23121642]

[138] Mounier C, Dumas V, Posner BI. Regulation of hepatic insulin-like growth factor-binding protein-1 gene expression by insulin: central role for mammalian target of rapamycin independent of forkhead box O proteins. Endocrinology 2006; 147(5): 2383-91.
[http://dx.doi.org/10.1210/en.2005-0902] [PMID: 16455781]

[139] Guarner F, Khan AG, Garisch J, *et al.* World Gastroenterology Organisation Global Guidelines. J Clin Gastroenterol 2012; 46(6): 468-81.
[http://dx.doi.org/10.1097/MCG.0b013e3182549092] [PMID: 22688142]

[140]   Altun HK, Yıldız EA. Prebiyotikler ve Probiyotiklerin Diyabet ile_Ilis,kisi. Turk J Life Sci 2017; 1: 149-56.

[141]   Karamali M, Eghbalpour S, Rajabi S, *et al.* Effects of Probiotic Supplementation on Hormonal Profiles, Biomarkers of Inflammation and Oxidative Stress in Women With Polycystic Ovary Syndrome: A Randomized, Double-Blind, Placebo-Controlled Trial. Arch Iran Med 2018; 21(1): 1-7. [PMID: 29664663]

[142]   Askari G, Shoaei T, Tehrani H, Heidari-Beni M, feizi A, Esmaillzadeh A. Effects of probiotic supplementation on pancreatic β-cell function and c-reactive protein in women with polycystic ovary syndrome: A randomized double-blind placebo-controlled clinical trial. Int J Prev Med 2015; 6(1): 27. [http://dx.doi.org/10.4103/2008-7802.153866] [PMID: 25949777]

[143]   Repaci A, Gambineri A, Pasquali R. The role of low-grade inflammation in the polycystic ovary syndrome. Mol Cell Endocrinol 2011; 335(1): 30-41. [http://dx.doi.org/10.1016/j.mce.2010.08.002] [PMID: 20708064]

[144]   Deepika MLN, Nalini S, Maruthi G, *et al.* Analysis of oxidative stress status through MN test and serum MDA levels in PCOS women. Pak J Biol Sci 2014; 17(4): 574-7. [http://dx.doi.org/10.3923/pjbs.2014.574.577] [PMID: 25911850]

[145]   Ostadmohammadi V, Jamilian M, Bahmani F, Asemi Z. Vitamin D and probiotic co-supplementation affects mental health, hormonal, inflammatory and oxidative stress parameters in women with polycystic ovary syndrome. J Ovarian Res 2019; 12(1): 5. [http://dx.doi.org/10.1186/s13048-019-0480-x] [PMID: 30665436]

[146]   Heshmati J, Farsi F, Yosaee S, *et al.* The Effects of Probiotics or Synbiotics Supplementation in Women with Polycystic Ovarian Syndrome: A Systematic Review and Meta-Analysis of Randomized Clinical Trials. Probiotics Antimicrob Proteins 2019; 11(4): 1236-47. [http://dx.doi.org/10.1007/s12602-018-9493-9] [PMID: 30547393]

[147]   Ahmadi S, Jamilian M, Karamali M, *et al.* Probiotic supplementation and the effects on weight loss, glycaemia and lipid profiles in women with polycystic ovary syndrome: A randomized, double-blind, placebo-controlled trial. Hum Fertil (Camb) 2017; 20(4): 254-61. [http://dx.doi.org/10.1080/14647273.2017.1283446] [PMID: 28142296]

[148]   Rashad NM, El-Shal AS, Amin AI, Soliman MH. Effects of probiotics supplementation on macrophage migration inhibitory factor and clinical laboratory feature of polycystic ovary syndrome. J Funct Foods 2017; 36: 317-24. [http://dx.doi.org/10.1016/j.jff.2017.06.029]

[149]   Ghanei N, Rezaei N, Amiri GA, Zayeri F, Makki G, Nasseri E. The probiotic supplementation reduced inflammation in polycystic ovary syndrome: A randomized, double-blind, placebo-controlled trial. J Funct Foods 2018; 42: 306-11. [http://dx.doi.org/10.1016/j.jff.2017.12.047]

[150]   Zhang F, Ma T, Cui P, *et al.* Diversity of the Gut Microbiota in Dihydrotestosterone-Induced PCOS Rats and the Pharmacologic Effects of Diane-35, Probiotics, and Berberine. Front Microbiol 2019; 10: 175. [http://dx.doi.org/10.3389/fmicb.2019.00175] [PMID: 30800111]

[151]   Zhang J, Sun Z, Jiang S, *et al.* Probiotic *Bifidobacterium lactis* V9 Regulates the Secretion of Sex Hormones in Polycystic Ovary Syndrome Patients through the Gut-Brain Axis. mSystems 2019; 4(2): e00017-19. [http://dx.doi.org/10.1128/mSystems.00017-19] [PMID: 31020040]

CHAPTER 2

Antibody Therapy as Alternative to Antibiotics

Manoj Lama[1,*]

[1] *Molecular Immunology Laboratory, Department of Zoology, University of Gour Banga, Malda - 732103, India*

Abstract: In the 1890s, Behring and Kitasato established the principle of serum therapy, which proved useful in treating infectious diseases. However, by the 1940s, serum therapy was abandoned mainly due to complications associated with the toxicity of heterologous sera and the introduction of more effective antibiotics. Although the availability of antibiotics had a tremendous impact on saving lives from infectious diseases, there was a rapid emergence of antibiotic resistance. As a result, an alternative therapy is being given due consideration. With the advent of antibody production technology, antibody therapy has gained interest as a promising treatment for emerging infectious diseases. Some monoclonal antibodies (mAbs) had already been approved for the treatment of certain infectious diseases. Many mAb candidates are currently in different phases of clinical testing for a variety of infectious pathogens. There is hope that antibody therapy may appear as a promising treatment option against infectious diseases in the near future.

Keywords: Antibacterials, Antibiotics, Antibiotic resistance, Antibody, Antibody therapy, Antifungals, B cells, Chimeric antibodies, Clinical trials, Complementarity-determining regions, Efficacy, Fragment crystallizable, Fragment for antigen binding, Humanized antibodies, Hybridoma technology, Infectious diseases, Monoclonal antibodies, Serum therapy, Toxins.

INTRODUCTION

The protection from the bacterial toxins mediated by the specific antibodies was first demonstrated by Behring and Kitasato in the 1980s [1]. This finding led to design the antibody-based therapies for different infectious diseases [2, 3]. The therapy was called 'serum therapy' as the antibodies were isolated from the serum of immunized animals or immune human donors. Although this form of therapy was proved effective against many infectious diseases, the use of animal serum had significant side effects, including immediate hypersensitive reactions and serum sickness, resulting in the accumulation of antigenantibody complex [1].

* **Corresponding author Manoj Lama:** Molecular Immunology Laboratory, Department of Zoology, University of Gour Banga, Malda - 732103, West Bengal, India. E-mail: manoj0071061@rediffmail.com

Tilak Saha, Manab Deb Adhikari and Bipransh Kumar Tiwary (Eds.)

Further, serum therapy was associated with toxicity, difficulty in administration, narrow specificity, lot-to-lot variation and expense. All these complications related to early serum therapy and the rise of the antibiotic age by the 1940s led to the abandonment of passive antibody therapy against bacterial infections.

By the mid-1900s, the impact of antibiotics to combat infectious diseases was remarkably astounding. After the commercialization of antibiotics, the mortality rate of infectious diseases declined dramatically below 1% in England [4]. The impact of the antibiotics was regarded as a 'medical miracle'. The role of antibiotics is not only noteworthy in saving lives from infectious diseases but also significant in the advancement of medicine and surgery [5]. However, the rapid emergence of antibiotic resistance and slow progress in the development of novel drugs by pharmaceutical industries challenge the extraordinary health benefits of antibiotics. This altogether resulted in a global crisis [6].

During the last few decades, antibodies as therapeutics have gained priority for treating various infectious diseases because of several reasons, viz. revolutionary advancement of antibody production technology, appearance of new pathogens, existence of drug-resistant microbes and increasing numbers of immune suppressed patients [7]. Antibodies are extremely versatile antimicrobial molecules produced by the B cells in response to infection or immunization. Several effector mechanisms, such as inhibition of attachment of microbes to the host cells, promotion of opsonization for phagocytosis, neutralization of viruses and toxins, activation of the complement system and antibody-dependent cellular cytotoxicity, are mediated by the antibodies [8].

The advent of hybridoma technology in 1975 for the production of monoclonal antibodies, followed by further development of antibody engineering, leading to the creation of fully humanized antibodies, offers antibody therapy as an alternative to antibiotics for the treatment of various infections. This chapter attempts to discuss serum therapy used in pre-antibiotic era, antibiotics and antibiotic resistance, overview of antibody, advancement of monoclonal antibody (mAb) production technology, and development of antibody-based therapies against infectious diseases and limitations of antibody-based therapies.

Antibody Therapy in Pre-Antibiotic Era

Serum Therapy

In the 1890s, Emil von Behring and Shibasaburo Kitasato demonstrated that serum from rabbits immunized with tetanus toxin could prevent tetanus in rabbits, and the same phenomenon was also shown for diphtheria toxin [9]. Emil von Behring was awarded the first Nobel Prize in Physiology or Medicine in 1901 for his discovery of serum therapy for diphtheria [10]. In fact, the immune sera contained specific antibodies which mediated therapeutic effects by promoting opsonization, neutralizing toxins, and/or triggering complement-mediated bacterial lysis [2]. In the pre-antibiotic era, many infectious diseases, including diphtheria, tetanus, scarlet fever, pneumococcal pneumonia and meningitis caused by *Neisseria meningitis* and *Haemophilus influezae* were treated by administration of immune animal sera [2]. Several controlled trials demonstrated an approximately 50% reduction in death rate in patients with pneumococcal pneumonia after the administration of type-specific serum [11]. Although the efficacy of serum therapy was uncertain in whooping cough, anthrax, dysentery (*Shigella dysenteriae*) and gas gangrene, human convalescent serum was effective against measles, which had a mortality rate of 6-7% in some populations [3].

Shortfall of Serum Therapy

Although serum therapy was a choice of treatment for various infectious diseases in the early 20th century, this treatment had serious side effects as an immune response was induced against the animal-derived antibodies. The most severe was serum sickness, a type of hypersensitivity response characterized by fever, chills, rashes, arthritis, and occasionally glomerulonephritis. Also because of the high specificity of the antibodies, separate immune sera had to be raised for different pathogens [2]. Another disadvantage of hyperimmune sera was being a polyclonal antibody preparation with undefined concentrations of multiple specific and non-specific antibodies. Therefore, it was challenging to standardize serum quality and ensure the efficacy of the therapeutic serum. Certainly, all these complications together with the discovery of antibiotics, led to the abandonment of serum therapy by the early 1940s [12].

Antibiotics and Antibiotic Resistance Crisis

Selman Waksman introduced the term 'antibiotic', referring to any small molecule, produced by a microbe, with antagonistic properties on the growth of other microbes [13]. Generally, antibiotics refer to antibacterial, however, these are differentiated as antibacterial, antifungal and antiviral for the action they exert on the group of microorganisms [14]. The modern era of antibiotics started with the discovery of penicillin by Sir Alexander Fleming in 1928 from a soil-inhabiting fungus *Penicillium notatum* [15, 16]. The discovery of penicillin was reported in 1929 [17], and its first clinical trial was conducted on humans in 1940 [14]. During the 1940–1960s, most antibiotic classes used today were discovered, and this period is known as the antibiotic golden age. The rate at which antibiotics were discovered during this period, it was believed that infectious diseases would soon be a controlled public health issue [18, 19]. Although the discovery of antibiotics transformed modern medicine and saved millions of lives [5], penicillin resistance soon became a substantial clinical problem, and by the 1950s, the advances of the prior decade were threatened [20]. However, confidence was gained by developing and discovering a new class of antibiotics [19, 20]. Unfortunately, the case of methicillin-resistant *Staphylococcus aureus* (MRSA) appeared for the first time in 1962 in the United Kingdom and in 1968 in the United States [19, 21]. Nevertheless, resistance to virtually all classes of antibiotics was developed [21].

Infections with antibiotic-resistant strains are widespread worldwide [22]. According to a survey conducted in 2011, above 60 percent of the participants were reported to have infections with pan-resistant untreatable bacterial strains in the previous year [20]. The occurrence of antibiotic-resistant bacterial strains has been described as a 'crisis', which could have catastrophic consequences [23]. Subsequently, there was a declaration from the Centers for Disease Control and Prevention (CDC) in 2013 that there is the end of the antibiotic era, and we are living in the post-antibiotic era. Further, World Health Organization (WHO) emphasized on terrible consequences of the widespread occurrence of antibiotic resistance [24]. Several factors, such as inappropriate usage of antibiotics, lack of novel drug discovery by the pharmaceutical companies because of the low economic incentives and challenging regulatory requirements, *etc.*, are responsible for the rapid development of antibiotic-resistant pathogens [6, 16, 25 - 28].

Overview of Antibody

Structurally, the antibody is composed of identical copies of two heavy and two light chains (Fig. **1**). Each chain is subdivided into domains, which are small units consisting of around 110 amino acids [29]. The N terminal domain amino acids are highly variable in both heavy and light chains and therefore known as variable regions. The other domains of the antibody chains are comparatively less variable and are known as constant regions. Constant regions are present in C terminal domains of both heavy and light chains. In of light chains, there is only one constant domain, while heavy chains contain three or four constant region domains (C_{H1}, C_{H2}, and C_{H3}) [30]. The molecular weight of each heavy chain (H) is 50 kDa, and each light chain (L) is 25 kDa. Comparative sequence analysis of different antibodies showed that a highly variable region is present within the variable region domain, known as hypervariable region (HVR) or complementarity-determining region (CDR). Each variable region consists of three CDR regions (CDR 1-3) comprised of 9-12 amino acids [31]. A complete antibody molecule is comprised of two distinct regions, Fab (fragment for antigen binding) and Fc (fragment crystallizable) regions. Fab works to identify and bind to an antigen, while Fc region is responsible for various effector functions. It is observed that enzymatic digestion separates two regions; when Fc region is cleaved from its counterpart Fab region, it instantly becomes crystallized, and for this reason, it is termed as fragment crystallizable [32]. The Fab and Fc region are joined together by another region known as the hinge region. This region contains two disulfide bonds which hold the entire structure together and provide flexibility to the Fab region for antigen binding [33]. Hinge region can be sub-classified into three distinct regions *viz.* core hinge, upper hinge and lower hinge regions. Movement and rotation of the Fab region on N terminal region are controlled by the upper hinge region, whereas a variable number of cysteine residues are present in the core hinge region. The lower hinge region allows the movement of the Fc region along with FcγR binding activity [34]. Hinge region also has protease activity (papain, pepsin) which allows the separation of Fab and Fc regions [35, 36].

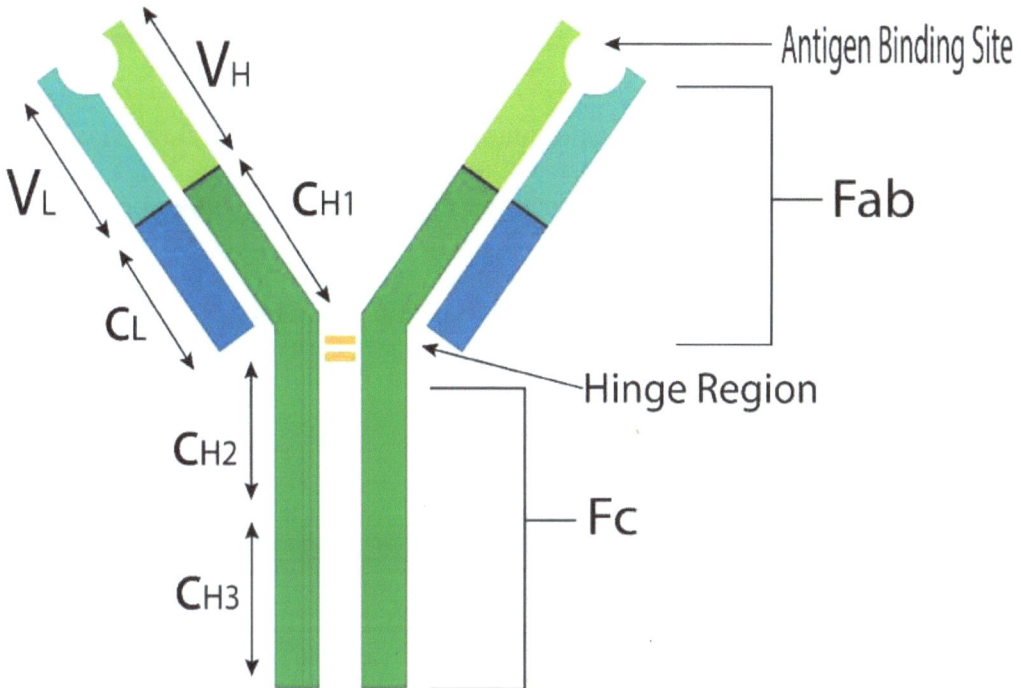

Fig. (1). Basic structure of an antibody. V_L – light chain variable domain; C_L – light chain constant domain; V_H – heavy chain variable domain; $C_H1/C_H2/C_H3$ – heavy chain constant domains 1, 2 and 3; Fab- antigen binding fragment; Fc – fragment crystallizable.

Advancement of Monoclonal Antibody (mAb) Production Technology

With the advent of mouse hybridoma technology by Kohler and Milstein in 1975, there was excitement for the production of monoclonal antibodies for therapy [37]. The monovalent antibodies, produced from a single B-lymphocyte clone, that bind to the particular epitope are called monoclonal antibodies [38]. Two important properties of antibodies viz. low toxicity as antibodies are endogenous proteins native to the body and their high specificity make them highly versatile therapeutic agents [12]. Monoclonal antibody therapy has become a versatile therapeutic alternative in the treatment of various diseases, including transplantation, oncology, autoimmune, cardiovascular and infectious diseases [39].

Unfortunately, the hope turned into disappointment when it became evident that the monoclonal antibody molecules as therapeutics encountered some serious problems. When the early murine mAbs were injected into patients, these were recognized as foreign molecules and were reacted upon by the host's immune mechanism. In addition, the biological efficacy of murine mAbs was severely restricted because of their failure to interact properly with components of the human immune system [40].

The production of recombinant antibodies by means of subcloning, random or directed mutagenesis and molecular evolution procedures ushered in the age of antibody engineering [41]. It was feasible to create chimeric antibodies with the application of antibody engineering. The chimeric antibodies were created by fusing murine variable domains with human constant domains [42]. This led to the development of a new generation of therapeutic candidates [43]. The chimeric antibodies possess 70% human segment containing fully human Fc portion, which makes them considerably less immunogenic in humans and allows them to interact with human effector cells and complement cascade [40].

Advances in antibody engineering techniques made it possible to further reduce the murine part of mAbs by replacing only the hypervariable region of a fully human antibody with the hypervariable region of a murine antibody using the complementarity determining region grafting approach [44]. The antibodies produced by this approach are called 'humanized' antibodies with 85-90% human part and are even less immunogenic than chimeric counterparts. The approved mAbs antibodies currently used are either humanized or chimeric [40]. Phage display is another method for the production of monoclonal antibodies, initially described by Smith in 1985 and further developed by other groups. It is based on the genetic engineering of bacteriophages (viruses that infect bacteria) and repeated rounds of antigen-guided selection and phage propagation [45]. The technique of antibody phage display (APD) allows *in vitro* selection of mAbs of virtually any specificity, enabling the production of recombinant reagents for research, clinical diagnostics and therapeutic use in humans. The first fully human APD-derived mAb is adalimumab [46].

Development of Antibody-Based Therapies Against Infectious Diseases

Monoclonal antibodies have the potential to address a wide range of infectious diseases. Certain mAbs designed currently against viral, bacterial and fungal pathogens are summarized in Table **1**.

Table 1. mAb candidates designed against various targets of infectious agents.

mAb Candidates	Target	Pathogen	References
PRO140	CCR5	HIV	[47]
Cenicriviroc	CCR2 & CCR5	HIV	[48]
CSJ148 (Combination of two mAbs, LJP538 & LJP539)	LJP538 – viral gB protein LJP539 – viral gH pentameric complex	HCMV	[53]
G7667 (Combination of two mAbs, MCMV5322A & MCMV3068A)	MCMV5322A - epitope on HCMVgH/GL MCMV3068A - pentameric gH complex	HCMV	[54, 55]
Palivizumab, Motavizumab & Motavizumab-YTE	F glycoprotein	RSV	[56, 57, 60]
MHAA4549A (also known as 39.29)	Conserved epitope on the stalk of influenza A HA	Influenza virus	[62]
VIS410	Group 1 & Group 2 HAs of influenza A viruses	Influenza virus	[65]
E6F6	HBsAg	HBV	[68]
SIgN-3C	Intact viral particles	Dengue virus	[71]
Visterra 513	E protein domain III (EDIII)	Dengue virus	[72, 73]
Z23 & Z3L1	Tertiary epitopes in envelope Protein domain I, II, or III	Zika virus	[76]
Bezlotoxumab	TcdB	*C. difficile*	[79]
MEDI4893	Alpha hemolysin	*S. aureus*	[80]
Human IgG1	Shiga toxin subunit	*S. aureus*	[81]
MEDI3902 (A bispecific Ab)	Fimbrial protein PcrV & Exopolysaccharide Ps1	*Pseudomomas aeruginosa*	[86]
18B7	Polysaccharide capsule	*Cryptococcus*	[95]
4F11	Kexin-like protein KEX1	*Pneumocystis*	[96]

(Table 1) cont.....

mAb Candidates	Target	Pathogen	References
E3	Glycoprotein 43 (gp43)	*Paracoccidioides brasiliensis*	[97]
Anti-gp70 mAb	Glycoprotein 70 (gp70)	*Paracoccidioides brasiliensis*	[98]

List of mAbs is not exhaustive.

Anti-viral mAb Candidates

Human Immunodeficiency Virus

Anti-HIV mAbs have been considered as a viable option for prophylactic and therapeutic treatment of HIV infection and AIDS. Currently, two mAbs PRO140 and Cenicriviroc, are in Phase III clinical trials. PRO140 mAb designed for targeting CCR5 (cysteine-cysteine chemokine receptor type 5) has entered Phase IIb/III trials [47]. The other mAb Cenicriviroc targets CCR2 and CCR5, and its potential for a number of indications, including HIV infection, has been investigated [48]. Recently, approval was given for mAb ibalizumab as a second-line treatment for HIV [49].

Human Cytomegalovirus

Human cytomegalovirus (HCMV) belongs to the herpes family. A lifelong latent infection in the host can be established by this virus with occasional reactivation [50]. HCMV is a highly complex virus having multiple antigens, such as glycoprotein B (gB) and the gH pentameric complex [50, 51]. Most of the antibodies that are generated against HCMV target the gB antigen. However, it has been shown that such antibodies alone do not have a strong neutralizing ability to control HCMV infection and reactivation [52]. Rather, the antibodies targeting the pentameric gH complex have been shown to be effective in neutralizing antiviral antibodies [51]. Therefore, mostly mAbs targeting the gH pentameric complex are being developed [52]. Currently, two antibodies targeting HCMV that are the combination of two mAbs are under clinical trials [53, 54]. CSJ148 is developed by the combination of two anti-HCMV human mAbs, LJP538 and LJP539. The targets bound by LJP538 and LJP539 mAbs are gB protein and gH pentameric complex of the HCMV [53]. Similarly, MCMV5322A and MCMV3068A have been combined to develop G7667. The target recognized by MCMV5322A is an epitope on HCMV gH/gL [55], while the pentameric gH complex is recognized by MCMV3068A [54].

Respiratory Syncytial Virus

In 1998, the Food and Drug Administration (FDA) approved Palivizumab against severe disease caused by RSV in children. F glycoprotein of the virus is targeted by Palivizumab [56, 57]. Motavizumab is another mAb, derived from the affinity matured palivizumab, which is tenfold more effective in F glycoprotein binding as compared to palivizumab [56, 58]. The results of the clinical studies demonstrated a significant decrease in RSV-associated hospitalization among motavizumab recipients compared to palivizumab recipients [59]. Further, another derivative, motaavizumab-YTE, has shown better results than motavizumab in terms of lower clearance and longer half-life [60].

Influenza Virus

Infections caused by the influenza virus are widespread, with generally mild illness. Antibody therapy has become a feasible option to tackle the menace of the influenza pandemic [61]. The plasmablast cell obtained from the influenza-vaccinated donor was cloned to develop MHAA4549A, which is also designated as 39.29. The target recognized by MHAA4549A is an epitope located on the stalk of influenza A HA. All the strains of human influenza A virus are effectively neutralized by this mAb [62]. Currently, MHAA4549A has been evaluated in clinical studies for hospitalized patients with severe influenza A infection [63, 64]. Further, VIS410 mAb developed by engineering human IgG1 mAb, targets the stem (or stalk) region of influenza A HA. This mAb has been shown to bind both group 1 and group 2 HAs of influenza A viruses [65]. The assessment of VIS410 with respect to its safety and tolerability in subjects with influenza infection has been made recently [66]. CR6261 is another mAb that neutralizes the influenza virus. This mAb was derived from the display libraries of human IgM(+) memory B cells derived from seasonal influenza vaccinees [67].

Hepatitis B virus

Serious liver ailment is caused by the infection of the hepatitis B virus [61]. In HBV infection, an elevated level of circulating HBsAg is thought to mask the immune response. E6F6 mAb developed to reduce the levels of circulating HBsAg in HBV-infected patients is currently under evaluation [68]. Furthermore, designing therapeutic mAbs targeting viral S protein has been spearheaded in recent times [69].

Dengue Virus

Dengue, a vector-borne disease, has a worldwide prevalence. Every year, millions of people are afflicted by dengue virus infection [70]. Antibody therapy is emerging as a treatment option for dengue. Several murine and human mAb candidates against dengue have been designed and evaluated. SIgN-3C mAb, when administered in mice after two days of infection, showed neutralizing effects on all serotypes and decreased viral load [71]. A panserotype anti-DENV humanized mAb, Visterra 513 (VIS513), which targets E protein domain III (EDIII), demonstrated a promising antiviral effect by neutralizing all serotypes of dengue virus [72, 73].

Ebola Virus

Efforts have been made to isolate potent neutralizing antibodies against the Ebola virus. The neutralizing antibodies targeting the Ebola virus surface glycoprotein (EBOV GP) were isolated using various methods [74]. The neutralizing antibodies were also isolated from the B cells of the convalescent individuals who were infected during the Ebola virus Zaire outbreak in 2014 [75]. The antibodies could be promising in the treatment of Ebola virus infection.

Zika Virus

Antibody-based therapy has also been proposed for Zika virus infection. Two neutralizing ZIKV-specific antibodies obtained from an infected individual were demonstrated to confer protection in mice exposed to the virus [76]. The rational antibody targets for ZIKV-specific therapy have been shown by the binding of Z23 and Z3L1 antibodies to epitopes in the envelope protein domain I, II, or III of Zika virus [76].

Anti-bacterial mAb Candidates

The targets exploited in devising antibody-based therapies for bacterial infections include the bacterial toxins, capsular polysaccharides and surface proteins [77]. However, only the antibodies targeting bacterial toxins particularly anthrax and *Clostridium* cytotoxin have been marketed so far [78].

Bezlotoxumab targets *Clostridium difficile* TcdB which is currently being approved for use in preventing the recurrence of *C. difficile* infection, although this was ineffective to cure active infection [79]. The results of the clinical study on MEDI4893 showed its potential in reaching levels in blood and nares sufficient for neutralizing *S. aureus* alpha-hemolysin to prevent invasion [80]. Human IgG1 mAb was generated in transgenic mice against Shiga toxin subunit. Prevention of the fatal systemic complications in piglets was observed following the administration of this mAb after the onset of diarrhea [81]. Other mAbs developed against *Pseudomonas aeruginosa* exotoxin A, *Clostridium perfringens* epsilon toxin and *Clostridium botulinum* neurotoxin have been demonstrated to have therapeutic potentials [82 - 84]. The promising result obtained from the clinical studies of mAb AR301 provides hope for successful antitoxin therapies [71].

Many bacterial surface proteins are mostly conserved across clinical strains and are considered as easier targets for designing therapeutic antibodies [85]. A bispecific antibody MEDI3902 designed for targeting *Pseudomonas aeruginosa* fimbrial protein PcrV and exopolysaccharide Ps1 was a successful endeavour [86]. This antibody showed improved survival and lung oxygenation with decreased bacterial burden and disease in rabbits with acute *P. aeruginosa* pneumonia [87]. Another antibody 514G3 was identified to have effective *in vivo* prophylaxis against MRSA bacteremia and acted in synergy with vancomycin to reduce lethality [88].

Bacterial polysaccharide such as lipopolysaccharide (LPS) and capsular polysaccharide (CPS) have been explored as potential target antigens for the development of antibody therapeutics. Antibodies generated against CPS have been shown to promote opsonophagocytosis of normally 'slippery' bacteria [89, 90]. Further, such antibodies have also been shown to directly affect bacterial metabolism [91]. However, designing mAbs for bacterial polysaccharide targets seems challenging as the polysaccharides are extremely variable [92, 93]. In addition, as evident from the response of *S. pneumonia* strains to vaccination, designing successful mAb therapy against a polysaccharide antigen may eventually lead to antigenic shift in the pathogens [91].

Anti-fungal mAb Candidates

Treatment for fungal infections has become a global challenge. The antibody therapies designed against the fungal pathogens include *Cryptococcus*, *Pneumocystis*, *Paracoccidioides* and *Candida* [94]. Clinical study of 18B7mAb developed against the polysaccharide capsule of *Cryptococcus* showed significantly decreased circulating cryptococcal antigen [95]. Unfortunately, further development of this mAb candidate was held back. Similarly, mAbs were also designed against *Pneumocystis* epitopes. The preclinical study of 4F11 mAb, which binds to *Pneumocystis* kexin-like protein KEX1, showed its potential to prevent *Pneumocystis* pneumonia in the susceptible mice kept together with the infected mice. The result shows the therapeutic potential of 4F11 mAb to deal with the *Pneumocystis* infection [96].

The major targets used for designing mAb therapeutics against *Paracoccidioides brasiliensis* include glycoprotein 43 (gp43) and gp70. Evaluation of anti-gp43 mAb E3 on *P. brasiliensis* infected mouse model demonstrated enhanced phagocytosis, increased IFN-γ production and reduction in fungal burden [97]. On the other hand, anti-gp70 mAb showed significantly reduced fungal colony-forming units and complete suppression of granuloma formation in the lungs [98]. The high morbidity and mortality in HIV and TB patients have often been associated with *Candida albicans*, an opportunistic fungal pathogen [99]. In a recent study, attempt has been made to clone antibody genes of cultured B cells derived from *C. albicans* infected patients. This study demonstrated that the antibodies generated had conferred protection in the mouse model of disseminated candidiasis by enhancing the opsonophagocytic activity of macrophages [100].

Limitations of Antibody-based Therapies

The limitations of antibodies as therapeutics are mainly related to cost, route of administration, selective specificity towards particular pathogen or serotype, storage, risk of infection, *etc.* The major limitation of passive antibody therapies is that it is very expensive. For instance, the cost of antibody prophylaxis for CMV infections comes around $4,000 to $9,000 per patient [101]. However, extensive use of antibodies as therapeutics would definitely lower down the cost.

Systemic administration is another disadvantage of antibody-based therapies. Antibody therapy administered orally seems unlikely to be effective except for enteric pathogens [102, 103]. Furthermore, the blood-brain barrier is a potential obstacle to antibody therapy for infections in the brain. Thus, alternative administration logistics or modifications of the antibody molecule are essential for facilitating blood-brain penetration [104].

Generally, the pathogenic microorganisms are antigenically variable, while the antibodies are highly specific to a particular pathogen/serotype. The highly specific nature of antibodies makes antibody therapy disadvantageous, particularly in dealing with mixed infections. For example, the failure of type-specific serum therapy was seen in mixed infection with multiple serotypes by *S. pneumoniae* [105]. However, the use of antibody cocktails (mixture of antibodies/mAbs of different isotypes directed against common antigenic types) helps deal successfully with mixed infections [11].

Furthermore, storage at the appropriate condition for maintaining the activity of immunoglobulins is another limitation of antibody-based therapies [7]. Although antibodies are considered to be protective effector molecules, however, all antibody responses to pathogens are not beneficial to the host. Sometimes antibody responses may exert detrimental effects on the host, for instance, some viral-specific antibodies have been shown to enhance the infection in the host cells [106].

CONCLUDING REMARKS

The rapid emergence of antibiotic-resistant pathogens coupled with the paucity of novel drug discovery invited the search for alternative therapy for the treatment of infectious diseases in particular. Technological advancement has enabled researchers to design fully human antibodies targeting specific antigens using *in vitro* and *in vivo* screening techniques. There has been rapid progress in designing mAbs against infectious agents with an ever-expanding knowledge of the expression and the roles of target antigens. However, identification of the suitable antigenic target, understanding roles of Fc receptor, isotype and other regions of the antibody in mediating protection, and further improvement in correlation between preclinical and clinical study results need to be addressed for developing antibody therapies as promising future therapeutics for infectious diseases.

CONSENT FOR PUBLICATION

Not applicable.

CONFLICT OF INTEREST

The author declares no conflict of interest, financial or otherwise.

ACKNOWLEDGEMENT

Declared none.

REFERENCES

[1]   Casadevall A, Dadachova E, Pirofski L. Passive antibody therapy for infectious diseases. Nat Rev Microbiol 2004; 2(9): 695-703.
        [http://dx.doi.org/10.1038/nrmicro974] [PMID: 15372080]

[2]   Casadevall A, Scharff MD. Serum therapy revisited: animal models of infection and development of passive antibody therapy. Antimicrob Agents Chemother 1994; 38(8): 1695-702.
        [http://dx.doi.org/10.1128/AAC.38.8.1695] [PMID: 7985997]

[3]   Casadevall A, Scharff MD. Return to the past: the case for antibody-based therapies in infectious diseases. Clin Infect Dis 1995; 21(1): 150-61.
        [http://dx.doi.org/10.1093/clinids/21.1.150] [PMID: 7578724]

[4]   Smith PW, Watkins K, Hewlett A. Infection control through the ages. Am J Infect Control 2012; 40(1): 35-42.
        [http://dx.doi.org/10.1016/j.ajic.2011.02.019] [PMID: 21783278]

[5]   Gould IM, Bal AM. New antibiotic agents in the pipeline and how they can help overcome microbial resistance. Virulence 2013; 4(2): 185-91.
        [http://dx.doi.org/10.4161/viru.22507] [PMID: 23302792]

[6]   Bartlett JG, Gilbert DN, Spellberg B. Seven ways to preserve the miracle of antibiotics. Clin Infect Dis 2013; 56(10): 1445-50.
        [http://dx.doi.org/10.1093/cid/cit070] [PMID: 23403172]

[7]   Kipriyanov SM. Recombinant antibodies in infectious disease 2004.
        [http://dx.doi.org/10.1517/13543776.14.2.135]

[8]   Heinzel FP. Antibodies. In: Mandell GL, Bennett JE, Dolin R, Eds. Principles and Practice of Infectious Diseases 1995; 36-57.

[9]   Winau F, Winau R. Emil von Behring and serum therapy. Microbes Infect 2002; 4(2): 185-8.
        [http://dx.doi.org/10.1016/S1286-4579(01)01526-X] [PMID: 11880051]

[10]   Graham BS, Ambrosino DM. History of passive antibody administration for prevention and treatment of infectious diseases. Curr Opin HIV AIDS 2015; 10(3): 129-34.
        [http://dx.doi.org/10.1097/COH.0000000000000154] [PMID: 25760933]

[11]   Casadevall A. Antibody-based therapies for emerging infectious diseases. Emerg Infect Dis 1996; 2(3): 200-8.
        [http://dx.doi.org/10.3201/eid0203.960306] [PMID: 8903230]

[12]   Chan CEZ, Chan AHY, Hanson BJ, Ooi EE. The use of antibodies in the treatment of infectious diseases. Singapore Med J 2009; 50(7): 663-72.
        [PMID: 19644620]

[13]   Clardy J, Fischbach MA, Currie CR. The natural history of antibiotics. Curr Biol 2009; 19(11): R437-41.
        [http://dx.doi.org/10.1016/j.cub.2009.04.001] [PMID: 19515346]

[14]   Russell AD. Types of antibiotics and synthetic antimicrobial agents. In: Denyer SP, Hodges NA, German SP, Eds. Hugo and Russell's Pharmaceutical Microbiology 7th ed. 2004; 152-86.
        [http://dx.doi.org/10.1002/9780470988329.ch10]

[15]   Sengupta S, Chattopadhyay MK, Grossart HP. The multifaceted roles of antibiotics and antibiotic resistance in nature. Front Microbiol 2013; 4: 47.
        [http://dx.doi.org/10.3389/fmicb.2013.00047] [PMID: 23487476]

[16]   Piddock LJV. The crisis of no new antibiotics—what is the way forward? Lancet Infect Dis 2012; 12(3): 249-53.
        [http://dx.doi.org/10.1016/S1473-3099(11)70316-4] [PMID: 22101066]

[17]   Aminov RI. A brief history of the antibiotic era: lessons learned and challenges for the future. Front Microbiol 2010; 1(134): 134.
        [http://dx.doi.org/10.3389/fmicb.2010.00134] [PMID: 21687759]

[18]   Livermore DM, Blaser M, Carrs O, *et al.* Discovery research: the scientific challenge of finding new antibiotics. J Antimicrob Chemother 2011; 66(9): 1941-4.
        [http://dx.doi.org/10.1093/jac/dkr262] [PMID: 21700626]

[19]   Saga T, Yamaguchi K. History of antimicrobial agents and resistant. Japan Med Assoc J 2009; 137: 103-8.

[20]   Spellberg B, Gilbert DN. The future of antibiotics and resistance: a tribute to a career of leadership by John Bartlett. Clin Infect Dis 2014; 59(2) (Suppl. 2): S71-5.
        [http://dx.doi.org/10.1093/cid/ciu392] [PMID: 25151481]

[21]   Centers for Disease Control and Prevention, Office of Infectious Disease. Antibiotic resistance threats in the United States, 2013. April 2013.

[22]   Golkar Z, Bagasra O, Pace DG. Bacteriophage therapy: a potential solution for the antibiotic resistance crisis. J Infect Dev Ctries 2014; 8(2): 129-36.
        [http://dx.doi.org/10.3855/jidc.3573] [PMID: 24518621]

[23]   Viswanathan VK. Off-label abuse of antibiotics by bacteria. Gut Microbes 2014; 5(1): 3-4.
        [http://dx.doi.org/10.4161/gmic.28027] [PMID: 24637595]

[24]   Michael CA, Dominey-Howes D, Labbate M. The antimicrobial resistance crisis: causes, consequences, and management. Front Public Health 2014; 2: 145.
        [http://dx.doi.org/10.3389/fpubh.2014.00145] [PMID: 25279369]

[25]   Wright GD. Something old, something new: revisiting natural products in antibiotic drug discovery.

Can J Microbiol 2014; 60(3): 147-54.
[http://dx.doi.org/10.1139/cjm-2014-0063] [PMID: 24588388]

[26]   Read AF, Woods RJ. Antibiotic resistance management. Evol Med Public Health 2014; 2014(1): 147.
[http://dx.doi.org/10.1093/emph/eou024] [PMID: 25355275]

[27]   Lushniak BD. Antibiotic resistance: a public health crisis. Public Health Rep 2014; 129(4): 314-6.
[http://dx.doi.org/10.1177/003335491412900402] [PMID: 24982528]

[28]   Gross M. Antibiotics in crisis. Curr Biol 2013; 23(24): R1063-5.
[http://dx.doi.org/10.1016/j.cub.2013.11.057] [PMID: 24501765]

[29]   Prabakaran P, Dimitrov DS. Human Antibody Structure and Function. Methods and Principles in Medicinal Chemistry 2017; 9(1): 51-84.
[http://dx.doi.org/10.1002/9783527699124.ch3]

[30]   Wu TT, Kabat EA. An analysis of the sequences of the variable regions of Bence Jones proteins and myeloma light chains and their implications for antibody complementarity. J Exp Med 1970; 132(2): 211-50.
[http://dx.doi.org/10.1084/jem.132.2.211] [PMID: 5508247]

[31]   Kabat EA, Wu TT, Bilofsky H. Unusual distributions of amino acids in complementarity-determining (hypervariable) segments of heavy and light chains of immunoglobulins and their possible roles in specificity of antibody-combining sites. J Biol Chem 1977; 252(19): 6609-16.
[http://dx.doi.org/10.1016/S0021-9258(17)39891-5] [PMID: 408353]

[32]   Poljak RJ, Amzel LM, Avey HP, Chen BL, Phizackerley RP, Saul F. Three-dimensional structure of the Fab' fragment of a human immunoglobulin at 2,8-A resolution. Proc Natl Acad Sci USA 1973; 70(12): 3305-10.
[http://dx.doi.org/10.1073/pnas.70.12.3305] [PMID: 4519624]

[33]   Hayashi Y, Miura N, Isobe J, Shinyashiki N, Yagihara S. Molecular dynamics of hinge-bending motion of IgG vanishing with hydrolysis by papain. Biophys J 2000; 79(2): 1023-9.
[http://dx.doi.org/10.1016/S0006-3495(00)76356-9] [PMID: 10920032]

[34]   Saphire EO, Stanfield RL, Max Crispin MD, *et al.* Contrasting IgG structures reveal extreme asymmetry and flexibility. J Mol Biol 2002; 319(1): 9-18.
[http://dx.doi.org/10.1016/S0022-2836(02)00244-9] [PMID: 12051932]

[35]   Porter RR. The hydrolysis of rabbit γ-globulin and antibodies with crystalline papain. Biochem J 1959; 73(1): 119-27.
[http://dx.doi.org/10.1042/bj0730119] [PMID: 14434282]

[36]   Turner MW, Bennich H. Subfragments from the Fc fragment of human immunoglobulin G. Isolation and physicochemical characterization. Biochem J 1968; 107(2): 171-8.
[http://dx.doi.org/10.1042/bj1070171] [PMID: 4171016]

[37]   Köhler G, Milstein C. Continuous cultures of fused cells secreting antibody of predefined specificity. Nature 1975; 256(5517): 495-7.
[http://dx.doi.org/10.1038/256495a0] [PMID: 1172191]

[38]   Little M, Kipriyanov SM, Le Gall F, Moldenhauer G. Of mice and men: hybridoma and recombinant antibodies 2000.
[http://dx.doi.org/10.1016/S0167-5699(00)01668-6]

[39]   Nissim A, Chernajovsky Y. Historical development of monoclonal antibody therapeutics. Handb Exp Pharmacol 2008; 181(181): 3-18.
[http://dx.doi.org/10.1007/978-3-540-73259-4_1] [PMID: 18071939]

[40]   Chames P, Van Regenmortel M, Weiss E, Baty D. Therapeutic antibodies: successes, limitations and hopes for the future. Br J Pharmacol 2009; 157(2): 220-33.
[http://dx.doi.org/10.1111/j.1476-5381.2009.00190.x] [PMID: 19459844]

[41]   Hoogenboom HR, Chames P. Natural and designer binding sites made by phage display technology. Immunol Today 2000; 21(8): 371-8.
[http://dx.doi.org/10.1016/S0167-5699(00)01667-4] [PMID: 10916139]

[42]   Neuberger MS, Williams GT, Mitchell EB, Jouhal SS, Flanagan JG, Rabbitts TH. A hapten-specific chimaeric IgE antibody with human physiological effector function. Nature 1985; 314(6008): 268-70.
[http://dx.doi.org/10.1038/314268a0] [PMID: 2580239]

[43]   Reichert JM, Rosensweig CJ, Faden LB, Dewitz MC. Monoclonal antibody successes in the clinic. Nat Biotechnol 2005; 23(9): 1073-8.
[http://dx.doi.org/10.1038/nbt0905-1073] [PMID: 16151394]

[44]   Jones PT, Dear PH, Foote J, Neuberger MS, Winter G. Replacing the complementarity-determining regions in a human antibody with those from a mouse. Nature 1986; 321(6069): 522-5.
[http://dx.doi.org/10.1038/321522a0] [PMID: 3713831]

[45]   Barbas CF. Phage display: a laboratory manual 2001.

[46]   Lee CMY, Iorno N, Sierro F, Christ D. Selection of human antibody fragments by phage display. Nat Protoc 2007; 2(11): 3001-8.
[http://dx.doi.org/10.1038/nprot.2007.448] [PMID: 18007636]

[47]   Thompson MA. The return of PRO 140, a CCR5-directed mAb. Curr Opin HIV AIDS 2018; 13(4): 346-53.
[http://dx.doi.org/10.1097/COH.0000000000000479] [PMID: 29708899]

[48]   Covino DA, Purificato C, Catapano L, *et al.* APOBEC3G/3A Expression in Human Immunodeficiency Virus Type 1-Infected Individuals Following Initiation of Antiretroviral Therapy Containing Cenicriviroc or Efavirenz. Front Immunol 2018; 9: 1839.
[http://dx.doi.org/10.3389/fimmu.2018.01839] [PMID: 30135687]

[49]   Markham A. Ibalizumab: First Global Approval. Drugs 2018; 78(7): 781-5.
[http://dx.doi.org/10.1007/s40265-018-0907-5] [PMID: 29675744]

[50]   Fu TM, An Z, Wang D. Progress on pursuit of human cytomegalovirus vaccines for prevention of congenital infection and disease. Vaccine 2014; 32(22): 2525-33.
[http://dx.doi.org/10.1016/j.vaccine.2014.03.057] [PMID: 24681264]

[51]   Freed DC, Tang Q, Tang A, *et al.* Pentameric complex of viral glycoprotein H is the primary target for potent neutralization by a human cytomegalovirus vaccine. Proc Natl Acad Sci USA 2013; 110(51): E4997-5005.
[http://dx.doi.org/10.1073/pnas.1316517110] [PMID: 24297878]

[52]   Ohlin M, Söderberg-Nauclér C. Human antibody technology and the development of antibodies against cytomegalovirus. Mol Immunol 2015; 67(2): 153-70.
[http://dx.doi.org/10.1016/j.molimm.2015.02.026] [PMID: 25802091]

[53]   Dole K, Segal FP, Feire A, *et al.* A first-in-human study to assess the safety and pharmacokinetics of monoclonal antibodies against human cytomegalovirus in healthy volunteers. Antimicrob Agents Chemother 2016; 60(5): 2881-7.
[http://dx.doi.org/10.1128/AAC.02698-15] [PMID: 26926639]

[54]   Ishida JH, Burgess T, Derby MA, *et al.* Phase 1 randomized, double-blind, placebo-controlled study of RG7667, an anticytomegalovirus combination monoclonal antibody therapy, in healthy adults. Antimicrob Agents Chemother 2015; 59(8): 4919-29.
[http://dx.doi.org/10.1128/AAC.00523-15] [PMID: 26055360]

[55]   Li B, Fouts AE, Stengel K, *et al.* In vitro affinity maturation of a natural human antibody overcomes a barrier to in vivo affinity maturation. MAbs 2014; 6(2): 437-45.
[http://dx.doi.org/10.4161/mabs.27875] [PMID: 24492299]

[56]   McLellan JS, Chen M, Kim A, Yang Y, Graham BS, Kwong PD. Structural basis of respiratory

syncytial virus neutralization by motavizumab. Nat Struct Mol Biol 2010; 17(2): 248-50.
[http://dx.doi.org/10.1038/nsmb.1723] [PMID: 20098425]

[57] Mejias A, Ramilo O. New options in the treatment of respiratory syncytial virus disease. J Infect 2015; 71 (Suppl. 1): S80-7.
[http://dx.doi.org/10.1016/j.jinf.2015.04.025] [PMID: 25922289]

[58] Wu H, Pfarr DS, Tang Y, *et al.* Ultra-potent antibodies against respiratory syncytial virus: effects of binding kinetics and binding valence on viral neutralization. J Mol Biol 2005; 350(1): 126-44.
[http://dx.doi.org/10.1016/j.jmb.2005.04.049] [PMID: 15907931]

[59] Carbonell-Estrany X, Simões EAF, Dagan R, *et al.* Motavizumab for prophylaxis of respiratory syncytial virus in high-risk children: a noninferiority trial. Pediatrics 2010; 125(1): e35-51.
[http://dx.doi.org/10.1542/peds.2008-1036] [PMID: 20008423]

[60] Robbie GJ, Criste R, Dall'Acqua WF, *et al.* A novel investigational Fc-modified humanized monoclonal antibody, motavizumab-YTE, has an extended half-life in healthy adults. Antimicrob Agents Chemother 2013; 57(12): 6147-53.
[http://dx.doi.org/10.1128/AAC.01285-13] [PMID: 24080653]

[61] Salazar G, Zhang N, Fu TM, An Z. Antibody therapies for the prevention and treatment of viral infections. NPJ Vaccines 2017; 2(1): 19.
[http://dx.doi.org/10.1038/s41541-017-0019-3] [PMID: 29263875]

[62] Gupta P, Kamath AV, Park S, *et al.* Preclinical pharmacokinetics of MHAA4549A, a human monoclonal antibody to influenza A virus, and the prediction of its efficacious clinical dose for the treatment of patients hospitalized with influenza A. MAbs 2016; 8(5): 991-7.
[http://dx.doi.org/10.1080/19420862.2016.1167294] [PMID: 27031797]

[63] Genentech. A. Study of MHAA4549A as Monotherapy for Acute Uncomplicated Seasonal Influenza A in Otherwise Healthy Adults. NCT02623322 (ClinicalTrials. gov, 2016.

[64] Genentech. A. Study of MHAA4549A in Combination with Oseltamivir Versus Oseltamivir in Participants with Severe Influenza A Infection. NCT02293863 (ClinicalTrials.gov, 2016.

[65] Naik G. Scientists' elusive goal: reproducing study results. Wall St J 2015; 258(130): A1.

[66] Visterra. A. Phase 2a Double-blind, Placebo-controlled Study to Assess the Safety and Tolerability of a Single Intravenous Dose of VIS410 in Subjects with Uncomplicated Influenza A Infection. NCT02989194 (ClinicalTrials.gov, 2016.

[67]] NIAID. Randomized, Double-Blind, Placebo-Controlled, Phase 2 Study in Healthy Volunteers to Evaluate the Efficacy and Safety of CR6261 in an H1N1 Influenza Healthy Human Challenge Model. NCT02371668 (ClinicalTrials.gov, 2016.

[68] Gao Y, Zhang TY, Yuan Q, Xia NS. Antibody-mediated immunotherapy against chronic hepatitis B virus infection. Hum Vaccin Immunother 2017; 13(8): 1768-73.
[http://dx.doi.org/10.1080/21645515.2017.1319021] [PMID: 28521640]

[69] Cerino A, Bremer CM, Glebe D, Mondelli MU. A Human Monoclonal Antibody against Hepatitis B Surface Antigen with Potent Neutralizing Activity. PLoS One 2015; 10(4)e0125704
[http://dx.doi.org/10.1371/journal.pone.0125704] [PMID: 25923526]

[70] Fibriansah G, Lok SM. The development of therapeutic antibodies against dengue virus. Antiviral Res 2016; 128: 7-19.
[http://dx.doi.org/10.1016/j.antiviral.2016.01.002] [PMID: 26794397]

[71] Xu M, Zuest R, Velumani S, *et al.* A potent neutralizing antibody with therapeutic potential against all four serotypes of dengue virus. NPJ Vaccines 2017; 2(1): 2.
[http://dx.doi.org/10.1038/s41541-016-0003-3] [PMID: 29263863]

[72] Ong EZ, Budigi Y, Tan HC, *et al.* Preclinical evaluation of VIS513, a therapeutic antibody against dengue virus, in non-human primates. Antiviral Res 2017; 144: 44-7.

[http://dx.doi.org/10.1016/j.antiviral.2017.05.007] [PMID: 28529000]

[73]   Budigi Y, Ong EZ, Robinson LN, *et al.* Neutralization of antibody-enhanced dengue infection by VIS513, a pan serotype reactive monoclonal antibody targeting domain III of the dengue E protein. PLoS Negl Trop Dis 2018; 12(2)e0006209
[http://dx.doi.org/10.1371/journal.pntd.0006209] [PMID: 29425203]

[74]   Zhang Q, Gui M, Niu X, *et al.* Potent neutralizing monoclonal antibodies against Ebola virus infection. Sci Rep 2016; 6(1): 25856.
[http://dx.doi.org/10.1038/srep25856] [PMID: 27181584]

[75]   Bornholdt ZA, Turner HL, Murin CD, *et al.* Isolation of potent neutralizing antibodies from a survivor of the 2014 Ebola virus outbreak. Science 2016; 351(6277): 1078-83.
[http://dx.doi.org/10.1126/science.aad5788] [PMID: 26912366]

[76]   Wang Q, Yang H, Liu X, *et al.* Molecular determinants of human neutralizing antibodies isolated from a patient infected with Zika virus. Sci Transl Med 2016; 8(369)369ra179
[http://dx.doi.org/10.1126/scitranslmed.aai8336] [PMID: 27974667]

[77]   Motley MP, Banerjee K, Fries BC. Monoclonal antibody-based therapies for bacterial infections. Curr Opin Infect Dis 2019; 32(3): 210-6.
[http://dx.doi.org/10.1097/QCO.0000000000000539] [PMID: 30950853]

[78]   Berry JD, Gaudet RG. Antibodies in infectious diseases: polyclonals, monoclonals and niche biotechnology. N Biotechnol 2011; 28(5): 489-501.
[http://dx.doi.org/10.1016/j.nbt.2011.03.018] [PMID: 21473942]

[79]   Wilcox MH, Gerding DN, Poxton IR, *et al.* MODIFY I and MODIFY II Investigators. Bezlotoxumab for prevention of recurrent Clostridium difficile infection. N Engl J Med 2017; 376(4): 305-17.
[http://dx.doi.org/10.1056/NEJMoa1602615] [PMID: 28121498]

[80]   Ruzin A, Wu Y, Yu L, *et al.* Characterisation of anti-alpha toxin antibody levels and colonisation status after administration of an investigational human monoclonal antibody, MEDI4893, against *Staphylococcus aureus* alpha toxin. Clin Transl Immunology 2018; 7(1)e1009
[http://dx.doi.org/10.1002/cti2.1009] [PMID: 29484186]

[81]   Sheoran AS, Chapman-Bonofiglio S, Harvey BR, *et al.* Human antibody against shiga toxin 2 administered to piglets after the onset of diarrhea due to Escherichia coli O157:H7 prevents fatal systemic complications. Infect Immun 2005; 73(8): 4607-13.
[http://dx.doi.org/10.1128/IAI.73.8.4607-4613.2005] [PMID: 16040972]

[82]   Fogle MR, Griswold JA, Oliver JW, Hamood AN. Anti-ETA IgG neutralizes the effects of Pseudomonas aeruginosa exotoxin A. J Surg Res 2002; 106(1): 86-98.
[http://dx.doi.org/10.1006/jsre.2002.6433] [PMID: 12127813]

[83]   McClain MS, Cover TL. Functional analysis of neutralizing antibodies against Clostridium perfringens epsilon-toxin. Infect Immun 2007; 75(4): 1785-93.
[http://dx.doi.org/10.1128/IAI.01643-06] [PMID: 17261609]

[84]   Smith TJ, Lou J, Geren IN, *et al.* Sequence variation within botulinum neurotoxin serotypes impacts antibody binding and neutralization. Infect Immun 2005; 73(9): 5450-7.
[http://dx.doi.org/10.1128/IAI.73.9.5450-5457.2005] [PMID: 16113261]

[85]   Visan L, Rouleau N, Proust E, Peyrot L, Donadieu A, Ochs M. Antibodies to PcpA and PhtD protect mice against *Streptococcus pneumoniae* by a macrophage- and complement-dependent mechanism. Hum Vaccin Immunother 2018; 14(2): 489-94.
[http://dx.doi.org/10.1080/21645515.2017.1403698] [PMID: 29135332]

[86]   Tabor DE, Oganesyan V, Keller AE, *et al.* Pseudomonas aeruginosa PcrV and Psl, the molecular targets of bispecific antibody MEDI3902, are conserved among diverse global clinical isolates. J Infect Dis 2018; 218(12): 1983-94.
[http://dx.doi.org/10.1093/infdis/jiy438] [PMID: 30016475]

[87]   Le HN, Quetz JS, Tran VG, *et al.* MEDI3902 correlates of protection against severe pseudomonas aeruginosa pneumonia in a rabbit acute pneumonia model. Antimicrob Agents Chemother 2018; 62(5)e02565-17
[http://dx.doi.org/10.1128/AAC.02565-17] [PMID: 29483116]

[88]   Varshney AK, Kuzmicheva GA, Lin J, *et al.* A natural human monoclonal antibody targeting Staphylococcus Protein A protects against Staphylococcus aureus bacteremia. PLoS One 2018; 13(1)e0190537
[http://dx.doi.org/10.1371/journal.pone.0190537] [PMID: 29364906]

[89]   Diago-Navarro E, Motley MP, Ruiz-Peréz G, *et al.* Novel, broadly reactive anticapsular antibodies against carbapenem-resistant Klebsiella pneumonia protect from infection. MBio 2018; 9(2)e00091-18
[http://dx.doi.org/10.1128/mBio.00091-18] [PMID: 29615497]

[90]   Kobayashi SD, Porter AR, Freedman B, *et al.* Antibody-mediated killing of carbapenem-resistant ST258 *Klebsiella pneumoniae* by human neutrophils. MBio 2018; 9(2)e00297-18
[http://dx.doi.org/10.1128/mBio.00297-18] [PMID: 29535199]

[91]   Doyle CR, Moon JY, Daily JP, Wang T, Pirofski L. A capsular polysaccharide-specific antibody alters *Streptococcus pneumoniae* gene expression during nasopharyngeal colonization of mice. Infect Immun 2018; 86(7)e00300-18
[http://dx.doi.org/10.1128/IAI.00300-18] [PMID: 29735523]

[92]   Mostowy RJ, Holt KE. Diversity-generating machines: genetics of bacterial sugar-coating. Trends Microbiol 2018; 26(12): 1008-21.
[http://dx.doi.org/10.1016/j.tim.2018.06.006] [PMID: 30037568]

[93]   Pennini ME, De Marco A, Pelletier M, *et al.* Immune stealth-driven O2 serotype prevalence and potential for therapeutic antibodies against multidrug resistant *Klebsiella pneumoniae.* Nat Commun 2017; 8(1): 1991.
[http://dx.doi.org/10.1038/s41467-017-02223-7] [PMID: 29222409]

[94]   Hooft van Huijsduijnen R, Kojima S, Carter D, *et al.* Reassessing therapeutic antibodies for neglected and tropical diseases. PLoS Negl Trop Dis 2020; 14(1)e0007860
[http://dx.doi.org/10.1371/journal.pntd.0007860] [PMID: 31999695]

[95]   Larsen RA, Pappas PG, Perfect J, *et al.* Phase I evaluation of the safety and pharmacokinetics of murine-derived anticryptococcal antibody 18B7 in subjects with treated cryptococcal meningitis. Antimicrob Agents Chemother 2005; 49(3): 952-8.
[http://dx.doi.org/10.1128/AAC.49.3.952-958.2005] [PMID: 15728888]

[96]   Gigliotti F, Haidaris CG, Wright TW, Harmsen AG. Passive intranasal monoclonal antibody prophylaxis against murine Pneumocystis carinii pneumonia. Infect Immun 2002; 70(3): 1069-74.
[http://dx.doi.org/10.1128/IAI.70.3.1069-1074.2002] [PMID: 11854184]

[97]   Buissa-Filho R, Puccia R, Marques AF, *et al.* The monoclonal antibody against the major diagnostic antigen of Paracoccidioides brasiliensis mediates immune protection in infected BALB/c mice challenged intratracheally with the fungus. Infect Immun 2008; 76(7): 3321-8.
[http://dx.doi.org/10.1128/IAI.00349-08] [PMID: 18458072]

[98]   de Mattos Grosso D, de Almeida SR, Mariano M, Lopes JD. Characterization of gp70 and anti-gp70 monoclonal antibodies in Paracoccidioides brasiliensis pathogenesis. Infect Immun 2003; 71(11): 6534-42.
[http://dx.doi.org/10.1128/IAI.71.11.6534-6542.2003] [PMID: 14573675]

[99]   Kim J, Sudbery P. Candida albicans, a major human fungal pathogen. J Microbiol 2011; 49(2): 171-7.
[http://dx.doi.org/10.1007/s12275-011-1064-7] [PMID: 21538235]

[100] Rudkin FM, Raziunaite I, Workman H, *et al.* Single human B cell-derived monoclonal anti-Candida antibodies enhance phagocytosis and protect against disseminated candidiasis. Nat Commun 2018; 9(1): 5288.

[http://dx.doi.org/10.1038/s41467-018-07738-1] [PMID: 30538246]

[101] Conti DJ, Freed BM, Gruber SA, Lempert N. Prophylaxis of primary cytomegalovirus disease in renal transplant recipients. A trial of ganciclovir vs immunoglobulin. Arch Surg 1994; 129(4): 443-7.
[http://dx.doi.org/10.1001/archsurg.1994.01420280121016] [PMID: 8154971]

[102] Barnes GL, Hewson PH, Mclellan JA, *et al.* A randomised trial of oral gammaglobulin in low-birt--weight infants infected with rotavirus. Lancet 1982; 319(8286): 1371-3.
[http://dx.doi.org/10.1016/S0140-6736(82)92496-5] [PMID: 6177981]

[103] Borowitz SM, Saulsbury FT. Treatment of chronic cryptosporidial infection with orally administered human serum immune globulin. J Pediatr 1991; 119(4): 593-5.
[http://dx.doi.org/10.1016/S0022-3476(05)82412-6] [PMID: 1919892]

[104] Bickel U, Yoshikawa T, Pardridge WM. Delivery of peptides and proteins through the blood–brain barrier. Adv Drug Deliv Rev 2001; 46(1-3): 247-79.
[http://dx.doi.org/10.1016/S0169-409X(00)00139-3] [PMID: 11259843]

[105] Bullowa JGM. The management of the pneumonias. New York: Oxford University Press: 1937; pp 283-298.

[106] Halstead SB. Immune enhancement of viral infection. Prog Allergy 1982; 31: 301-64.
[PMID: 6292921]

CHAPTER 3

Cationic Amphiphiles as Antimicrobial Agents

Sovik Dey Sarkar[1] and **Chirantan Kar**[1,*]

[1] *Amity Institute of Applied Sciences, Amity University Kolkata, Kolkata 700135, India*

Abstract: Numerous antimicrobial peptides (AMP) obtained from natural sources are currently tested in clinical or preclinical settings for treating infections triggered by antimicrobial-resistant bacteria. Several experiments with cyclic, linear and diastereomeric AMPs have proved that the geometry, along with the chemical properties of an AMP, is important for the microbiological activities of these compounds. It is understood that the combination of the hydrophobic and hydrophilic nature of AMPs is crucial for the adsorption and destruction of the bacterial membrane. However, the application of AMPs in therapeutics is still limited due to their poor pharmacokinetics, low bacteriological efficacy and overall high manufacturing costs. To overcome these problems, a variety of newly synthesized cationic amphiphiles have recently appeared, which imitate not only the amphiphilic nature but also the potent antibacterial activities of the AMPs with better pharmacokinetic properties and lesser *in vitro* toxicity. Thus, amphiphiles of this new genre have enough potential to deliver several antibacterial molecules in years to come.

Keywords: Amphiphiles, Antibacterial, Antibiotic, Bacterial cell, Cationic amphiphiles, Cytotoxicity, Drug, Gram-negative bacteria, Gram-positive bacteria, Hydrophobicity, Mammalian cell, Membrane penetration, Membrane permeability, Peptide, Positively charged, Quaternary ammonium salt, Steroid, Surfactants.

INTRODUCTION

The major reasons behind the research to find a newer class of antibacterial drug molecules are the surge in antibiotic-resistant bacteria and the decreasing rate of innovation of new antibacterial drugs [1], especially in to fight against multidrug-resistant bacteria [2 - 4]. Self-healing from bacterial disease in higher order organism do not depend completely on enzymatic inhibitor, rather, higher-order organism generates wide-range antimicrobial peptides, which may target the lysis of cellular membrane leading to cell death. Disruption of membranes and cell walls is one of the most necessary steps in killing bacteria, and hence they serve

* **Corresponding author Chirantan Kar:** Amity Institute of Applied Sciences, Amity University Kolkata, Kolkata 700135, India; E-mail: ckar@kol.amity.edu

Tilak Saha, Manab Deb Adhikari and Bipransh Kumar Tiwary (Eds.)

as suitable targets for the development of new antimicrobial agents or antibiotics. Recently, various naturally occurring as well as synthetically developed antimicrobial molecule has been reported for interrupting different stages of the formation of the bacterial cell wall. Among these new reports, there are beta-lactams and glycolipid antibiotic moenomycin A, which can permanently prevent the biosynthetic step of peptidoglycan trans peptization. There are also vancomycin-like glycopeptides that can entirely obstruct the trans-peptization step [5 - 8]. Although there are several examples of antibiotics that can target bacterial cell walls, there are only a few known for targeting bacterial membranes. Some significant advantages of developing such antibiotics are:

1. Especially for infection caused by slow-growing or dormant bacteria, bacterial membrane targeting is a better option compared to treating with the current range of clinically available antibiotics [9].

2. Bacterial membranes are hidden structures compared to the cell wall, disruption of the bacterial membrane will cause rapid killing of the bacteria, and hence it will be difficult for them to evolve against such attacks. Thus, membrane-targeting antibiotics can be used for a prolonged time without facing any resistance from the bacteria [10].

3. Although bacterial cell permeation poses a major challenge for designing antibiotics to aim intracellular targets, bacterial membrane targeting antibiotics does not need cell permeability [11].

Nevertheless, for developing an effective but safe membrane targeting antibiotics, the molecule must have the ability to differentiate between mammalian and bacterial cell membranes. The inability to elude such non-selective membrane damage could cause cytotoxicity towards eukaryotic cells and hence can be a major challenge for developing membrane-targeting antibiotics. The distinctive compositional feature of the bacterial membrane and mammalian cell membrane heightens the possibilities of developing membrane-selective antimicrobial agents. Here are some of the striking differences between the mammalian and bacterial cell membranes:

(A) In the case of Gram-positive bacterial cell membrane, teichoic acid is one of the major constituents and in the case of Gram-negative bacteria, lipopolysaccharides are one of the key components. In mammalian cell membranes, lipopolysaccharides and teichoic acids do not exist.

(B) Due to the presence of a comparatively low percentage of negatively charged lipids in the mammalian cell membrane, the negative charges on it are much weaker compared to the bacterial cell membrane [12 - 15].

(C) Neutral proteins like phosphatidylcholine and sphingomyelin are localized on the outer leaflet of the mammalian cell membrane, and the inner leaflet consists of negatively charged lipids, such as phosphatidylserine and phosphatidylinositol [16, 17]. Since the attractive force working between the positively charged amphiphilic molecules and oppositely (negative) charged membrane is an electrostatic interaction and is dependent on the distance between the opposite charges, the presence of the negative charges on the inner leaflet of the mammalian cell membrane significantly minimizes this force of attraction compared to the bacterial cell membrane where the negative charges are on the outer surface.

The above-mentioned differences in the structure and composition between the bacterial cell membrane and mammalian cell membrane help in designing several membranes, targeting cationic antimicrobial molecules.

In this chapter, we will discuss the structural aspects and working principles of some significant amphiphilic molecules and the current challenges in using these molecules as antimicrobial agents.

SYNTHETIC CATIONIC PEPTIDES AS ANTIMICROBIAL AGENTS

Antimicrobial peptides perform a major function in the human immune system and are related to diseases like Crohn's disease and Morbus Kostmann. Molecules such as AMPs, which can target cell membranes, are advantageous over the general method of drug activity as they may offer a variety of active structures while facing lesser resistance [18]. Apart from their various benefits, naturally occurring antimicrobial peptides also have a few drawbacks, which should be taken care of before applying them widely as a replacement for already well-known antibiotics. Naturally occurring antimicrobial peptides have some significant pharmacokinetic deficiencies, including low metabolic stability, poor bioavailability and formulation difficulties. These shortcomings are mainly due to several amide linkages between the constituent amino acids and their size [19]. Overcoming these shortfalls is the reason behind the recent interest in the design and development of their partially or fully synthetic analogues. In various versions of these synthetic analogues, the cationic and hydrophobic properties of the naturally occurring molecules are retained, and the cationic charges in the molecules help to increase the long-range repulsive forces and overcome the short-range hydrophobic interactions.

Amphiphilic Helices

Among various antimicrobial peptides designed, the alpha-helical peptides are one of the most heavily investigated. The secondary structure of these synthetically prepared positively charged peptide amphiphiles is fully dependent on their environment, which controls their activity. Folding of the peptide molecule to yield an alpha-helical structure separates the cationic charges and the hydrophobic amino acids, giving it a complete amphiphilic structure. Interaction with negatively charged phospholipid can boost the formation of helical structure; this was further corroborated by CD spectra which show these amphiphilic peptide molecules attain a more structured form in their target membrane but remain in an unorganized form in simple buffer solution [19]. The alpha-helical structure of the peptides can also be disrupted by the replacement of some key amino acids (L-enantiomers) with the corresponding D-enantiomers; studies with these modified compounds show that some pre-folded structures may insert into the neutral membranes leading to hemolysis [20]. This fact is further supported by studies with V681. V681 forms a stable alpha-helical structure in hydrophobic and aqueous solutions [21]. A modification of this peptide was made by rupturing the hydrophobic region in the structure with the inclusion of a polar lysine molecule which disrupts the alpha helix in buffer solution; this variant is found to have unaltered antimicrobial activity with several times decreased hemolytic activity (Table **1**). The opposite effect has also been observed in AMP RTA3, which has polar amino acid residue on its hydrophobic face. Replacement of the polar residue with a hydrophobic one in RTA-R5L protrudes the hemolytic effect by strengthening amphiphilic alpha helix formation [22]. Breaking the alpha helix structure of V681 by incorporating specific amino acids residues with the corresponding D-isomers has resulted in the formation of V13VD and V13KD with a comparable antimicrobial property of that of V681 but lesser hemolytic effect [23]. The researcher has also replaced the peptide linkages of these amphiphiles with peptoid linkage, which is also found to disrupt the helical structure. Peptoids are just like peptides; the only difference is in the position of the side chain of the amino acid. In peptide, the side chain is attached to the alpha carbon, whereas in peptoid, it is attached to the amide nitrogen; due to this structural feature, peptoids also alter the hydrophobic moment. Some recent studies done by replacing the amino acid residues with their peptoid variants show slightly lesser antimicrobial effect, but the hemolytic effect is reduced significantly [24].

Charge and Hydrophobicity

Antimicrobial proteins also have another set of significant characteristics: Cationic charge and hydrophobicity, which regulate the overall amphiphilic character of the molecule. Recent studies also suggest that the biological activity of the AMPs is vastly dependent on the amphiphilic character of the molecule. This is because the bacterial membrane is generally negatively charged, hence it is essential that the amphiphilic molecule have a positively charged head group which may easily interact with the bacterial membrane through electrostatic force of attraction. The hydrophobic chain is responsible for interacting with the hydrophobic core of the bacterial membrane, which results in insertion into the membrane leading to lysis, cytoplasm leakage and overall rupture of the bacterial cell [25].

Table 1. Comparison of alpha helical stability with AMP activity.

Peptide	Sequence	MIC (μM)		MHC (μM)	Refs.
		Gram-Positive cells	Gram-Negative cells		
Melittin	GIGAVLKVLTTGLPALISWIKRKRQQ-NH$_2$	2 (EC)	0.5 (SA)	0.78	[24]
Melittin peptoid F	GIGAVNfKVLTTGNfPALISWNfKRKRQQ-NH$_2$	4 (EC)	2 (SA)	>100	[24]
RTA	RPAFRKAAFRVMRACV-NH$_2$	4 (PA)	Not Performed	>1,300	[22]
RTA-F4W/ R5L	RPAWLKAAFRVMRACV-NH$_2$	4 (PA)	Not Performed	~7	[21]
V681	Ac-KWKSFLKTFKSAVKTVLHTALKAISS-NH$_2$	2.1 (EC)	Not Performed	5.2	[21]
V681-V13K	Ac-KWKSFLKTFKSAKKTVLHTALKAISS-NH$_2$	0.89 (EC)	Not Performed	>88.6	[21]
V681-V13VD	Ac-KWKSFLKTFKSAVDKTVLHTALKAISS-NH	1.1 (EC)	Not Performed	22.2	[23]
D-V681-V13K	Ac-K$_D$W$_D$S$_D$F$_D$L$_D$K$_D$T$_D$F$_D$K$_D$S$_D$A$_D$K$_D$K$_D$T$_D$V$_D$L$_D$ H$_D$T$_D$A$_D$L$_D$K$_D$A$_D$I$_D$S$_D$S$_D$-NH$_2$	1.1 (EC)	Not Performed	>100	[23]

EC=*Escherichia coli*; PA= *Pseudomonas aeruginosa*; SA=*Staphylococcus aureus*; MHC= minimal hemolytic concentration; Nf,= H5C6-CH2-NH-CH2-COOH; Ac= acetyl.

Although it is a well-known concept that increasing the hydrophobicity of amphiphiles will increase the antimicrobial activity, it is sometimes associated with the drawback of elevated hemolytic activity. For instance, when two leucine residues of buforin-II 10 are replaced with more hydrophobic hexafluoro-leucine (Fig. **1**), it was found that the antimicrobial activity has significantly increased with no significant variation in hemolytic activity (Table **2**) [3]. In contrast, a similar kind of alteration in amino acid residues of magainin 2 has resulted in variants, which show a significant increase in hemolytic activity but without changing the antimicrobial activity [26]. Hence, there must be a range of hydrophobicity where the effectiveness is highest. This is further corroborated by a similar experiment done by varying the hydrophobic character of V68 [27]. It is also found that amphiphiles which are not hydrophobic enough are neither hemolytic nor antimicrobial in nature, whereas increasing the hydrophobicity too much may increase the antimicrobial activity but will also elevate the hemolytic effect to a significant level (Table **2**). In some cases, too much hydrophobicity of the amphiphiles may also decrease the antimicrobial property due to the self-aggregation of the hydrophobic chains [21, 28].

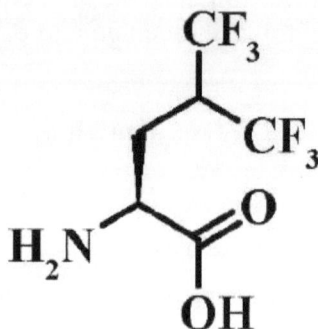

Fig. (1). Structure of Hexafluoro-leucine.

Increasing the positive charge on amphiphiles may also enhance bactericidal activity. Positive charges in antimicrobial peptides can be increased by including amino acid residues like histidine, arginine or lysine in the peptide chain. As mentioned earlier, a positive charge enhances the electrostatic attraction with the bacterial cell membrane without affecting any other possible interaction with zwitterionic lipids present in mammalian cell membranes [29, 30]. Similar to the hydrophobicity, increasing the positive charge density too much will not increase the antibacterial effect indefinitely. Methodically enhancing the cationic charge of V681 shows that after a threshold charge density, the antibacterial activity doesn't change any farther, but the hemolytic effect increases dramatically (Table **2**) [31]. This may be due to the enhanced electrostatic force of attraction between the negative membrane potential of the mammalian cells and the positive charge of the amphiphiles [32]. In another case, dimerization of penetratin, a widely known cell-penetrating peptide, has formed an antimicrobial peptide with comparable antimicrobial activity but several times enhanced hemolytic activity (Table 2). Although the dimerized form of penetratin has similar CD spectra as that of penetratin, only the dimer is capable of disrupting liposomes (artificial bacterial membrane) made up of phosphatidyl ethanolamine and phosphatidyl glycerol. This indicates that even with a similar structure, a longer dimerized form may show a distinct window of activity [30]. These observations also show that delicate changes in the peptide structure of AMPs may induce novel characteristics in the overall activity spectrum of the amphiphiles, leading to interactions with multiple targets [33].

Table 2. Effect of hydrophobicity and positive charge on AMP activity.

Peptide	Sequence	MIC (µM)		MHC (µM)	Refs.
		Gram-Positive cells	Gram-Negative cells		
Buforin II 10	FPVGRVHRLLRK-H	>173 (EC)	>173 (BS)	270 (HC$_{50}$)	[26]
Buforin II 10-L9L$_f$/L10L$_f$	FPVGRVHRL$_f$L$_f$RK-H	23 (EC)	5.9 (BS)	>235 (HC$_{50}$)	[26]
Magainin 2	GIGKFLHAAKKFAKAFVAEIMNS-NH2	1.0 (EC)	1.0 (BS)	70.6 (HC$_{50}$)	[26]
Magainin 2-L6L$_f$/I20L$_f$ 2	GIGKFL$_f$HAAKKFAKAFVAEL$_f$MNS-NH$_2$	0.92 (EC)	0.92 (BS)	7.4 (HC$_{50}$)	[26]
Magainin 2-L6L$_f$/A9L$_f$/A13L$_f$/I20L$_f$	GIGKFL$_f$HAL$_f$KKFL$_f$KAFL$_f$AEL$_f$MNS-NH$_2$	13 (EC)	3.3 (BS)	3.6 (HC$_{50}$)	[26]
Penetratin	RQIKIWFQNRRMKWKK-NH$_2$	2 (EC)	1 (SA)	>200	[75]
Dual penetratin	(RQIKIWFQNRRMKWKK)$_2$K-NH$_2$	2 (EC)	1 (SA)	25	[75]
V$_{681}$-L6A/L21A	Ac-KWKSFAKTFKSAKKTVLHTAAKAISS-NH$_2$	168 (PA)	Not Performed	336	[27]
V$_{681}$-A12L/A20L/A23L	Ac-KWSFLKTFKSLKKTVLHTLLKLISS-NH$_2$	168 (PA)	Not Performed	1.3	[27]
V$_{681}$-V13K/T15K/T19K	Ac-KWKSFLKTFKSAKKKVLHKALKAISS-NH$_2$	2.8 (EC)	11 (BS)	<2.8	[21]
D-V$_{681}$-V13K	Ac-K$_D$W$_D$S$_D$F$_D$L$_D$K$_D$T$_D$F$_D$K$_D$S$_D$A$_D$K$_D$K$_D$T$_D$V$_D$L$_D$ H$_D$T$_D$A$_D$L$_D$K$_D$A$_D$I$_D$S$_D$S$_D$-NH$_2$	5.3 (PA)	Not Performed	88.6	[23]

EC=*Escherichia coli*; PA=*Pseudomonas aeruginosa*; SA=*Staphylococcus aureus*; BS=*Bacillus subtilis*; MHC= minimal hemolytic concentration; Lf=(CF3)2-CH-CH2-CH(NH2)-COOH; Ac= acetyl.

CATIONIC STEROIDAL AMPHIPHILES AS ANTIMICROBIAL AGENTS

Similar to the peptide amphiphiles synthesized by coupling of alpha amino acids, amide polymers can also be obtained from the combination of beta-amino acids. Like the alpha variant, they are also capable of forming facially amphiphilic conformation. These new molecules, synthesized by Liu *et al.* [34] and Gellman *et al.* [35], have hydrophobic groups at one face and multiple cationic groups on the opposite face. These examples of membrane-active antibiotics derived from sources other than alpha amino acids suggest that the antimicrobial activity of the molecule depends on the nature of the charged and hydrophobic groups in the secondary structure and not on the component monomeric units. Likewise, as several other non-peptide molecules can be synthesized to provide the same disposition of charged and hydrophobic groups (amphiphilicity), a similar kind of membrane activity to that of AMPs should also be possible. Following this track, Kahne and co-workers reported amphiphilic derivatives of cholic acid, which can be used as a transfection agent [36]. These compounds were synthesized from cholic acid with attached carbohydrate units, the hydrocarbon skeleton of the steroid molecule forms the hydrophobic face, and the carbohydrate molecules act as the polar face A in Fig. (**2**), the amphiphilic nature of these molecules is proved to enhance the transmission of polar molecules through eukaryotic cell membranes. Although, the antibacterial activity of the above-mentioned compounds has not been reported, several other steroidal amphiphilic compounds mimicking the facial amphiphilic nature of widely used antibiotic polymyxin B have been reported [37 - 41]. These positively charged (cationic) steroid molecules were found to have comparable antibacterial activities to those of many AMPs. Due to the similar kind of amphiphilic nature see B in Fig. (**2**), the mechanism of action of these molecules is also similar to that of AMPs and can efficiently permeabilize the bacterial membrane [42, 43] for example, see B in Fig. (**2**), it is not surprising that they appear to share common mechanisms of action [41].

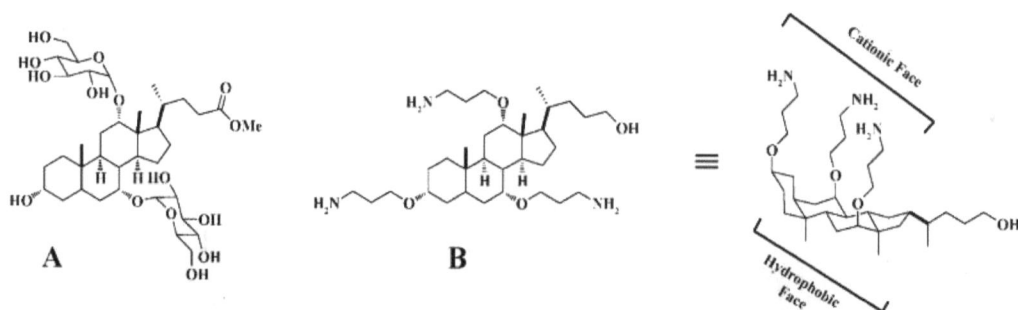

Fig. (2). Structures of facial amphiphiles A and B, showing the hydrophobic and cationic face.

Structural Classes of Cationic Steroid Amphiphiles

Cationic steroid amphiphiles are divided into two major classes polymyxin mimics and squalamine and its mimics. Although there are various derivatives prepared from each of the classes have been prepared but here we will discuss only some selected antibiotics and their corresponding activities. Molecules falling into the polymyxin mimics group has three propyl amine groups attached to the steroid nucleus *via* ether linkage *e.g.,* B in Fig. (**3**) [39, 41]. Several different molecules are designed based on this structure by varying the side chain at C24 by introducing hydrophobic chains and polyamines, which is found to influence the antibacterial activity of the compounds (Fig. **3**). Squalamine (Fig. **3**). is a naturally occurring cationic steroidal amphiphile obtained from the dogfish shark [44]. This molecule has been extensively mimicked by Regen *et al.* by the synthesis of a series of new compounds [45, 46]. They have interchanged the position of the sulphate group and the polyamine sidechain, in some cases, they have also completely removed the sulfate group (compounds E and F of Fig. (**3**). In spite of the difference between the structure of polymyxin and squalamine, it has been found that squalamine mimics can take a facially amphiphilic conformation in the presence of bacterial membranes by keeping all the polyamine chains at one face of the molecule [45].

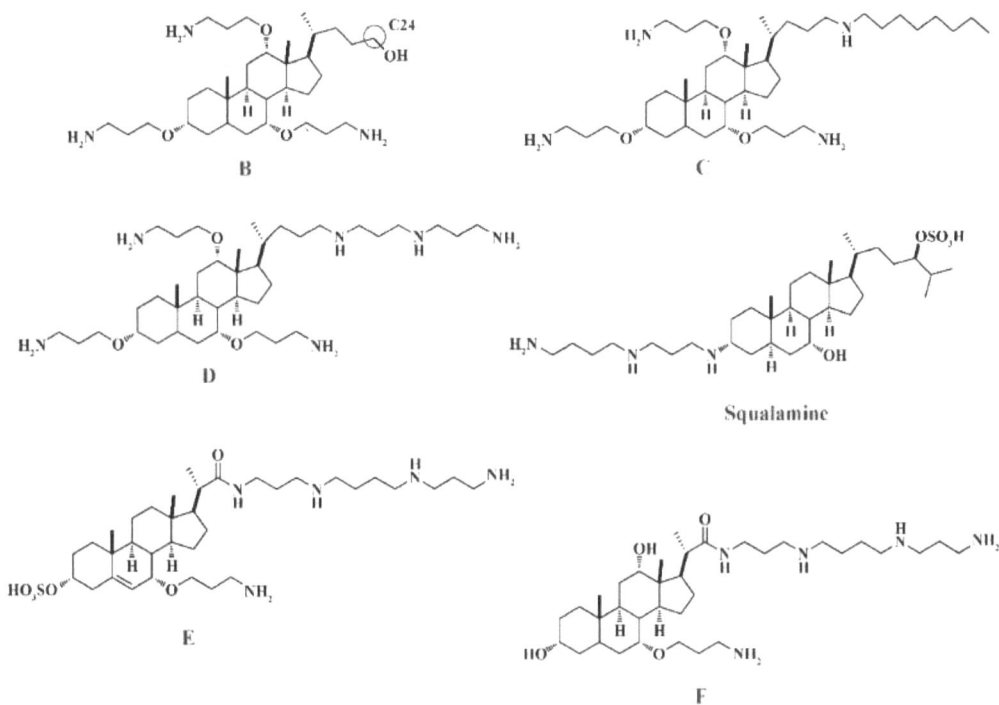

Fig. (3). Structure of various cationic steroidal amphiphiles and squalamine.

Membrane Permeabilizing Property of Cationic Steroid Amphiphiles

The outer membrane of Gram-negative bacteria can work as an effective shield for hydrophobic molecules. Due to this fact, many antimicrobial compounds which work effectively in Gram-positive bacteria are relatively inactive for Gram-negative bacteria. Cross-bridging between lipid A molecules through divalent cations such as magnesium and calcium forms the permeability barrier for the outer membrane of bacteria [47]. Therefore, any molecule which can either bind with the lipid or the divalent cations can disrupt the permeability barrier of the outer membrane. For instance, well-known metal ion binders such as EDTA can sensitize Gram-negative bacteria towards hydrophobic antibiotics [47].

A shorter version of polymyxin B known as polymyxin B nonapeptide [48] can bind to lipid A and can sensitize Gram-negative bacteria, although the compound alone doesn't display any antibacterial properties. Cationic steroid amphiphiles can permeabilize the bacterial membrane in multiple ways; in some cases, it kills the bacteria so rapidly that it is difficult to study the permeabilization activity. Several steroidal amphiphiles are synthesized with varying hydrophobic chains at C24; comprehensive studies study with these variants revealed that the hydrophobicity of the probe is necessary for the antibacterial property but not essential for permeabilizing the membrane [39]. The hydrophobic chain helps the probe to move through the outer membrane to the cytoplasmic membrane of the cell. The permeabilizing abilities of cationic steroid amphiphiles can make the outer membrane of Gram-negative vulnerable to hydrophobic antibiotics. Hence, in the future, steroid-based cationic amphiphiles can improve the arsenal in the fight against infections caused by Gram-negative bacteria.

Mechanism of Action of Cationic Steroid Amphiphiles

As the amphiphilic morphology of cationic steroid amphiphiles and antimicrobial peptides are similar, so it is expected that their mechanism of action will also be similar. To prove this fact, comparative studies between some selected steroidal amphiphiles (B and C in Fig. (**3**). and amphiphilic peptide magainin have been performed. These studies using fluorescence probes further confirmed that both kinds of antibiotic amphiphiles are membrane-active, and their activity is quite rapid in nature. It was also noticed that steroidal amphiphilic antibiotics are much active at a concentration lower than the antimicrobial peptide molecule. Furthermore, it was also found that their mechanistic aspects are also similar [41]. Antimicrobial peptides have two action mechanisms, namely 'barrel-stave model' and 'carpet model' [49]. In carpet model, the cationic peptide molecules get absorbed throughout the cell surface due to ionic interactions; when a critical concentration of peptide molecules is reached, the whole cover of the membrane is removed (Fig. **4**). In the barrel-stave model, the peptide molecules accumulate in a bundle as staves of a barrel to form a pore-like structure on the membrane. Most of the antimicrobial peptides which may form an alpha helix, such as magainins and cecropins, act *via* the carpet model. Due to the similar kind of activity observed for both, cationic steroid amphiphiles and these peptide molecules, it is believed that the antimicrobial activity of the steroidal amphiphiles is also taking place *via* the same kind of carpet model.

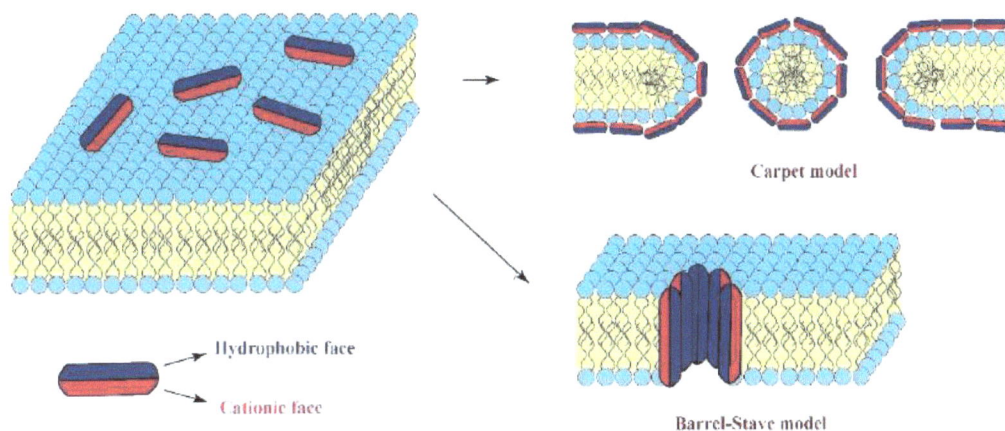

Fig. (4). Mechanism of action of cationic amphiphiles.

QUATERNARY AMMONIUM COMPOUNDS AS ANTIMICROBIAL AGENTS

Quaternary ammonium compounds (QACs) (Fig. **5**) are so far the most useful antiseptics and disinfectants that have ever been invented. Most of the other amphiphilic compounds with free amine groups mentioned in this chapter are neutral as a compound but acquire a positive charge after entering the physiological environment. On the other hand, quaternary ammonium compounds have permanent positive charges on the nitrogen atoms. Quaternary ammonium compounds (QACs) possess surface-active properties, detergency, and antimicrobial properties, including activity against bacteria, fungi, molds and viruses. The first reports of quaternary ammonium compounds with biocidal activity were reported in 1916 [50], and since then, they are being used extensively. The quaternary ammonium compounds are synthesized by the nucleophilic substitution reaction of tertiary amines by alkyl halides, benzyl halides, or similar compounds.

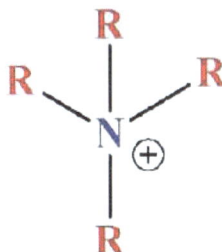

Fig. (5). The general structure of QAC.

Each of the -R group represents a covalently bonded alkyl group, and the X represents the anion, which is usually halides (chlorides and bromides; iodide salts exhibit decreased solubility) [51]. The functional portion of the molecule varies in structure and chain length. The R groups may be saturated or unsaturated, cyclic or acyclic, branched or non-branched, and aromatic or substituted aromatics. And typically, the alkyl chain lengths vary from C-8 to C-22. QACs usually have very high solubility in water but their high water solubility can be dramatically decreased by anionic substances. For example, traditional anionic fatty acids found in soaps can precipitate them by causing an ionic attraction between the positively and negatively charged groups. One of the major drawbacks with typical quaternary ammonium compounds is the lack of specificity, and they show antimicrobial activity by disrupting the bio-membranes without any discrimination [52, 53], which means they also rupture the cell membrane regardless of cell type, hence showing strong toxic side effects. Although recent researches have been made to focus on the specificity of the antimicrobial activity of QACs by incorporating biocompatible groups.

Structure-activity Relationship

QACs were found to be notably effective against several of the ESKAPE bacteria (*Enterococcus faecium, Staphylococcus aureus, Klebsiella pneumonia, Acinetobacter baumannii, Pseudomonas aeruginosa* and *Enterobacter* species) [54]. It is generally accepted that lipid bilayer structures of cell membranes are principal targets for this type of compound. Similar to all the other cationic amphiphiles, the QACs also work by electrostatic interaction between the positively charged quaternary ammonium head and the negatively charged phospholipids membrane of the bacterial cell, which is followed by permeation of the QAC side chains into the lipid bilayer leading to leakage of cytoplasmic material, eventually destroying the target cell [55]. As QACs target the bacterial cell membranes non-specifically, they are considered broad-spectrum antibiotics. However, it was found that they exhibit significantly increased activity against Gram-positive bacteria [56]. This is because Gram-positive bacteria contain a single phospholipid cellular membrane and a thicker cell wall comprised of peptidoglycan, whereas Gram-negative bacteria are covered by two cellular membranes and a rather thin peptidoglycan layer. It is due to the presence of this second membrane that QACs and other membrane-targeting antiseptics like QACs tend to show decreased activity.

Effect of Charge Density of the Cationic Head Groups

The positive charge density of amphiphilic molecules is dependent on various structural constraints, like counter anion, the structure of the positively charged head groups, the functional group present in the head group, the acid dissociation constant (pK_a) of head groups, the total number of head groups, *etc.*

The counter anions selected for positively charged surfactants are normally halide ions, such as chloride, bromide or iodide. Extensive experimental studies show that the antimicrobial capability of any cationic surfactants is dependent on the counter anions [57]. The reason behind this is the dissociation ability of the halide anions, which is different for each halide and provides the cationic head groups with unique charge density, which is proportional to the order of dissociation ability of the halide anions *i.e.* if the dissociation ability of the counter-ions are high, higher will be the cationic charge density, and of course, a high cationic charge density will result in an efficient attraction with the oppositely charged bacterial membrane. However, iodide anions as counter anion generally tend to decrease the solubility of the surfactant, resulting in a compromised antimicrobial property. This decrease in solubility also increases the value of minimum inhibitory concentration (MIC). Hence, in most cases, chloride and bromides are used as the counter-ion.

The design of the head group for cationic surfactants is dependent on the functionality of the head groups, the acidic dissociation constant (pK_a) of head groups, the number of head groups, *etc.* For example, it was observed that the antimicrobial activity of dodecyl-pyridinium amphiphile against *E. coli* can be increased by the inclusion of electron donors, like as amino and methyl groups; on the other hand, carboxyl and carbamoyl groups, which are known electron withdrawer can reduce the activity [58]. The electron donor as a substituent pushes the electron towards the pyridine nitrogen, which increases the electron-donating ability of the pyridine nitrogen, and when it forms a pyridinium cation, better charge separation takes place, resulting in a more (comparatively) positively charged pyridinium species. The higher cationic charge density favors greater electrostatic attraction with the negatively charged phospholipid membrane of bacteria, increasing the effectivity of the compound. On the other hand, electron-withdrawing groups show the opposite effect and lead to a weakening of the attraction with bacterial membrane. It has also been proved that the increase in antimicrobial capacity of the dodecylpyridinium iodide is directly related to the increase of pK_a value of the pyridine unit. The pK_a value of the pyridine unit is also proportional to the positive charge density in their head group. In case of amino acid-based cationic amphiphiles, the cationic charge is largely dependent on the extent of protonation of the amino groups, hence amphiphiles with pKa value higher than 9 will act as a better antimicrobial agent (at the physiological condition) than an amphiphile with pK_a value lower than 7, especially against Gram-negative bacteria [59]. This pK_a dependency is true for any cationic surfactants.

As the number of charged headgroups significantly increases the antimicrobial activity of cationic amphiphiles thus, increasing the number of charged head groups is one popular way to increase the effectivity of QACs [52, 60]. It was found that the inclusion of multiple head-groups in cationic surfactants greatly enhances antimicrobial activity. This is because of the increase in the charge density, which leads to an increase in the extent of interaction with the bacterial membrane and a decrease in MIC. An increase in the number of the cationic head group also significantly decreases the time to need kill the bacteria.

Effect of Lipophilic Tails

The antimicrobial efficiency of a cationic amphiphile is generally not dependent on the length of the alkyl chain. The antimicrobial activity increases steadily up to a maximum, and then again starts to decrease. This is often called "cut-off effect" [61, 62]. Generally, cationic amphiphiles with carbon chain lengths 12 to 14 show optimal antimicrobial activity. The cut-off effect was conveniently explained by "Free volume hypothesis". The amphiphilic polar groups will interact with lipid polar groups and their chains will orient parallel to the lipid hydrocarbon chains. After that, during the insertion of the hydrophobic alkyl chain, these molecules are capable of forming a free volume under their lipophilic chain, which is possible due to their asymmetric conical structure. This free volume increases the membrane permeability and hence enhances the antimicrobial activity of the amphiphile. In the case of a short alkyl chain, the free volume created will be larger owing to larger conical asymmetry, but the interaction with the lipid part of the bacterial cell membrane will be weak due to the low membrane partition coefficient. Consequently, at a given amphiphile concentration, the total free volume created in the bacterial membrane will be smaller. If the chain length of the amphiphiles is increased, the similarity with the bacterial lipid chain (as bacterial lipids are also long-chain lipids) will also increase, resulting in a considerable enhancement of the partition coefficient but with a very long chain length of the free volume created will be close to zero. Thus, amphiphilic molecules with alkyl chain lengths between these two extremes can make a good combination of substantial partition coefficient with an optimum total free volume in the membrane.

Apart from the length of the alkyl chain, the antimicrobial activity of QACs also depend on the number of chains, therefore, a Gemini surfactant with two alkyl chain is more efficient in killing bacteria than the corresponding single-chain analogue [59, 63]. One such example is the cationic double-chained Gemini surfactant di-N-alkyldimethylammonium halides [64, 65]. The reason behind this is the enhanced hydrophilic-hydrophilic interaction between the two alkyl chains and the lipid membrane of the bacteria, which enables a better insertion of the chain into the membrane. This means amphiphiles with multiple chain lengths can rupture the bacterial cell membrane more effectively and be considered efficient anti-microbials.

The inclusion of aromatic groups in the quaternary ammonium substituent of the QACs increases the hydrophobicity of the amphiphiles, which leads to better insertion into the lipid layer, effectively increasing the activity of the compound as an antimicrobial agent. For example, benzalkonium bromide (BAB) bearing a benzyl group on the head group region shows much greater antimicrobial activity than its corresponding non-aromatic analogue DTAB (Dodecyl tri methyl ammonium bromide) [66]. This is due to the presence of the aromatic phenyl ring, which is highly hydrophobic in nature and therefore enhances the interaction of the amphiphiles with the bacterial cell membrane. The elaborate relationship of antimicrobial activity with aromatic groups of cationic amphiphiles like QACs was shown by a group of scientists [67]; according to them, when the aromatic core is kept the same, the antimicrobial activity is dependent on the alkyl chain length. The optimal chain length is hexyl and octyl for both Gram-positive and Gram-negative bacteria. For the QACs with smaller aromatic core, the antimicrobial efficiency increases linearly with the alkyl chain length and touches a maximum at decyl and dodecyl, respectively. On the other hand, if the aromatic core is changed, keeping the alkyl chain length constant, an interesting aromatic group-dependent effect is observed. It was shown that when the alkyl group is shorter (hexyl to octyl), MIC change follows the order anthracene>naphthalene>benzene, *i.e.*, with the decreasing size of the aromatic cores, the activity greatly decreases. Especially for QACs where the alkyl chain length is short, the aromatic core controls antimicrobial activity, whereas for QACs long alkyl chain, the antimicrobial activity doesn't change in a significant amount with respect to their non-aromatic, non-cyclic analogues, rather it was seen that upon increasing the alkyl chain, the antimicrobial activity depends more on the alkyl chain than the aromatic substituent in the alkyl group.

Effect of Self-assembly

If the MIC value of a particular compound is lesser than its CMC, then instead of the self-assembly of that compound, the monomers exert the antimicrobial effect. In such cases, the antimicrobial property of amphiphiles shows a non-linear relationship with their CMC value and presents a cut-off effect, similar to their dependency on the alkyl chain length. For example, this effect was seen in a gemini QAC derived from bis-(2-dimethylaminoethyl) glutarate [68], where it showed increased activity against *E. coli* and *S. aureus* up to a maximum at $1/9^{th}$ of its CMC and again it decreases afterwards.

If MIC value is higher than its CMC, then the amphiphiles self-assemblies exert antimicrobial activity [69]. Self-assembly formation can significantly increase the cationic charge density and local concentration, which in turn enhances the binding ability of the molecule with the cell membrane and increases the accumulation process of the individual molecule on the bacterial surface. An increase in the lipophilic tail length decreases the CMC, which means now the amphiphile molecules will be able to form self-assemblies at a much lower concentration, thereby decreasing the MIC. The morphology of the self-assemblies can also vary the effectivity of the quaternary ammonium surfactants towards the target. Another thing should be noted for these amphiphiles, which show antimicrobial activity by forming aggregates that it is the monomeric form of the surfactant, which is actually inserted into the lipid bilayer. So there, the self-assemblies should not be too stable, or the concentration of the monomeric form will be too low to effectively rupture the cell wall.

The Structure-cytotoxicity Relationship

Due to the unspecific membrane targeting ability of the QACs, their toxicity towards mammalian cells is a very important thing to consider. Although QACs are very effective as an antimicrobial agent, most typical QACs are bounded by their cytotoxic effects. Similar to their antimicrobial activity, the cytotoxicity of quaternary ammonium surfactants is also linearly proportional to the hydrophobicity of the alkyl tail and positive charge density of the head-group, therefore generally, Gemini surfactants are more cytotoxic than their corresponding single-chain analogues [70]. As the number of alkyl chains in Gemini surfactant is more, so it generates a higher amount of hydrophobicity and enables better interaction with the cell membrane. For different Gemini surfactants of the same spacer length, the hydrophobicity depends on the alkyl chain length and hence the cytotoxicity increases with increasing the same. Similarly, for Gemini surfactants with the same alkyl chain, the cytotoxicity will increase with increasing the spacer length, as increasing spacer length will also increase the hydrophobicity. Increasing the spacer length also decreases the cationic charge densities of Gemini surfactants and makes them weaker to interact with cells. Due to the above reasons, quaternary ammonium Gemini surfactants are highly cytotoxic even when present at a smaller concentration, which limits their application as an antimicrobial agent. Two parameters that are widely used for evaluating cytotoxicity are "half maximal inhibitory concentration" (IC_{50}) and the concentration to induce 50% hemolysis (HC_{50}). Where high IC_{50} or HC_{50} value means low cytotoxicity.

The most convenient way to reduce the cytotoxicity is by the introduction of biocompatible groups/moieties like amino acids, amide, and ester or using natural product-derived QACs [71 - 74]. Although biocompatible groups are effective to reduce cytotoxicity as a compromise, sometimes those groups which reduce the cytotoxicity can also reduce the effectivity of the QACs as antimicrobial agents, and as also mentioned before, we can further decrease the toxic effect for those QACs, which shows inhibitory effect *via* the formation of self-assemblies by the introduction of hydrophobic groups which increases the hydrophobicity of the tails and decreases the MIC.

CONCLUDING REMARKS AND FUTURE PERSPECTIVE

Due to the recent surge in numerous antibiotic-resistant bacteria, scientists are showing keen interest in finding new antimicrobial agents. The search for new drug molecules has sparked an interest in developing positively charged amphiphilic molecules with antimicrobial properties. This new genre of drug molecules will definitely broaden the arsenal of antibiotics that could be used in the fight against infectious diseases. As discussed throughout the chapter, the major challenge scientists face while working with cationic amphiphiles is their toxicity to mammalian cells, and due to this reason, very few of these drugs are approved for clinical use. However, new families of cationic amphiphiles are recently reported, which are less harmful to the eukaryotic cells and show a broad range of activity towards multiple antibiotic-resistant bacteria. In the near future, more cationic amphiphiles will undergo clinical trial and will provide a series of new antibiotics for safer internal and external use.

CONSENT FOR PUBLICATION

Not applicable.

CONFLICT OF INTEREST

The authors declare no conflict of interest, financial or otherwise.

ACKNOWLEDGEMENT

Declared none.

REFERENCES

[1]   Song JH. What's new on the antimicrobial horizon? Int J Antimicrob Agents 2008; 32 (Suppl. 4): S207-13.
        [http://dx.doi.org/10.1016/S0924-8579(09)70004-4] [PMID: 19134521]

[2]   Mulvey MR, Simor AE. Antimicrobial resistance in hospitals: How concerned should we be? CMAJ 2009; 180(4): 408-15.
        [http://dx.doi.org/10.1503/cmaj.080239] [PMID: 19221354]

[3]   Naarmann N, Bilgiçer B, Meng H, Kumar K, Steinem C. Fluorinated interfaces drive self-association of transmembrane α helices in lipid bilayers. Angew Chem Int Ed 2006; 45(16): 2588-91.

[http://dx.doi.org/10.1002/anie.200503567] [PMID: 16532504]

[4]   Otter JA, French GL. Nosocomial transmission of community-associated meticillin-resistant Staphylococcus aureus: An emerging threat. Lancet Infect Dis 2006; 6(12): 753-5.
[http://dx.doi.org/10.1016/S1473-3099(06)70636-3] [PMID: 17123892]

[5]   Healy VL, Lessard IAD, Roper DI, Knox JR, Walsh CT. Vancomycin resistance in enterococci: reprogramming of the d-Ala–d-Ala ligases in bacterial peptidoglycan biosynthesis. Chem Biol 2000; 7(5): R109-19.
[http://dx.doi.org/10.1016/S1074-5521(00)00116-2] [PMID: 10801476]

[6]   Ge M, Chen Z, Russell H. Vancomycin derivatives that inhibit peptidoglycan biosynthesis without binding D-Ala-D-Ala. Science 1999; 284(5413): 507-11.

[7]   Yuan Y, Fuse S, Ostash B, Sliz P, Kahne D, Walker S. Structural analysis of the contacts anchoring moenomycin to peptidoglycan glycosyltransferases and implications for antibiotic design. ACS Chem Biol 2008; 3(7): 429-36.
[http://dx.doi.org/10.1021/cb800078a] [PMID: 18642800]

[8]   Fuse S, Tsukamoto H, Yuan Y, *et al.* Functional and structural analysis of a key region of the cell wall inhibitor moenomycin. ACS Chem Biol 2010; 5(7): 701-11.
[http://dx.doi.org/10.1021/cb100048q] [PMID: 20496948]

[9]   Hurdle JG, O'Neill AJ, Chopra I, Lee RE. Targeting bacterial membrane function: An underexploited mechanism for treating persistent infections. Nat Rev Microbiol 2011; 9(1): 62-75.
[http://dx.doi.org/10.1038/nrmicro2474] [PMID: 21164535]

[10]   Daugelavičius R, Bakiene̊ E, Bamford DH. Stages of polymyxin B interaction with the Escherichia coli cell envelope. Antimicrob Agents Chemother 2000; 44(11): 2969-78.
[http://dx.doi.org/10.1128/AAC.44.11.2969-2978.2000] [PMID: 11036008]

[11]   Nikaido H. Prevention of drug access to bacterial targets: permeability barriers and active efflux. Science 1994; 264(5157): 382-8.
[http://dx.doi.org/10.1126/science.8153625] [PMID: 8153625]

[12]   Virtanen JA, Cheng KH, Somerharju P. Phospholipid composition of the mammalian red cell membrane can be rationalized by a superlattice model. Proc Natl Acad Sci USA 1998; 95(9): 4964-9.
[http://dx.doi.org/10.1073/pnas.95.9.4964] [PMID: 9560211]

[13]   Epand RF, Pollard JE, Wright JO, Savage PB, Epand RM. Depolarization, bacterial membrane composition, and the antimicrobial action of ceragenins. Antimicrob Agents Chemother 2010; 54(9): 3708-13.
[http://dx.doi.org/10.1128/AAC.00380-10] [PMID: 20585129]

[14]   Zhang YM, Rock CO. Membrane lipid homeostasis in bacteria. Nat Rev Microbiol 2008; 6(3): 222-33.
[http://dx.doi.org/10.1038/nrmicro1839] [PMID: 18264115]

[15]   van Meer G, Voelker DR, Feigenson GW. Membrane lipids: where they are and how they behave. Nat Rev Mol Cell Biol 2008; 9(2): 112-24.
[http://dx.doi.org/10.1038/nrm2330] [PMID: 18216768]

[16]   Ohvo-Rekilä H, Ramstedt B, Leppimäki P, Slotte JP. Cholesterol interactions with phospholipids in membranes. Prog Lipid Res 2002; 41(1): 66-97.
[http://dx.doi.org/10.1016/S0163-7827(01)00020-0] [PMID: 11694269]

[17]   Dowhan W, Bogdanov M. Functional roles of lipids in membranes. New Comprehensive Biochemistry. Elsevier 2002; 36: pp. 1-35.

[18]   Zasloff M. Antimicrobial peptides in health and disease.
[http://dx.doi.org/10.1056/NEJMe020106]

[19]   Klocek G, Schulthess T, Shai Y, Seelig J. Thermodynamics of melittin binding to lipid bilayers. Aggregation and pore formation. Biochemistry 2009; 48(12): 2586-96.

[http://dx.doi.org/10.1021/bi802127h] [PMID: 19173655]

[20]   Dathe M, Wieprecht T. Structural features of helical antimicrobial peptides: their potential to modulate activity on model membranes and biological cells. Biochim Biophys Acta Biomembr 1999; 1462(1-2): 71-87.
[http://dx.doi.org/10.1016/S0005-2736(99)00201-1] [PMID: 10590303]

[21]   Chen Y, Mant CT, Farmer SW, Hancock REW, Vasil ML, Hodges RS. Rational design of α-helical antimicrobial peptides with enhanced activities and specificity/therapeutic index. J Biol Chem 2005; 280(13): 12316-29.
[http://dx.doi.org/10.1074/jbc.M413406200] [PMID: 15677462]

[22]   Hawrani A, Howe RA, Walsh TR, Dempsey CE. Origin of low mammalian cell toxicity in a class of highly active antimicrobial amphipathic helical peptides. J Biol Chem 2008; 283(27): 18636-45.
[http://dx.doi.org/10.1074/jbc.M709154200] [PMID: 18434320]

[23]   Chen Y, Vasil AI, Rehaume L, et al. Comparison of biophysical and biologic properties of α-helical enantiomeric antimicrobial peptides. Chem Biol Drug Des 2006; 67(2): 162-73.
[http://dx.doi.org/10.1111/j.1747-0285.2006.00349.x] [PMID: 16492164]

[24]   Zhu WL, Song YM, Park Y, et al. Substitution of the leucine zipper sequence in melittin with peptoid residues affects self-association, cell selectivity, and mode of action. Biochim Biophys Acta Biomembr 2007; 1768(6): 1506-17.
[http://dx.doi.org/10.1016/j.bbamem.2007.03.010] [PMID: 17462584]

[25]   Herzog IM, Fridman M. Design and synthesis of membrane-targeting antibiotics: from peptides- to aminosugar-based antimicrobial cationic amphiphiles. MedChemComm 2014; 5(8): 1014-26.
[http://dx.doi.org/10.1039/C4MD00012A]

[26]   Meng H, Kumar K. Antimicrobial activity and protease stability of peptides containing fluorinated amino acids. J Am Chem Soc 2007; 129(50): 15615-22.
[http://dx.doi.org/10.1021/ja075373f] [PMID: 18041836]

[27]   Chen Y, Guarnieri MT, Vasil AI, Vasil ML, Mant CT, Hodges RS. Role of peptide hydrophobicity in the mechanism of action of α-helical antimicrobial peptides. Antimicrob Agents Chemother 2007; 51(4): 1398-406.
[http://dx.doi.org/10.1128/AAC.00925-06] [PMID: 17158938]

[28]   Sarig H, Rotem S, Ziserman L, Danino D, Mor A. Impact of self-assembly properties on antibacterial activity of short acyl-lysine oligomers. Antimicrob Agents Chemother 2008; 52(12): 4308-14.
[http://dx.doi.org/10.1128/AAC.00656-08] [PMID: 18838600]

[29]   Matsuzaki K. Control of cell selectivity of antimicrobial peptides. Biochim Biophys Acta Biomembr 2009; 1788(8): 1687-92.
[http://dx.doi.org/10.1016/j.bbamem.2008.09.013]

[30]   Zhu WL, Lan H, Park IS, et al. Design and mechanism of action of a novel bacteria-selective antimicrobial peptide from the cell-penetrating peptide Pep-1. Biochem Biophys Res Commun 2006; 349(2): 769-74.
[http://dx.doi.org/10.1016/j.bbrc.2006.08.094] [PMID: 16945333]

[31]   Jiang Z, Vasil AI, Hale JD, Hancock REW, Vasil ML, Hodges RS. Effects of net charge and the number of positively charged residues on the biological activity of amphipathic α-helical cationic antimicrobial peptides. Biopolymers 2008; 90(3): 369-83.
[http://dx.doi.org/10.1002/bip.20911] [PMID: 18098173]

[32]   Stewart KM, Horton KL, Kelley SO. Cell-penetrating peptides as delivery vehicles for biology and medicine. Org Biomol Chem 2008; 6(13): 2242-55.
[http://dx.doi.org/10.1039/b719950c] [PMID: 18563254]

[33]   Pütsep K, Carlsson G, Boman HG, Andersson M. Deficiency of antibacterial peptides in patients with morbus Kostmann: An observation study. Lancet 2002; 360(9340): 1144-9.

[http://dx.doi.org/10.1016/S0140-6736(02)11201-3] [PMID: 12387964]

[34]   Liu D, DeGrado WF. De novo design, synthesis, and characterization of antimicrobial β-peptides. J Am Chem Soc 2001; 123(31): 7553-9.
[http://dx.doi.org/10.1021/ja0107475] [PMID: 11480975]

[35]   Porter EA, Weisblum B, Gellman SH. Mimicry of host-defense peptides by unnatural oligomers: Antimicrobial β-peptides. J Am Chem Soc 2002; 124(25): 7324-30.
[http://dx.doi.org/10.1021/ja0260871] [PMID: 12071741]

[36]   Walker S, Sofia MJ, Kakarla R, *et al.* Cationic facial amphiphiles: A promising class of transfection agents. Proc Natl Acad Sci USA 1996; 93(4): 1585-90.
[http://dx.doi.org/10.1073/pnas.93.4.1585] [PMID: 8643675]

[37]   Li C, Peters AS, Meredith EL, Allman GW, Savage PB. Design and synthesis of potent sensitizers of Gram-negative bacteria based on a cholic acid scaffolding. J Am Chem Soc 1998; 120(12): 2961-2.
[http://dx.doi.org/10.1021/ja973881r]

[38]   Li C, Budge LP, Driscoll CD, Willardson BM, Allman GW, Savage PB. Incremental conversion of outer-membrane permeabilizers into potent antibiotics for Gram-negative bacteria. J Am Chem Soc 1999; 121(5): 931-40.
[http://dx.doi.org/10.1021/ja982938m]

[39]   Atiq-ur-Rehman , Li C, Budge LP, Street SE, Savage PB. Preparation of amino acid-appended cholic acid derivatives as sensitizers of Gram-negative bacteria. Tetrahedron Lett 1999; 40(10): 1865-8.
[http://dx.doi.org/10.1016/S0040-4039(99)00075-1]

[40]   Guan Q, Li C, Schmidt EJ, *et al.* Preparation and characterization of cholic acid-derived antimicrobial agents with controlled stabilities. Org Lett 2000; 2(18): 2837-40.
[http://dx.doi.org/10.1021/ol0062704] [PMID: 10964378]

[41]   Ding B, Guan Q, Walsh JP, *et al.* Correlation of the antibacterial activities of cationic peptide antibiotics and cationic steroid antibiotics. J Med Chem 2002; 45(3): 663-9.
[http://dx.doi.org/10.1021/jm0105070] [PMID: 11806717]

[42]   Li C, Lewis MR, Gilbert AB, *et al.* Antimicrobial activities of amine- and guanidine-functionalized cholic acid derivatives. Antimicrob Agents Chemother 1999; 43(6): 1347-9.
[http://dx.doi.org/10.1128/AAC.43.6.1347] [PMID: 10348750]

[43]   Schmidt EJ, Boswell JS, Walsh JP, *et al.* Activities of cholic acid-derived antimicrobial agents against multidrug-resistant bacteria. J Antimicrob Chemother 2001; 47(5): 671-4.
[http://dx.doi.org/10.1093/jac/47.5.671] [PMID: 11328782]

[44]   Moore KS, Wehrli S, Roder H, *et al.* Squalamine: An aminosterol antibiotic from the shark. Proc Natl Acad Sci USA 1993; 90(4): 1354-8.
[http://dx.doi.org/10.1073/pnas.90.4.1354] [PMID: 8433993]

[45]   Deng G, Dewa T, Regen SL. A synthetic ionophore that recognizes negatively charged phospholipid membranes. J Am Chem Soc 1996; 118(37): 8975-6.
[http://dx.doi.org/10.1021/ja961269e]

[46]   Kikuchi K, Bernard EM, Sadownik A, Regen SL, Armstrong D. Antimicrobial activities of squalamine mimics. Antimicrob Agents Chemother 1997; 41(7): 1433-8.
[http://dx.doi.org/10.1128/AAC.41.7.1433] [PMID: 9210661]

[47]   Vaara M. Agents that increase the permeability of the outer membrane. Microbiol Rev 1992; 56(3): 395-411.
[http://dx.doi.org/10.1128/mr.56.3.395-411.1992] [PMID: 1406489]

[48]   Vaara M, Vaara T. Sensitization of Gram-negative bacteria to antibiotics and complement by a nontoxic oligopeptide. Nature 1983; 303(5917): 526-8.
[http://dx.doi.org/10.1038/303526a0] [PMID: 6406904]

[49]   Shai Y. Mechanism of the binding, insertion and destabilization of phospholipid bilayer membranes by α-helical antimicrobial and cell non-selective membrane-lytic peptides. Biochim Biophys Acta Biomembr 1999; 1462(1-2): 55-70.
[http://dx.doi.org/10.1016/S0005-2736(99)00200-X] [PMID: 10590302]

[50]   Jacobs WA, Heidelberger M, Amoss HL. THE BACTERICIDAL PROPERTIES OF THE QUATERNARY SALTS OF HEXAMETHYLENETETRAMINE. J Exp Med 1916; 23(5): 569-76.
[http://dx.doi.org/10.1084/jem.23.5.569] [PMID: 19868008]

[51]   Xie X, Cong W, Zhao F, *et al.* Synthesis, physiochemical property and antimicrobial activity of novel quaternary ammonium salts. J Enzyme Inhib Med Chem 2018; 33(1): 98-105.
[http://dx.doi.org/10.1080/14756366.2017.1396456] [PMID: 29148294]

[52]   Colomer A, Pinazo A, Manresa MA, *et al.* Cationic surfactants derived from lysine: effects of their structure and charge type on antimicrobial and hemolytic activities. J Med Chem 2011; 54(4): 989-1002.
[http://dx.doi.org/10.1021/jm101315k] [PMID: 21229984]

[53]   Manaargadoo-Catin M, Ali-Cherif A, Pougnas JL, Perrin C. Hemolysis by surfactants — A review. Adv Colloid Interface Sci 2016; 228: 1-16.
[http://dx.doi.org/10.1016/j.cis.2015.10.011] [PMID: 26687805]

[54]   Boucher HW, Talbot GH, Bradley JS, *et al.* Bad bugs, no drugs: no ESKAPE! An update from the Infectious Diseases Society of America. Clin Infect Dis 2009; 48(1): 1-12.
[http://dx.doi.org/10.1086/595011] [PMID: 19035777]

[55]   Denyer SP. Mechanisms of action of antibacterial biocides. Int Biodeterior Biodegradation 1995; 36(3-4): 227-45.
[http://dx.doi.org/10.1016/0964-8305(96)00015-7]

[56]   Salton MRJ. Lytic agents, cell permeability, and monolayer penetrability. J Gen Physiol 1968; 52(1): 227-52.
[http://dx.doi.org/10.1085/jgp.52.1.227] [PMID: 19873623]

[57]   Łudzik K, Kustrzepa K, Kowalewicz-Kulbat M, *et al.* Antimicrobial and cytotoxic properties of bisquaternary ammonium bromides of different spacer length. J Surfactants Deterg 2018; 21(1): 91-9.
[http://dx.doi.org/10.1002/jsde.12005]

[58]   Okazaki K, Manabe Y, Maeda T, Nagamune H, Kourai H. MANABE Y, MAEDA T, NAGAMUNE H, KOURAI H. Quantitative structure-activity relationship of antibacterial dodecylpyridinium iodide derivatives. Biocontrol Sci 1996; 1(1): 51-9.
[http://dx.doi.org/10.4265/bio.1.51]

[59]   Pinazo A, Manresa MA, Marques AM, Bustelo M, Espuny MJ, Pérez L. Amino acid–based surfactants: New antimicrobial agents. Adv Colloid Interface Sci 2016; 228: 17-39.
[http://dx.doi.org/10.1016/j.cis.2015.11.007] [PMID: 26792016]

[60]   Haldar J, Kondaiah P, Bhattacharya S. Synthesis and antibacterial properties of novel hydrolyzable cationic amphiphiles. Incorporation of multiple head groups leads to impressive antibacterial activity. J Med Chem 2005; 48(11): 3823-31.
[http://dx.doi.org/10.1021/jm049106l] [PMID: 15916434]

[61]   Balgavý P, Devínsky F. Cut-off effects in biological activities of surfactants. Adv Colloid Interface Sci 1996; 66: 23-63.
[http://dx.doi.org/10.1016/0001-8686(96)00295-3] [PMID: 8857708]

[62]   Devínsky F, Kopecka-Leitmanová A, Šeršeň F, Balgavý P. Cut-off effect in antimicrobial activity and in membrane perturbation efficiency of the homologous series of N,N-dimethylalkylamine oxides. J Pharm Pharmacol 2011; 42(11): 790-4.
[http://dx.doi.org/10.1111/j.2042-7158.1990.tb07022.x] [PMID: 1982303]

[63]   Zhang S, Ding S, Yu J, Chen X, Lei Q, Fang W. Antibacterial activity, *in vitro* cytotoxicity, and cell

cycle arrest of gemini quaternary ammonium surfactants. Langmuir 2015; 31(44): 12161-9.
[http://dx.doi.org/10.1021/acs.langmuir.5b01430] [PMID: 26474336]

[64]   Leclercq L, Lubart Q, Dewilde A, Aubry JM, Nardello-Rataj V. Supramolecular effects on the antifungal activity of cyclodextrin/di-n-decyldimethylammonium chloride mixtures. Eur J Pharm Sci 2012; 46(5): 336-45.
[http://dx.doi.org/10.1016/j.ejps.2012.02.017] [PMID: 22406295]

[65]   Zana R, Benrraou M, Rueff R. Alkanediyl-.alpha.omega.-bis(dimethylalkylammonium bromide) surfactants. 1. Effect of the spacer chain length on the critical micelle concentration and micelle ionization degree. Langmuir 1991; 7(6): 1072-5.
[http://dx.doi.org/10.1021/la00054a008]

[66]   Daoud NN, Dickinson NA, Gilbert P. Antimicrobial activity and physico-chemical properties of some alkyldimethylbenzylammonium chlorides. Microbios 1983; 37(148): 73-85.
[PMID: 6413825]

[67]   Ghosh C, Manjunath GB, Akkapeddi P, *et al.* Small molecular antibacterial peptoid mimics: the simpler the better! J Med Chem 2014; 57(4): 1428-36.
[http://dx.doi.org/10.1021/jm401680a] [PMID: 24479371]

[68]   Pavlíková-Mořická M, Lacko I, Devínsky F, Masárová L, Mlynarčík D. Quantitative relationships between structure and antimicrobial activity of new "Soft" bisquaternary ammonium salts. Folia Microbiol (Praha) 1994; 39(3): 176-80.
[http://dx.doi.org/10.1007/BF02814644] [PMID: 7995599]

[69]   Fukushima K, Liu S, Wu H, *et al.* Supramolecular high-aspect ratio assemblies with strong antifungal activity. Nat Commun 2013; 4(1): 2861.
[http://dx.doi.org/10.1038/ncomms3861] [PMID: 24316819]

[70]   Almeida JAS, Faneca H, Carvalho RA, Marques EF, Pais AACC. Dicationic alkylammonium bromide gemini surfactants. Membrane perturbation and skin irritation. PLoS One 2011; 6(11): e26965.
[http://dx.doi.org/10.1371/journal.pone.0026965] [PMID: 22102870]

[71]   Piecuch A, Obłąk E, Guz-Regner K. Antibacterial activity of alanine-derived gemini quaternary ammonium compounds. J Surfactants Deterg 2016; 19(2): 275-82.
[http://dx.doi.org/10.1007/s11743-015-1778-3] [PMID: 26949329]

[72]   Zhou C, Wang F, Chen H, *et al.* Selective antimicrobial activities and action mechanism of micelles self-assembled by cationic oligomeric surfactants. ACS Appl Mater Interfaces 2016; 8(6): 4242-9.
[http://dx.doi.org/10.1021/acsami.5b12688] [PMID: 26820390]

[73]   Fatma N, Panda M, Kabir-ud-Din , Beg M. Ester-bonded cationic gemini surfactants: Assessment of their cytotoxicity and antimicrobial activity. J Mol Liq 2016; 222: 390-4.
[http://dx.doi.org/10.1016/j.molliq.2016.07.044]

[74]   Joyce MD, Jennings MC, Santiago CN, Fletcher MH, Wuest WM, Minbiole KPC. Natural product-derived quaternary ammonium compounds with potent antimicrobial activity. J Antibiot (Tokyo) 2016; 69(4): 344-7.
[http://dx.doi.org/10.1038/ja.2015.107] [PMID: 26577453]

[75]   Zhu WL, Shin SY. Antimicrobial and cytolytic activities and plausible mode of bactericidal action of the cell penetrating peptide penetratin and its lys-linked two-stranded peptide. Chem Biol Drug Des 2009; 73(2): 209-15.
[http://dx.doi.org/10.1111/j.1747-0285.2008.00769.x] [PMID: 19207423]

CHAPTER 4

Amphiphilic Nanocarriers to Fight Against Pathogenic Bacteria

Amit Sarder[1,*] and **Chanchal Mandal**[1]

[1] *Biotechnology and Genetic Engineering Discipline, Khulna University, Khulna-9208, Bangladesh*

Abstract: The emergence and expansion of antibiotic resistance in pathogenic bacteria have become a global threat to both humans and animals. Immense use, overuse and misuse of antibiotics over several decades have increased the frequencies of resistance in pathogenic bacteria and resulted in significant medical problems. To fight against the widespread drug-resistant pathogenic bacteria has become a terrific challenge for the modern healthcare system. The major challenges to fight against pathogenic bacteria involve long-term antibiotic therapy with combinations of drugs. The abundance of resistance mechanisms in pathogenic bacteria has compelled many therapeutic antibiotics to become ineffective. As a result, the elimination of drug-resistant pathogenic bacteria requires a judicious strategy. The advent of nanotechnology has unveiled a new horizon in the field of nanomedicine. Nanoparticle-based techniques have the potential to overcome the challenges faced by traditional antimicrobials. In this way, self-assembling amphiphilic molecules have emerged as a fascinating technique to fight against pathogenic bacteria because of their ability to function as nanocarriers of bactericidal agents and interact and disrupt bacterial membranes. Nanocarrier-based drug delivery systems can mitigate toxicity issues and the adverse effects of high antibiotic doses. The focus of this chapter is to discuss various amphiphilic nanocarriers and their roles and possibilities in fighting against pathogenic bacteria.

Keywords: Amphiphile, Bacteria, Copolymers, Cubosome, Dendrimer, Drug Delivery, Hydrogel, Hydrophilic, Hydrophobic, Lipidated Peptide Amphiphile, Liposome, Micelle, Multidrug Resistance, Nanocarrier, Niosome, Pathogen, Peptide Amphiphile, Polymerosome, Self-assembly, Synthetic Amphiphile.

INTRODUCTION

The emergence of bacterial-resistant strains is a global health concern [1 - 3]. Antibiotics that control bacterial infections in human and animal has been corroded continuously by the emergence of drug resistance. This consequence is

* **Corresponding author Amit Sarder:** Biotechnology and Genetic Engineering Discipline, Khulna University, Khulna-9208, Bangladesh. E-mail: sarder_amit@yahoo.com

Tilak Saha, Manab Deb Adhikari and Bipransh Kumar Tiwary (Eds.)

the outcome of the substantial changes in the microbial environment due to the widespread use of antibiotics [4]. It is assumed that if antibiotic resistance increases at the present rate, by 2050, bacterial infections will cause around 10 million deaths annually which is more than the number of deaths caused by cancer presently. Due to this reason, "The Centers for Disease Control and Prevention" has mentioned in recent times that the globe is very close to entering the "post-antibiotic era" where more people will die from infections of bacteria than cancer [3, 5]. There is a factual possibility that slight injuries and common infections can cause death in the 21st century [3].

To fight against the widespread drug-resistant pathogenic bacteria has become a terrific challenge for the modern healthcare system. The abundance of resistance mechanisms in pathogenic bacteria has compelled many therapeutic antibiotics to become ineffective. As a result, the elimination of drug-resistant pathogenic bacteria requires a judicious strategy [6]. Thus, designing and developing antibiotics that can resist pathogenic bacterial resistance is very important. But, progression in developing them has become slow. In fact, the discovery and development of antibiotics are declining while antibiotic resistance in pathogenic bacteria is rising. Unfortunately, the new classes of antibiotics introduced in the early 1960s have not yet made a major impact. The global market of antibiotics is still dominated by the previously discovered classes of antibiotics [1, 3, 7, 8].

Several classes of antibiotic-resistant pathogens have emerged as major threats [7]. In fact, six antibiotic-resistant pathogenic bacterial species, namely *Enterococcus faecium*, *Staphylococcus aureus*, *Klebsiella pneumoniae*, *Acinetobacter baumannii*, *Pseudomonas aeruginosa* and *Enterobacter* species termed ESKAPE are considered as a great threat for human health [3, 9]. Again, methicillin-resistant *Staphylococcus aureus* [MRSA], multidrug-resistant [MDR] and pandrug-resistant [PDR] Gram-negative bacteria, MDR and extensively drug-resistant [XDR] *Mycobacterium tuberculosisetc.* are also emerging as a major threat. The increasing prevalence of MRSA also increases the risk of vancomycin-resistant *Staphylococcus aureus* [VRSA] [7]. Another common example of MDR bacteria encountered presently includes *Escherichia coli* [10].

The advent of nanotechnology has unveiled a new horizon in the field of nanomedicine. The progression of nanotechnology has allowed the synthesis of nanoparticles that can be assembled into various complex nanostructures. This self-assembling property is regarded as critical for the formation of nanostructures [11, 12]. Nanoparticle-based techniques have the potential to overcome the challenges faced by traditional antimicrobials [5]. In this way, self-assembling amphiphilic molecules have emerged as a fascinating technique to fight against pathogenic bacteria because of their ability to function as nanocarriers of bactericidal agents and also to interact and disrupt bacterial membranes [1]. Nanoparticles are likely to have an increased tendency to interact with pathogenic bacterial cells due to their larger surface area [6].

SELF-ASSEMBLY AND AMPHIPHILICITY

Self-assembly induced by the amphiphilic property is of much importance for creating functionality. In fact, one of the major driving forces for self-assembly is amphiphilicity [13]. The process of self-assembly is controlled by various factors like electrostatic interactions, hydrophobic interactions, van der Waals forces, intermolecular hydrogen bonds or intramolecular hydrogen bonds, *etc.* Altering the hydrophobic region length or the charge of the hydrophilic region also affects the self-assembling peptide morphology [14, 15]. The hydrophobic and electrostatic interactions are the main factors that help the peptide amphiphiles [PAs] to be self-assembled [16]. Amphiphilic molecules containing both polar and nonpolar elements tend to lessen unfavourable interactions with an aqueous environment through aggregation. In this process, the hydrophobic moieties of the amphiphilic molecules persist as shielded, whereas the hydrophilic domains get exposed. Various kinds of structures, from bilayer structures to micellar aggregates can be created based on the parameters like concentration, amphiphile geometry, *etc.* Examples of common amphiphiles include dialkylated molecules [*e.g.* phospholipids], single-chain surfactants [*e.g.*, fatty acids] *etc.* Amphiphilic behavior is also observed in peptides and proteins. Sometimes, a direct relationship is observed between the functions and the amphiphilic property of peptides or proteins [13].

Hamley IW has reported that there are two main classes of amphiphilic peptides. The first class of amphiphilic peptides includes pure peptides consisting of amphiphilic properties with both hydrophobic and hydrophilic residues. On the other hand, the second class of amphiphilic peptides includes peptides modified by attaching hydrophobic lipid chains termed PAs. Hamley IW differentiated between amphiphilic peptides and PAs in this way, as PAs are a subset of the amphiphilic peptides but not *vice versa* [17].

PEPTIDE-BASED AMPHIPHILES

We have already mentioned that different types of peptide-based amphiphiles have been reported, like amphiphilic peptides and PAs [18]. Amphiphilic peptides contain both hydrophobic and hydrophilic regions along their lengths [12]. Amphiphilic peptides are made up of hydrophobic and hydrophilic amino acids in which a charged head is attached to a non-charged tail. On the other hand, PAs, one kind of synthetic surfactant, is made up of one or more alkyl chains coupled to a peptide moiety. In an aqueous solution, PAs can assemble into nanofibres structurally like cylindrical micelles in which alkyl tails bury into the fiber core, whereas the hydrophilic oligopeptides persist on the surface of the fiber [18]. Thus, PAs can be defined as amphiphilic structures containing a hydrophilic head group in which a bioactive sequence is integrated and a hydrophobic tail which helps to align the head group, helps to drive the process of self-assembly and also helps to induce various conformations. Again, linking peptides to the hydrophobic tails helps them to be self-assembled into biomimetic films and also to attain secondary and tertiary conformations [19 - 21]. PAs combine the functional properties of bioactive peptides with the structural properties of amphiphilic surfactants [22]. PAs under conditions like pH, temperature, and ionic strength have the propensity to be self-assembled into high-aspect-ratio nanostructures [23 - 25]. PAs have attained much attention due to their capability to self-assemble into various nanostructures. Self-assembly is the capability to be associated with orderly three-dimensional structures by noncovalent interactions through a bottom-up approach [23]. The self-assembling properties are controlled by the amphiphilic nature of PAs because of the incorporation of a lipid chain to the epitope of peptide, which can participate in secondary structures like β-sheets. Spontaneously PAs can self-assemble into a diverse array of structures like ribbons, bilayers, vesicles, micelles, nanofibres, nanotubes *etc* [14, 16, 17, 19, 23, 24].

PAs self-assemble into highly ordered cylindrical supramolecular nanostructures. In these structures, the hydrophobic moieties persist in the interior space of the nanostructure, and therefore, the hydrophilic peptide sequences are displayed on the surface of the nanostructures. Thus, these supramolecular cylindrical nanostructures are different from the ordinary cylindrical micelles consisting of less ordered hydrophilic and hydrophobic compartments [16, 25, 26]. One type of PAs can self-assemble into one-dimensional nanofibres with cylindrical geometry with the possibility of being highly bioactive and is generally regarded as a great choice of interest in drug delivery [22, 27]. It has been found that micelle-forming peptide amphiphiles [PAs] have excellent antimicrobial activity against both Gram-positive and Gram-negative pathogenic bacteria [1]. The use of micelle as a drug delivery system offers some advantages like improved drug solubility, sustained release of encapsulated molecules, ability to target cells of interest to reduce the side effects, increased effectiveness, *etc* [28].

PAs are an excellent platform for antimicrobial action. The antimicrobial activity of PAs is dependent on various parameters like length of the alkyl tail, hydrophobicity of PAs, morphology of the pathogens, *etc*. PAs can increase the permeability of the cellular membrane of pathogens and disrupt the integrity of the membrane, which ultimately leads to cell lysis and death. Thus, acting as nanocarriers, PAs can contribute to developing synergistic antibacterial therapies

[1]. PAs are also considered potential therapeutic agents. They can transport hydrophobic drugs to a specific site and be biodegraded and metabolized into amino acids and lipids. In this way, PAs can also be easily removed from the kidneys. The hydrophobic tail increases bioavailability as it can travel across the cellular membrane. On the other hand, the peptide epitope can target a specific cell *via* receptor-ligand complex [23].

The peptide moiety on the surface of the nanostructures enables various biological and chemical functions. A dense hydrocarbon-like microenvironment is created within an aqueous gel by the self-assembling of PAs into nanostructures. This environment created by the self-assembled PA nanostructures makes them an ideal candidate for the delivery of hydrophobic molecules *in vivo*. Moreover, PA nanostructures with bioactive sequences show properties like pore formation in the membranes or the capability to reach a biological target in a selective way [26]. PA-based agents are biodegradable, biocompatible and demonstrate various biological actions [1, 29]. Other advantages of PAs include tunable bioactivity, low cost, chemical diversity, high drug loading capacity, specific targeting, *etc*. Thus, PAs are excellent drug delivery agents [16]. PAs can demonstrate their activity in two ways. They can demonstrate their activity either in a constructive way [*e.g.*, cell signal transduction] or in a destructive way by destroying the integrity of cellular membranes [13].

CLASSIFICATIONS OF PEPTIDE AMPHIPHILES [PAS]

Various classes of PAs include true PAs consisting only of both polar and nonpolar amino acids, hydrophilic peptides linked to hydrophobic lipid alkyl chains and peptide-based block copolymers [13, 19, 23].

Peptide Amphiphiles Consisting only of Amino Acids

Peptides and proteins that perform specific biological functions in membranes like membrane fusion, pore formation or signal transductions are needed to be amphiphilic [13]. PAs merely made of amino acids are organized into amphipathic sequences. When properly folded, amphipathic sequences contain both hydrophobic and hydrophilic domains. The peptides fold into different helices and sheets in order to allow the hydrophobic domains to interact with the interior surface of the membrane and the hydrophilic domains to get exposed to the aqueous environment. Again, in order to optimize the interactions with the surrounding environments, PAs can also self-assemble into lipid bilayers. An effective antimicrobial mode of action of PAs includes peptide-mediated permeabilization of biological membranes. But, unfortunately, most of the natural PAs don't possess suitable characteristics to be used as antibiotics. That's why it is important to understand the self-assembling properties and interactions with the biological membrane of all-amino acid-based PA model compounds [13, 30].

Hydrophilic Peptides Linked to Hydrophobic Lipid Alkyl Chains

Another class of PAs includes hydrophilic peptides linked to hydrophobic lipid alkyl chains [13, 23]. Lipidated PAs occurs naturally and have some definite biological functions. Examples of lipidated PAs include tyrosine kinase, guanine nucleotide peptides, *etc*. Lipidated PAs possess two main regions. The first region is a hydrophobic alkyl lipid chain linked to a hydrophilic peptide sequence which forms the head group. The aggregation of the hydrophobic tails governs the self-assembly in an aqueous solution. The peptide epitope can play crucial biological roles like signal transduction, cell adhesion, cell growth, *etc* [23]. It has been found that lipidated PAs demonstrate a broad range of activities against pathogenic bacteria like MDR *Pseudomonas aeruginosa,* MDR *Staphylococcus epidermidis, etc.* The length of the alkyl chain tail of lipopeptides is the key factor for interactions with the biological membrane [1].

Lipidated Peptides with Single Alkyl Chains

Lipidated peptides [*e.g.*, members of Src and Ras family proteins] play a crucial role in biological signal transduction pathways. It is assumed that lipid groups are associated with protein-lipid and protein-protein interactions and are believed to act as anchors to biological membranes. Additionally, N-terminally alkylated peptides can bind to iron and these peptides are found to be self-assembled into micelle, which in the presence of Fe [III] ions, can transform into vesicles. It has been found that a special group of lipopeptides with cyclic peptide head groups [*e.g.*, surfactins] possess antibacterial, antiviral or even anti-tumoural properties [13, 31].

Lipidated Peptides with Multiple Alkyl Chains

Hydrophobicity in the lipidated peptides can be increased by adding double or triple-chain alkyl moieties to the C terminus or N-terminus of lipidated peptides. Effective low molecular weight carriers can be prepared from the triple-chain lipopeptides. Again, the assembly and folding of peptides are affected by the addition of hydrophobic double alkyl chains [13, 32]. It has been found that dipeptides containing N-terminus double-chain hydrophobic moieties can interact specifically with water-soluble dipeptides [13].

Peptide-based Block Copolymers

Amphiphilic block copolymers are made up of blocks of dissimilar monomers, which have different polarities and affinities for aqueous solutions [33]. In peptide-based block copolymers, polymers are made up of amino acids along with hybrid systems connecting peptides with synthetic polymers. For self-assembled materials, this kind of block polymer serves as larger-sized building blocks [13]. Again, due to the hybrid characteristics, amphiphilic block copolymers are regarded as ideal drug delivery agents. Amphiphilic block copolymers must fulfill some criteria to serve as a drug delivery agent. These criteria include high drug loading capacity, controlled drug release, non-toxicity, smaller size for vascular permeability, *etc* [33].

On contrary to the above-mentioned classification of PAs, Dasgupta A and Das D have classified PAs as amphiphilic peptides, lipidated PAs and a new class of supramolecular PA conjugates. They have defined amphiphilic peptides as a class of PAs made up of amino acids only and consisting of both hydrophobic and hydrophilic amino acids. In lipidated PAs, one or two lipid groups are attached at the C-terminal of the amino acids of proteins. Again, they have claimed that supramolecular PA conjugates are the combination of conventional amphiphiles with supramolecular chemistry [34].

SYNTHETIC AMPHIPHILES

Synthetic amphiphiles have come out as a great bactericidal agent due to their membrane-directed mechanism of action and their high tendency to interact with the cells of bacteria [6]. Antimicrobial Peptides [AMPs] are an important component of the innate immune system and demonstrate a broad range of antimicrobial properties [30, 35]. The mode of action of AMPs includes penetrating and initiating physical lysis of bacterial membrane, which results in cell death by leakage of essential ions, nutrients and cytoplasmic components [1, 36]. Thus, AMP mimetic synthetic amphiphiles have emerged as an interesting class of therapeutic antibacterial because of their easy process of synthesis, structural tunability, tendency to interact with and disrupt membranes and proteolytic resistance [6, 29].

Bacterial biofilm formation, a virulence determinant related to many pathogenic bacterial infections, is of great concern because it is associated with chronic infections as well as device-associated infections [1, 6]. In general, dynamic communities of immobile microbes attached to a solid surface are termed biofilms [11, 37]. Bacterial infections are mostly due to their growth as biofilm and not because of their planktonic mode of growth. Biofilm-forming bacteria are more recalcitrant because of the anaerobic and acidic conditions in the biofilms and the low penetration and accumulation of antibiotics in the biofilms [38]. Biofilm formation can also increase bacteria resistance to antibiotics [1]. Bacterial strains from genera of *Staphylococcus aureus* and *Pseudomonas aeruginosa* are regarded as biofilm formers [35]. Other examples of biofilm-forming bacteria include *Moraxella catarrhalis, Streptococcus viridans, Proteus mirabilis, Klebsiella pneumonia, Escherichia coli, Enterococcus faecalis,etc* [11]. As the action of synthetic amphiphiles is initiated through biological membrane interactions leading to the substantial disruption of membranes, it is anticipated that synthetic amphiphiles hold remarkable potential in preventing bacterial biofilm formation. Again, because of the low molecular weight and smaller size, these synthetic amphiphiles are likely to have some therapeutic and pharmacological advantages, such as effective passage through biological membranes, higher biological fluid solubility and increased tissue distribution [6].

Based on the cell wall structure, bacteria are classified into Gram-positive and Gram-negative bacteria. The cell wall of Gram-positive bacteria possesses a thick peptidoglycan layer and a cytoplasmic membrane beneath it. On the other hand, Gram-negative bacteria possess a cytoplasmic membrane followed by a thin layer of peptidoglycan which is further surrounded by a lipid bilayer of lipopolysaccharides [5, 30]. In particular, Gram-negative bacteria are very difficult to kill because the outer membrane of Gram-negative bacteria serves as a strong permeability barrier in the transportation of bactericidal agents. Thus, the outer membrane in Gram-negative bacteria allows them to resist the action of bactericidal agents and makes many therapeutic antibiotics ineffective. Very often, Gram-negative bacteria also possess multiple efflux pumps and target modifying enzymes, which makes them more challengeable to kill [6, 39 - 41]. Being a membrane-acting agent, synthetic amphiphiles have the potential to permeabilize the outer membrane and enable the increased uptake of therapeutic antibiotics. Thus, synthetic amphiphiles have the capacity to increase their bactericidal efficacy [6].

The concerns that need to be resolved in the context of the therapeutic applications of synthetic amphiphiles involve the insufficient release of drug, after administration non-specific localization of drugs and reduction in the effective concentration of drugs due to the less stability and low solubility. In this regard, for effective entrapment and controlled release of synthetic amphiphiles, the development of nanocarriers can be a feasible strategy. The development of nanocarriers using nanomaterials provides some noteworthy advantages like increased solubility of therapeutic agents, increased drug release, improved half-life, targeted delivery, *etc.* [6].

NANOCARRIERS

Eradication of intracellular infections remains challenging because of the poor penetration of antibiotics. In fact, the intracellular localization provides a favorable niche for bacteria and also helps them to keep themselves protected from the host immune system as well as from the action of antibiotics. Thus, new strategies are needed to elude these problems. In this perspective, nanocarriers loaded with antibiotics appear to be a promising approach. Nanocarriers can encapsulate and conjugate antibiotics to deliver them intracellularly in order to treat intracellular infections [42]. Nanocarriers can encapsulate and deliver not only drugs but also compounds like enzymes, toxins, nutraceuticals, dyes *etc* [43]. Furthermore, the efficacy of antimicrobials can be increased using nanocarriers by successfully transporting them to the target site. The use of nanoparticles as a drug delivery system can also provide increased retention time and reduced non-specific distribution of antimicrobials [5]. Nanoparticles have a broad range of antimicrobial activities which can inhibit bacterial growth and also the growth of drug-resistant bacteria. Nanoparticles has the ability to cross the blood-brain barrier and can be utilized in the treatment of brain infections [44]. Cell-specific targeting can also be done by attaching drugs to designed nanocarriers. A drug can be transported to a specific place of action, and side effects can also be minimized. As a result, the accumulation of drugs at the target site is increased, and the required doses of drugs can also be lowered. Furthermore, a drug can be protected from degradation by attaching the drug to nanocarriers. Because of the smaller dimension, nanocarriers can cross the blood-brain barrier and can function at the cellular level [45].

Examples of commonly used amphiphilic nanocarriers include liposome, niosome, polymersome, cubosome, dendrimer, micelle, hydrogel, *etc.* (Table **1**) [43, 46 - 50]. The characteristic features, advantages and disadvantages of these amphiphilic nanocarriers are discussed below:

Table 1. Examples of amphiphilic nanocarriers with their remarkable features, advantages and the pathogens against which the nanocarriers can be used.

Amphiphilic Nanocarriers	Remarkable Features	Advantages	Susceptible Bacteria	References
Liposome	Bilayer phospholipid vesicle	Biodegradability, bio compatibility, low immunogenicity, controlled drug release, selective drug delivery, ability to entrap both hydrophobic and hydrophilic molecules *etc.*	*Mycobacterium tuberculosis, Salmonella enterica* serovar *Typhi*	[43, 51, 52, 72]

Amphiphilic Nanocarriers	Remarkable Features	Advantages	Susceptible Bacteria	References
Niosome	Nonionic surfactant vesicles	Biodegradability, biocompatibility, non-immunogenicity, low cost, less toxicity, chemical stability, controlled drug release, ability to carry both hydrophobic and hydrophilic molecules *etc.*	*Brucella neotomae, Brucella ovis, Staphylococcus aureus*	[43, 51, 54, 56, 57]
Polymersome	Polypeptide-based block copolymer vesicle	Colloidal stability, biodegradability, capability to encapsulate a broad range of antimicrobial agents, tunable membrane properties, controlled drug release, ability to respond to outer stimuli *etc.*	*Porphyromonas gingivalis, Escherichia coli*	[42, 46, 58, 59]
Cubosome	Liquid crystalline cubic phase nanoparticles	Ordered structure, compatible with both water-soluble and water-insoluble drugs, *etc.*	*Escherichia coli, Staphylococcus aureus, Pseudomonas aeruginosa, Acinetobacter baumannii*	[9, 43, 47, 60, 61]
Dendrimer	Immensely branched structured	Increase the stability and solubility of the drug, allows grafting of various functional groups, allows control of size, flexibility and topology *etc.*	*Pseudomonas aeruginosa, Mycobacterium avium, Mycobacterium xenopi, Mycobacterium intracellulare*	[38, 48, 62, 63, 73]
Micelle	Biodegradable colloidal system	Biodegradability, protects the drug from degradation, allows simultaneous delivery of various therapeutic drugs *etc.*	*Escherichia coli, Staphylococcus aureus, Pseudomonas aeruginosa, Mycobacterium tuberculosis*	[64 - 66, 74, 75]
Hydrogel	Three-dimensional networks of polymers	Sustained drug release, bioadhesive properties, biocompatibility, non-toxicity, biodegradability, responsiveness to various environmental stimuli, encapsulation of high amount of water, *etc.*	*Bacillus subtilis, Staphylococcus aureus, Escherichia coli, Klebsiella pneumonia, Staphylococcus saprophyticus, Enterococcus spp.*	[50, 67, 69 - 71]

Liposomes

Liposomes, small vesicles made up of amphiphilic phospholipid bilayer, are one of the promising amphiphilic nanocarriers [43, 51]. They are one of the most widely used nanocarriers as drug delivery agents [52]. Properties like biodegradability, biocompatibility and low immunogenicity have made liposomes as well-established drug carriers. Liposomes can entrap both hydrophobic and hydrophilic molecules, which serves as shielding and prevents degradation. It has been found that positively charged liposomes are more cytotoxic than neutral liposomes and anionic liposomes. In blood serum, positively charged liposomes also have a shorter half-life than neutral liposomes and anionic liposomes [43]. Liposomes are also reported to improve selective drug delivery and also to increase the therapeutic index of antimicrobial drugs. Liposomes are very suitable to carry drugs to the site of infection where the pathogenic bacteria reside [51]. They are also efficient in transdermal drug delivery, which can lessen the amount of drug permeation through the skin. Thus, it can increase the systemic concentration of the drug [53]. One of the promising applications of liposomes is to deliver the drug through implants. Liposomes can be immobilized into biodegradable implants for controlled drug release for the treatment of infections or even cancers. Implants are inserted at the target site or close to the target site [43]. But major problems associated with liposomes are low chemical stability during administration, high cost, *etc*. Upon storage, liposomes have the tendency to lose structural configuration and thus can result in leakage of encapsulated payloads [51, 54, 55].

Niosomes

Niosomes, composed of nonionic surfactant vesicles and cholesterol bilayers, can be used to carry both hydrophobic and hydrophilic molecules for the controlled release of drugs. It also increases the efficacy of the drugs. Niosomes have attained much attention due to their properties like biodegradability, biocompatibility and non-immunogenicity. Other advantages of niosomes include low cost, easy storage, easy handling, less toxicity and chemical stability. Thus, niosomes are very good drug delivery agents with good efficacy to fight against pathogenic bacteria [43, 51, 56, 57]. The permeability of small ions is much higher in niosomes which makes them attractive drug delivery agent [56]. Niosomes have the capability to encapsulate multiple drugs [57]. Though niosomes appear to be alike in physical properties to liposomes, they can lessen the problems associated with liposomes like chemical instability and high cost [54]. Despite these advantages, the disadvantages of niosomes include fusion and aggregation of vesicles [43].

Polymersomes

Polymersomes are one kind of self-assembled vesicles made up of polypeptide-based amphiphilic block copolymers which can effectively encapsulate a broad range of molecules. These spherical vesicles are called pepsomes [42, 46, 58]. It is possible to load hydrophilic bioactive molecules into the aqueous core of polymersomes and hydrophobic bioactive molecules to be loaded into the membrane bilayer of polymersomes [58]. The nano-aggregates of polymersomes provide numerous advantages like colloidal stability, biodegradability, capability to encapsulate a broad range of antimicrobial agents, large compartments to encapsulate biofunctional molecules, tunable membrane properties, controlled drug release, ability to respond to outer stimuli *etc*. Again, the vast aqueous interior core of polymersomes provides a suitable site to load proteins. Because of the thicker membrane of polymersomes, they have high membrane stability and low membrane permeability. By changing the species and lengths of copolymer blocks, it is possible to modulate the properties of the membranes of polymersomes [46]. It is possible to encapsulate water-soluble drugs into the aqueous core of polymersomes, and also, the lipophilic and amphiphilic drugs can be loaded into the membranes of polymersomes [59]. In this way, polymersomes are considered as more potential drug delivery agents as they can be loaded with both hydrophobic and hydrophilic antimicrobial agents [38]. Despite these remarkable advantages, polymersomes have some drawbacks also. The tough membrane of polymersomes may inhibit the fast release of drugs at the target site and thus can reduce the therapeutic efficacy of the drugs [46].

Cubosomes

Cubosomes can be defined as nanoparticles made up of liquid crystalline phase with cubic crystallographic symmetry created by the self-assembled amphiphilic molecules or by the surfactant-like molecules. Although cubosomes are made up of nanoparticles, they are not solid in nature [47]. Dispersion of a liquid crystalline cubic phase in the presence of a stabilizer is needed in order to form cubosomes [60]. However, cubosomes have unique characteristics in that they have high solid-like viscosities due to their bicontinuous structure. Thus, they can be treated as solid nanoparticles because of their high viscosity. As cubosomes can create biological lipids, they have the capability to solubilize biological molecules like proteins and possess a tortuous structure [47]. Cubosomes are considered great drug delivery agents for a wide range of drugs. Due to ordered structure and alternating hydrophobic and hydrophilic domains, they have unique delivery features, and they are compatible with both water-soluble and water-insoluble drugs [60]. AMP LL-37 is a 37 amino acid-long amphiphilic peptide from the human cathelicidin family. It has been found when cathelicidin AMP

LL-37 is loaded with cubosome; they collectively demonstrate antimicrobial activity [60, 61].

Dendrimers

Dendrimers are synthetic macromolecules of nanoscopic dimensions with immensely branched and star-shaped structures and demonstrate a broad range of efficacy against various planktonic bacteria. It is made up of several layers of branches positioned around a central core [38, 48, 62]. During the synthesis of dendrimers, the size, flexibility and topology of dendrimers can be heavily controlled. Dendrimers are prepared by repetition of a sequence of reactions emerging from a core molecule in order to develop repeating branched units. This allows the grafting of various functional groups on the outer shell that can interact with other molecules. Dendrimers are regarded as complex molecules of antimicrobial drugs. Dendrimers as drug delivery agents can efficiently increase the stabilization and solubility of drugs. However, one major limitation of polycationic dendrimers includes cytotoxicity. The toxicity of cationic dendrimers to eukaryotic cells has also been reported. The mechanism of cytotoxicity includes disruption of the integrity of cell membranes and an increase in the intracellular reactive oxygen species [ROS], which have the capability to induce apoptosis [63]. But, the cytotoxicity of dendrimers depends on the core and is highly influenced by the surface nature of dendrimers. The drugs can be attached to the internal surface of dendrimers or physically absorbed on the surface of dendrimers [45].

Micelles

PAs can self-assemble into various micellar structures, which can provide a multivalent functional peptide [49]. Polymeric micelle is a self-assembling biodegradable colloidal system. It is formed by a core-shell structure and created by the self-assembly of the amphiphilic copolymer. The hydrophobic core has the capability to encapsulate hydrophobic drugs, and the hydrophilic outer shell makes the drug more soluble. When the drug is loaded with the micelle, it protects the drug from degradation [64, 65]. The shape of the micelle is controlled by the size, conformation, hydrophilicity *etc*. The peptide secondary structure of the micelle is very important both in terms of functionality and the shape of the micelle. In many cases, the stability of the micelle is important because the stability determines whether the micelle will reach its target intact or it will be disintegrated to release its contents [49]. Block copolymers can self-assemble into micelles in selective solvents. They can be nanoreactors for silver nanoparticle synthesis to ensure stability, and solubility in appropriate solvents and to prevent them from aggregation [66]. In fact, polymeric micelles are a great candidate for growing silver nanoparticles and to stabilize them. They also offer advantages to simultaneously deliver various therapeutic drugs. The silver nanoparticles embedded micelle shell has the capability to damage the bacterial membrane structure [65].

Hydrogels

Hydrogels, a class of soft materials, can be defined as three-dimensional networks of polymers crosslinked by either physical interactions or covalent bonds consisting of large amounts of water by hydrophilic groups which are hydrated in an aqueous environment and which are similar to the natural extracellular matrix [ECM] [50, 67 - 69]. Hydrogels have become an excellent platform for delivering antimicrobial agents because of their porous structure and tailored functionality [68]. Hydrogels encapsulated with a bactericidal agent are promising in the treatment of surgical site infections. The drugs are generally released from the hydrogel *via* hydrogel degradation or passive degradation [70]. The main advantages of hydrogels include local drug delivery, sustained release of drugs, bioadhesive properties, high biocompatibility, controlled biodegradability, high oxygen permeability, non-toxicity, tunable properties, versatility to fabrication, responsiveness to various environmental stimuli, encapsulation of high amounts of water *etc*. These events also increase bioavailability and minimize bactericidal agents' cytotoxicity [50, 69 - 71]. Hydrogels provide a high surface area-t--volume ratio. Hydrogels also provide structural controllability, which makes it possible for hydrogels to selectively release loaded drugs at target sites [69].

DRUG RESISTANCE

The emergence and expansion of antibiotic resistance in pathogenic bacteria have become a global threat to both humans and animals [39, 76, 77]. Immense use, overuse and misuse of antibiotics over the several decades have increased the frequencies of resistance in pathogenic bacteria and resulted in significant medical problems. The antibiotic resistance mechanism in bacteria not only modifies their ability to survive against antibiotics but also changes their interactions with the environment and the host [2, 41, 77 - 79]. Antibiotics function upon bacteria by inhibiting cell wall synthesis as well as interfering with the synthesis of key proteins, DNA and RNA. But, bacteria have the intrinsic capability to evolve through genetic mutations and horizontal gene transfer [5]. In reality, along with intrinsic resistance and genetic mutations, antibiotic resistance in bacteria can be developed by several other mechanisms like poor penetration of antibiotics into the bacteria or antibiotic efflux, modifications of the antibiotic target by post-translational modifications, inactivation of antibiotics by hydrolysis or by other modification processes, *etc* [10, 39].

Again, MDR has become an emergent crisis for the treatment of infectious diseases [11]. In fact, the treatment of infections has been compromised by the emergence of MDR bacteria. MDR bacteria has now become a global problem and crossed international borders affecting members of all socio-economic classes extensively [10]. Undeniably, antibiotic resistance in pathogenic bacteria is as life-threating as cancer in the number of cases and probable outcomes. The human and economic cost created by antibiotic resistance is also immense. For example, in 2007 in Europe, infections caused by MDR bacteria were around 4,00,000, which caused around 25,000 deaths [39, 76]. In the United States, antibiotic-resistant bacteria infect more than 2 million people annually, along with around 23,000 direct deaths [39]. Drug-resistant pathogens like *Enterococci, Klebsiella pneumoniae, Staphylococci, Pseudomonas, etc.*, have created prolonged illness and increased the risk of death. These drug-resistant pathogens, also called as "super bugs", have emerged due to the intrinsic resistance to antibiotics [1, 8]. In fact, they have reacted to the man-made assault and adapted themselves to the changing environment by developing resistance. This adaptation is remarkably quick, and the indiscriminate use of antibiotics in humans, veterinary medicine and in the field of agriculture helped these superbugs to attain resistance by horizontal gene transfer and genetic mutations [1, 2, 4, 38].

ROLE OF AMPHIPHILIC NANOCARRIERS TO FIGHT AGAINST PATHOGENIC AND MULTIDRUG-RESISTANT [MDR] BACTERIA

Numerous diseases are caused by pathogenic and MDR bacteria. A brief discussion on some of the selected bacterial diseases caused by the pathogenic and MDR bacteria and the role of amphiphilic nanocarriers in to fight against that pathogenic and MDR bacteria are discussed here:

Tuberculosis

Tuberculosis, commonly known as TB, is caused by a Gram-positive pathogenic bacteria called *Mycobacterium tuberculosis* [80]. In 1993, tuberculosis was declared a public health emergency by the World Health Organization [WHO] [52]. Exposure to *Mycobacterium tuberculosis* does not essentially develop an active infection, and most of the people show symptom-free latent infection. But, susceptible individuals and individuals with poor immunity may develop an active infection. Being a Gram-positive bacteria, the thick cell wall of *Mycobacterium tuberculosis* serves as a strong permeability barrier, making them resistant to various antimicrobial agents. Furthermore, *Mycobacterium tuberculosis* may reside within the macrophages of the lungs. The treatment of *Mycobacterium tuberculosis* infections requires a long-term antibiotic therapy with a combination of drugs associated with side effects. These may contribute to side effects and lead to low patient compliance. These can also help to develop drug resistance [52, 80]. Liposomes have great potential as a drug delivery agent in the treatment of tuberculosis. The use of liposomes as a drug delivery agent increases the therapeutic index of anti-tuberculosis drugs. The administration of liposomes loaded with anti-tuberculosis drugs through inhalation has several benefits high local concentrations of drugs in the alveolar macrophages, reduction in the adverse effects and in frequency of administration. Thus, liposomes are regarded as attractive nanocarrier systems for anti-tuberculosis drugs and have great potential for pulmonary tuberculosis treatment [52].

Typhoid Fever

Typhoid fever is caused by *Salmonella enterica* serovar *Typhi (S. Typhi)*. This bacterium is ingested through contaminated water and food, and then it enters into the intestinal epithelial cells. After that, it resides and grows within the macrophages and later on spreads to the liver and spleen. The treatment approach to *S. Typhi* involves the use of antibiotics, but resistance to antibiotics for this pathogen has already been reported [80]. Antibiotic ciprofloxacin has been used in the treatment of *Salmonella* infections, including individuals with typhoid fever and chronic typhoid carriers. It has been found that liposome-incorporated ciprofloxacin showed superior activity than aqueous ciprofloxacin. The superiority of ciprofloxacin in *Salmonella* infections is most likely because of the intracellular penetration of ciprofloxacin [72].

Brucellosis

Brucellosis is one kind of bacterial infection caused by the *Brucella* species, which is a nonmotile Gram-negative coccobacillus. It is endemic in many countries of Asia, like China, India, Iran, Iraq, Saudi Arabia, *etc.Brucella* species reside as a cluster in the cytoplasm and don't create spores or capsules. Two *Brucella* species, namely *Brucella neotomae* and *Brucella ovis* have been reported to infect sheep and rats, respectively, but have not been reported yet to infect humans. The conventional treatment of brucellosis doesn't help to eradicate this disease. Niosomes have created a novel approach to treat brucellosis. Niosomes interact with the cell membranes of bacteria by fusion and unloading the encapsulated drug directly into or on the bacterial cell. Thus, niosomal levofloxacin can be a great approach to treat brucellosis. Encapsulation of levofloxacin also increases its antimicrobial activity several folds [56].

Food Poisoning, Nosocomial Infection, Endocarditis and Toxic Shock Syndrome

Staphylococcus aureus, an opportunistic pathogen, is responsible for a wide range of infections like food poisoning and nosocomial infections. It is also responsible for life-threating diseases like endocarditis and toxic shock syndrome. The emergence of MRSA made treatment of *Staphylococcus aureus* infections more challenging. Ciprofloxacin is used as an alternative to vancomycin for the treatment of MRSA infections. Unfortunately, 89% of MRSA infections are resistant to ciprofloxacin. Thus, treatment of ciprofloxacin-resistant MRSA is difficult with common antibiotics. Moreover, biofilm formation is a major factor in the development of MDR *Staphylococcus aureus*. But, niosome encapsulated ciprofloxacin is a good drug delivery system with biofilm inhibitory efficacy and improved antimicrobial activity against ciprofloxacin-resistant MRSA [57].

Periodontitis

Periodontitis is a common oral disease characterized by inflammation of the periodontium, which leads to progressive destruction of the tooth and, in some cases, may lead to the loss of the tooth. *Porphyromonas gingivalis*, a Gram-negative anaerobic bacterium, plays a crucial role in the initiation and progression of this disease. Although *Porphyromonas gingivalis* is susceptible to antibiotics, they can escape antibiotic action by residing within gingival keratinocytes. Polymersomes are efficient drug delivery agents for antibiotics that generally don't have access to host cells. Thus, polymersomes can be used as drug delivery agents to deliver antibiotics within monolayers of keratinocytes in order to kill *Porphyromonas gingivalis* [59]. Furthermore, it has been found that vancomycin-loaded polymersomes demonstrate a stronger impact on the membrane disruption of MRSA compared to free vancomycin [55]. Again, inhibition of *Escherichia coli* by the silver nanoparticles-embedded polymersomes has also been reported [58].

Miscellaneous

Antimicrobial peptide (AMP) LL-37, a peptide from the cathelicidin family, demonstrates a broad range of antibacterial activity. It is sensitive to proteolytic degradation by bacterial elastase enzyme, which limits the therapeutic use of AMP LL-37. But, cubosomes efficiently protect AMP LL-37 from proteolysis. Furthermore, cubosomes loaded with AMP LL-37 show increased antibacterial effects against *Escherichia coli*, *Staphylococcus aureus* and *Pseudomonas aeruginosa*. In this way, cubosomes provide the potential to protect peptide drugs from degradation [60, 61]. It is important to note that the drug delivery mechanism of cubosome and how they interact with bacterial cells is not clearly understood. But, Boge *et al.* reported that cubosomes loaded with cathelicidin AMP LL-37 distorted the membrane of the bacteria. It has been found that the contact between AMP LL-37 loaded cubosome and bacteria modulates bacteria in a lethal way. Thus, peptide-loaded cubosome collectively can exert antimicrobial activity [60]. Furthermore, it has been reported that phytantriol-based cubosomes have bactericidal activity against lipopolysaccharide-deficient, polymyxin-resistant *Acinetobacter baumannii*. Thus, cubosomes can be a novel strategy to fight against lipopolysaccharide-deficient Gram-negative bacteria [9].

Due to the synergistic effect, dendrimers and antibiotics demonstrate satisfactory antibacterial activity. Antibacterial activity of ammonium and amine-terminated carbosilane dendrimers has been reported. Furthermore, it has been found that cationic dendrimers are potential candidates for the development of antibacterial drugs. The antibacterial mechanism of action of dendrimers is related to the permeability effect of the cellular surface of bacteria. An increase in the cell permeability of *Pseudomonas aeruginosa* has been reported after incubation with various poly propyleneimine dendrimers [48, 63]. Several environmental nontuberculous mycobacteria like *Mycobacterium avium, Mycobacterium xenopi, Mycobacterium intracellulare, etc.*, are opportunistic human bacteria. Several species of mycobacteria cause subcutaneous and dermal infections in humans. It has been reported that dendritic amphiphiles can be a good platform to develop anti-mycobacterial drugs. But, the mechanism of action of dendritic amphiphiles can be species-specific [73]. Again, the mechanism of action of host defense peptides [HDPs] mimicking lipidated dendrimers demonstrates selective and potent antibacterial activity against both Gram-positive and Gram-negative bacteria as well as MDR strains of bacteria. It has been found that amphiphilicity is required for these broad spectra of antibacterial activity against Gram-positive, Gram-negative and MDR bacteria. These dendrimers not only demonstrate antibacterial activity against planktonic bacteria but also efficiently inhibit biofilm formation [62].

Many micelles have been prepared in the treatment of bacterial infections with increased efficacy, solubility and bioavailability. For instance, poly[D,L-lactide--co-poly[glycolide] [PLGA], a hydrophobic copolymer, has been developed to encapsulate hydrophobic drugs. It has been reported that when azithromycin is loaded with PLGA, the antibacterial activity of azithromycin has improved with its increasing distribution to phagocytic cells [64]. The micelles created by the self-assembly of amphiphilic poly[vinyl alcohol]-b-poly[acrylonitrile] block copolymers embedded with silver nanoparticles demonstrate a strong bactericidal activity against *Escherichia coli, Staphylococcus aureus* and *Pseudomonas aeruginosa* [66].

We already have mentioned that *Mycobacterium tuberculosis* is the causative agent of tuberculosis and rifampicin is a choice of drug in the treatment of tuberculosis. The high hydrophobicity of rifampicin is a major issue in using this drug which requires a specifically designed drug delivery system. Tripodo *et al.* designed two micelles, namely inulin functionalized with vitamin E [INVITE] and succinylated derivative of INVITE [INVITESA], for carrying the anti-tuberculosis drug rifampicin. It has been found that the delivery of rifampicin by INVITESA demonstrated higher bactericidal activity with respect to INVITE [74]. Again, the antibacterial activity of cationic micelles against *Escherichia coli* has also been reported. Cationic micelles kill *Escherichia coli* through two-step processes. Firstly, by the electrostatic interaction between micelles and the surface of *Escherichia coli*, cationic micelles disrupt the outer membrane integrity of *Escherichia coli*. Secondly, through the hydrophobic interaction between the hydrocarbon chains of the surfactants and the lipid of *Escherichia coli,* cationic micelles disintegrate the *Escherichia coli* inner membrane. These events eventually lead to bacterial death [75].

Peptide-based hydrogels have demonstrated significant antibacterial activity against both Gram-positive and Gram-negative bacteria like *Bacillus subtilis, Staphylococcus aureus, Escherichia colietc.* Moreover, hydrogels are proteolytically stable [67, 71]. Again, the inherent antibacterial activity of a peptide-based *β*-hairpin hydrogel has been reported [81]. The antibiotic ciprofloxacin has a broad spectrum of antimicrobial activity against both Gram-positive and Gram-negative bacteria and is used in the treatment of skin and eye infections. Ciprofloxacin-loaded self-assembled hydrogel demonstrates a broad range of antimicrobial activity against *Klebsiella pneumonia, Staphylococcus aureus* and *Escherichia coli*. Gentamicin is another traditional broad-spectrum antibiotic used in the treatment of skin infections and wounds. Posadowska *et al.* prepared a drug delivery system consisting of gentamicin-loaded poly[lactide-c--glycolide] embedded in gellan gum hydrogel, which demonstrates antibacterial activity against *Staphylococcus saprophyticus*. Recently, vancomycin-resistant *Enterococcus* has been discovered. It has been found that using hydrogels as a drug delivery system can protect and improve the effectiveness of vancomycin [69].

CONCLUSION

One of the major challenges to fighting against pathogenic bacteria involves long-term antibiotic therapy with combinations of drugs. Side effects also arise depending on the toxicity and the duration of the drug exposure. Another major challenge is to deliver enough drugs to the site of infections. Again, many antibiotics have a shorter half-life, which requires a frequent and large volume of doses. The most important challenge to fight against pathogenic bacteria is the development of antibiotic resistance to pathogenic bacteria [80]. Nanocarrier-based drug delivery systems can mitigate toxicity issues and the adverse effects associated with high antibiotic doses [5]. Encapsulation of antibiotics into nanocarriers has offered some advantages like the alleviation of side effects, enhanced drug solubility, reduction in the frequency of antibiotics administration, increased antimicrobial activity to fight against pathogenic bacteria, *etc.* Although complete eradication of bacteria in the infected site has not been completely possible yet with the advent of nanocarriers, it has definitely contributed to improve the well-being of patients [42].

CONSENT FOR PUBLICATION

Not applicable.

CONFLICT OF INTEREST

The authors declare no conflict of interest, financial or otherwise.

ACKNOWLEDGEMENT

Declared none.

REFERENCES

[1]   Rodrigues de Almeida N, Han Y, Perez J, Kirkpatrick S, Wang Y, Sheridan MC. Design, synthesis, and nanostructure-dependent antibacterial activity of cationic peptide amphiphiles. ACS Appl Mater Interfaces 2019; 11(3): 2790-801.
[http://dx.doi.org/10.1021/acsami.8b17808] [PMID: 30588791]

[2]   Odonkor ST, Addo KK. Bacteria resistance to antibiotics: recent trends and challenges. Int J Biol Med Res 2011; 2(4): 1204-10.

[3]   Melander RJ, Melander C. The challenge of overcoming antibiotic resistance: An adjuvant approach? ACS Infect Dis 2017; 3(8): 559-63.
[http://dx.doi.org/10.1021/acsinfecdis.7b00071] [PMID: 28548487]

[4]   Sköld O. Resistance to trimethoprim and sulfonamides. Vet Res 2001; 32(3-4): 261-73.
[http://dx.doi.org/10.1051/vetres:2001123] [PMID: 11432417]

[5]   Gupta A, Mumtaz S, Li CH, Hussain I, Rotello VM. Combatting antibiotic-resistant bacteria using nanomaterials. Chem Soc Rev 2019; 48(2): 415-27.
[http://dx.doi.org/10.1039/C7CS00748E] [PMID: 30462112]

[6]   Uday SP, Thiyagarajan D, Goswami S, Adhikari MD, Das G, Ramesh A. Amphiphile-mediated enhanced antibiotic efficacy and development of a payload nanocarrier for effective killing of pathogenic bacteria. J Mater Chem B Mater Biol Med 2014; 2(35): 5818-27.
[http://dx.doi.org/10.1039/C4TB00777H] [PMID: 32262025]

[7]   Fischbach MA, Walsh CT. Antibiotics for emerging pathogens. Science 2009; 325(5944): 1089-93.
[http://dx.doi.org/10.1126/science.1176667] [PMID: 19713519]

[8]   Li J, Nation RL, Turnidge JD, *et al.* Colistin: the re-emerging antibiotic for multidrug-resistant Gram-negative bacterial infections. Lancet Infect Dis 2006; 6(9): 589-601.

[http://dx.doi.org/10.1016/S1473-3099(06)70580-1] [PMID: 16931410]

[9]   Lai X, Ding Y, Wu CM, *et al.* Phytantriol-based cubosome formulation as an antimicrobial against lipopolysaccharide-deficient Gram-negative bacteria. ACS Appl Mater Interfaces 2020; 12(40): 44485-98.
[http://dx.doi.org/10.1021/acsami.0c13309] [PMID: 32942850]

[10]   Alekshun MN, Levy SB. Molecular mechanisms of antibacterial multidrug resistance. Cell 2007; 128(6): 1037-50.
[http://dx.doi.org/10.1016/j.cell.2007.03.004] [PMID: 17382878]

[11]   Franci G, Falanga A, Galdiero S, *et al.* Silver nanoparticles as potential antibacterial agents. Molecules 2015; 20(5): 8856-74.
[http://dx.doi.org/10.3390/molecules20058856] [PMID: 25993417]

[12]   Santoso SS, Vauthey S, Zhang S. Structures, function and applications of amphiphilic peptides. Curr Opin Colloid Interface Sci 2002; 7(5-6): 262-6.
[http://dx.doi.org/10.1016/S1359-0294(02)00072-9]

[13]   Löwik DWPM, van Hest JCM. Peptide based amphiphiles. Chem Soc Rev 2004; 33(4): 234-45.
[http://dx.doi.org/10.1039/B212638A] [PMID: 15103405]

[14]   Meng Q, Kou Y, Ma X, *et al.* Tunable self-assembled peptide amphiphile nanostructures. Langmuir 2012; 28(11): 5017-22.
[http://dx.doi.org/10.1021/la3003355] [PMID: 22352406]

[15]   Miravet JF, Escuder B, Segarra-Maset MD, *et al.* Self-assembly of a peptide amphiphile: transition from nanotape fibrils to micelles. Soft Matter 2013; 9(13): 3558-64.
[http://dx.doi.org/10.1039/c3sm27899a]

[16]   Fan T, Yu X, Shen B, Sun L. Peptide self-assembled nanostructures for drug delivery applications. Journal of Nanomaterials 2017; 2017
[http://dx.doi.org/10.1155/2017/4562474]

[17]   Hamley IW. Self-assembly of amphiphilic peptides. Soft Matter 2011; 7(9): 4122-38.
[http://dx.doi.org/10.1039/c0sm01218a]

[18]   Deng M, Yu D, Hou Y, Wang Y. Self-assembly of peptide-amphiphile C12-Abeta(11-17) into nanofibrils. J Phys Chem B 2009; 113(25): 8539-44.
[http://dx.doi.org/10.1021/jp904289y] [PMID: 19534562]

[19]   Kokkoli E, Mardilovich A, Wedekind A, Rexeisen EL, Garg A, Craig JA. Self-assembly and applications of biomimetic and bioactive peptide-amphiphiles. Soft Matter 2006; 2(12): 1015-24.
[http://dx.doi.org/10.1039/b608929a] [PMID: 32680204]

[20]   Palladino P, Castelletto V, Dehsorkhi A, Stetsenko D, Hamley IW. Conformation and self-association of peptide amphiphiles based on the KTTKS collagen sequence. Langmuir 2012; 28(33): 12209-15.
[http://dx.doi.org/10.1021/la302123h] [PMID: 22834769]

[21]   Paramonov SE, Jun HW, Hartgerink JD. Self-assembly of peptide-amphiphile nanofibers: the roles of hydrogen bonding and amphiphilic packing. J Am Chem Soc 2006; 128(22): 7291-8.
[http://dx.doi.org/10.1021/ja060573x] [PMID: 16734483]

[22]   Cui H, Webber MJ, Stupp SI. Self-assembly of peptide amphiphiles: From molecules to nanostructures to biomaterials. Biopolymers 2010; 94(1): 1-18.
[http://dx.doi.org/10.1002/bip.21328] [PMID: 20091874]

[23]   Dehsorkhi A, Castelletto V, Hamley IW. Self☐assembling amphiphilic peptides. J Pept Sci 2014; 20(7): 453-67.
[http://dx.doi.org/10.1002/psc.2633] [PMID: 24729276]

[24]   Dehsorkhi A, Castelletto V, Hamley IW, Adamcik J, Mezzenga R. The effect of pH on the self-assembly of a collagen derived peptide amphiphile. Soft Matter 2013; 9(26): 6033-6.

[http://dx.doi.org/10.1039/c3sm51029h]

[25]   Guler MO, Claussen RC, Stupp SI. Encapsulation of pyrene within self-assembled peptide amphiphile nanofibers. J Mater Chem 2005; 15(42): 4507-12.
[http://dx.doi.org/10.1039/b509246a]

[26]   Accardo A, Tesauro D, Mangiapia G, Pedone C, Morelli G. Nanostructures by self-assembling peptide amphiphile as potential selective drug carriers. Biopolymers 2007; 88(2): 115-21.
[http://dx.doi.org/10.1002/bip.20648] [PMID: 17154288]

[27]   Lee OS, Stupp SI, Schatz GC. Atomistic molecular dynamics simulations of peptide amphiphile self-assembly into cylindrical nanofibers. J Am Chem Soc 2011; 133(10): 3677-83.
[http://dx.doi.org/10.1021/ja110966y] [PMID: 21341770]

[28]   Feiner-Gracia N, Buzhor M, Fuentes E, Pujals S, Amir RJ, Albertazzi L. Micellar stability in biological media dictates internalization in living cells. J Am Chem Soc 2017; 139(46): 16677-87.
[http://dx.doi.org/10.1021/jacs.7b08351] [PMID: 29076736]

[29]   van den Heuvel M, Baptist H, Venema P, van der Linden E, Löwik DWPM, van Hest JCM. Mechanical and thermal stabilities of peptide amphiphile fibres. Soft Matter 2011; 7(20): 9737-43.
[http://dx.doi.org/10.1039/c1sm05642e]

[30]   Matsuzaki K. Why and how are peptide–lipid interactions utilized for self-defense? Magainins and tachyplesins as archetypes. Biochim Biophys Acta Biomembr 1999; 1462(1-2): 1-10.
[http://dx.doi.org/10.1016/S0005-2736(99)00197-2] [PMID: 10590299]

[31]   Hinterding K, Alonso-Díaz D, Waldmann H. Organic synthesis and biological signal transduction. Angew Chem Int Ed 1998; 37(6): 688-749.
[http://dx.doi.org/10.1002/(SICI)1521-3773(19980403)37:6<688::AID-ANIE688>3.0.CO;2-B]
[PMID: 29711371]

[32]   Prass W, Ringsdorf H, Bessler W, Wiesmüller KH, Jung G. Lipopeptides of the N-terminus of *Escherichia coli* lipoprotein: synthesis, mitogenicity and properties in monolayer experiments. Biochim Biophys Acta Biomembr 1987; 900(1): 116-28.
[http://dx.doi.org/10.1016/0005-2736(87)90283-5] [PMID: 3297144]

[33]   Wright ER, Conticello VP. Self-assembly of block copolymers derived from elastin-mimetic polypeptide sequences. Adv Drug Deliv Rev 2002; 54(8): 1057-73.
[http://dx.doi.org/10.1016/S0169-409X(02)00059-5] [PMID: 12384307]

[34]   Dasgupta A, Das D. Designer peptide amphiphiles: self-assembly to applications. Langmuir 2019; 35(33): 10704-24.
[http://dx.doi.org/10.1021/acs.langmuir.9b01837] [PMID: 31330107]

[35]   Chung PY, Khanum R. Antimicrobial peptides as potential anti-biofilm agents against multidrug-resistant bacteria. J Microbiol Immunol Infect 2017; 50(4): 405-10.
[http://dx.doi.org/10.1016/j.jmii.2016.12.005] [PMID: 28690026]

[36]   Ma Z, Yang J, Han J, *et al.* Insights into the antimicrobial activity and cytotoxicity of engineered α-helical peptide amphiphiles. J Med Chem 2016; 59(24): 10946-62.
[http://dx.doi.org/10.1021/acs.jmedchem.6b00922] [PMID: 28002968]

[37]   Geilich BM, Gelfat I, Sridhar S, van de Ven AL, Webster TJ. Superparamagnetic iron oxide-encapsulating polymersome nanocarriers for biofilm eradication. Biomaterials 2017; 119: 78-85.
[http://dx.doi.org/10.1016/j.biomaterials.2016.12.011] [PMID: 28011336]

[38]   Liu Y, Shi L, Su L, *et al.* Nanotechnology-based antimicrobials and delivery systems for biofilm-infection control. Chem Soc Rev 2019; 48(2): 428-46.
[http://dx.doi.org/10.1039/C7CS00807D] [PMID: 30601473]

[39]   Blair JMA, Webber MA, Baylay AJ, Ogbolu DO, Piddock LJV. Molecular mechanisms of antibiotic resistance. Nat Rev Microbiol 2015; 13(1): 42-51.
[http://dx.doi.org/10.1038/nrmicro3380] [PMID: 25435309]

[40]   Butler MS, Cooper MA. Antibiotics in the clinical pipeline in 2011. J Antibiot (Tokyo) 2011; 64(6): 413-25.
[http://dx.doi.org/10.1038/ja.2011.44] [PMID: 21587262]

[41]   Balakrishna R, Wood SJ, Nguyen TB, *et al.* Structural correlates of antibacterial and membrane-permeabilizing activities in acylpolyamines. Antimicrob Agents Chemother 2006; 50(3): 852-61.
[http://dx.doi.org/10.1128/AAC.50.3.852-861.2006] [PMID: 16495242]

[42]   Abed N, Couvreur P. Nanocarriers for antibiotics: A promising solution to treat intracellular bacterial infections. Int J Antimicrob Agents 2014; 43(6): 485-96.
[http://dx.doi.org/10.1016/j.ijantimicag.2014.02.009] [PMID: 24721232]

[43]   Salim M, Minamikawa H, Sugimura A, Hashim R. Amphiphilic designer nano-carriers for controlled release: from drug delivery to diagnostics. MedChemComm 2014; 5(11): 1602-18.
[http://dx.doi.org/10.1039/C4MD00085D]

[44]   Liu L, Xu K, Wang H, *et al.* Self-assembled cationic peptide nanoparticles as an efficient antimicrobial agent. Nat Nanotechnol 2009; 4(7): 457-63.
[http://dx.doi.org/10.1038/nnano.2009.153] [PMID: 19581900]

[45]   Wilczewska AZ, Niemirowicz K, Markiewicz KH, Car H. Nanoparticles as drug delivery systems. Pharmacol Rep 2012; 64(5): 1020-37.
[http://dx.doi.org/10.1016/S1734-1140(12)70901-5] [PMID: 23238461]

[46]   Zhao L, Li N, Wang K, Shi C, Zhang L, Luan Y. A review of polypeptide-based polymersomes. Biomaterials 2014; 35(4): 1284-301.
[http://dx.doi.org/10.1016/j.biomaterials.2013.10.063] [PMID: 24211077]

[47]   Spicer P. Cubosome processing: industrial nanoparticle technology development. Chem Eng Res Des 2005; 83(11): 1283-6.
[http://dx.doi.org/10.1205/cherd.05087]

[48]   Polcyn P, Jurczak M, Rajnisz A, Solecka J, Urbanczyk-Lipkowska Z. Design of antimicrobially active small amphiphilic peptide dendrimers. Molecules 2009; 14(10): 3881-905.
[http://dx.doi.org/10.3390/molecules14103881] [PMID: 19924036]

[49]   Trent A, Marullo R, Lin B, Black M, Tirrell M. Structural properties of soluble peptide amphiphile micelles. Soft Matter 2011; 7(20): 9572-82.
[http://dx.doi.org/10.1039/c1sm05862b]

[50]   Chang H, Li C, Huang R, Su R, Qi W, He Z. Amphiphilic hydrogels for biomedical applications. J Mater Chem B Mater Biol Med 2019; 7(18): 2899-910.
[http://dx.doi.org/10.1039/C9TB00073A]

[51]   Akbari V, Abedi D, Pardakhty A, Sadeghi-Aliabadi H. Ciprofloxacin nano-niosomes for targeting intracellular infections: An *in vitro* evaluation. J Nanopart Res 2013; 15(4): 1556.
[http://dx.doi.org/10.1007/s11051-013-1556-y]

[52]   Pinheiro M, Lúcio M, Lima JLFC, Reis S. Liposomes as drug delivery systems for the treatment of TB. Nanomedicine (Lond) 2011; 6(8): 1413-28.
[http://dx.doi.org/10.2217/nnm.11.122] [PMID: 22026379]

[53]   Mohamad EA, Fahmy HM. Niosomes and liposomes as promising carriers for dermal delivery of *Annona squamosa* extract. Braz J Pharm Sci 2020; 56: e18096.
[http://dx.doi.org/10.1590/s2175-97902019000318096]

[54]   Fang JY, Hong CT, Chiu WT, Wang YY. Effect of liposomes and niosomes on skin permeation of enoxacin. Int J Pharm 2001; 219(1-2): 61-72.
[http://dx.doi.org/10.1016/S0378-5173(01)00627-5] [PMID: 11337166]

[55]   Walvekar P, Gannimani R, Salih M, Makhathini S, Mocktar C, Govender T. Self-assembled oleylamine grafted hyaluronic acid polymersomes for delivery of vancomycin against methicillin

resistant Staphylococcus aureus (MRSA). Colloids Surf B Biointerfaces 2019; 182: 110388.
[http://dx.doi.org/10.1016/j.colsurfb.2019.110388] [PMID: 31369955]

[56]   Khan S, Akhtar MU, Khan S, Javed F, Khan AA. **Nanoniosome☐encapsulated levoflaxicin as an antibacterial agent against Brucella**. J Basic Microbiol 2020; 60(3): 281-90.
[http://dx.doi.org/10.1002/jobm.201900454] [PMID: 31856360]

[57]   Mirzaie A, Peirovi N, Akbarzadeh I, *et al.* Preparation and optimization of ciprofloxacin encapsulated niosomes: A new approach for enhanced antibacterial activity, biofilm inhibition and reduced antibiotic resistance in ciprofloxacin-resistant methicillin-resistance *Staphylococcus aureus.* Bioorg Chem 2020; 103: 104231.
[http://dx.doi.org/10.1016/j.bioorg.2020.104231] [PMID: 32882442]

[58]   Geilich BM, van de Ven AL, Singleton GL, Sepúlveda LJ, Sridhar S, Webster TJ. Silver nanoparticle-embedded polymersome nanocarriers for the treatment of antibiotic-resistant infections. Nanoscale 2015; 7(8): 3511-9.
[http://dx.doi.org/10.1039/C4NR05823B] [PMID: 25628231]

[59]   Wayakanon K, Thornhill MH, Douglas CWI, *et al.* Polymersome☐mediated intracellular delivery of antibiotics to treat *Porphyromonas gingivalis* ☐infected oral epithelial cells. FASEB J 2013; 27(11): 4455-65.
[http://dx.doi.org/10.1096/fj.12-225219] [PMID: 23921377]

[60]   Boge L, Browning KL, Nordström R, *et al.* Peptide-loaded cubosomes functioning as an antimicrobial unit against *Escherichia coli.* ACS Appl Mater Interfaces 2019; 11(24): 21314-22.
[http://dx.doi.org/10.1021/acsami.9b01826] [PMID: 31120236]

[61]   Boge L, Hallstensson K, Ringstad L, *et al.* Cubosomes for topical delivery of the antimicrobial peptide LL-37. Eur J Pharm Biopharm 2019; 134: 60-7.
[http://dx.doi.org/10.1016/j.ejpb.2018.11.009] [PMID: 30445164]

[62]   Gide M, Nimmagadda A, Su M, *et al.* Nano-sized lipidated dendrimers as potent and broad-spectrum antibacterial agents. Macromol Rapid Commun 2018; 39(24): 1800622.
[http://dx.doi.org/10.1002/marc.201800622] [PMID: 30408252]

[63]   Wrońska N, Majoral JP, Appelhans D, Bryszewska M, Lisowska K. Synergistic effects of anionic/cationic dendrimers and levofloxacin on antibacterial activities. Molecules 2019; 24(16): 2894.
[http://dx.doi.org/10.3390/molecules24162894] [PMID: 31395831]

[64]   Wei W, Li S, Xu H, *et al.* MPEG-PCL copolymeric micelles for encapsulation of azithromycin. AAPS PharmSciTech 2018; 19(5): 2041-7.
[http://dx.doi.org/10.1208/s12249-018-1009-0] [PMID: 29675667]

[65]   Huang F, Gao Y, Zhang Y, *et al.* Silver-decorated polymeric micelles combined with curcumin for enhanced antibacterial activity. ACS Appl Mater Interfaces 2017; 9(20): 16880-9.
[http://dx.doi.org/10.1021/acsami.7b03347] [PMID: 28481077]

[66]   Bryaskova R, Pencheva D, Kyulavska M, Bozukova D, Debuigne A, Detrembleur C. Antibacterial activity of poly(vinyl alcohol)-b-poly(acrylonitrile) based micelles loaded with silver nanoparticles. J Colloid Interface Sci 2010; 344(2): 424-8.
[http://dx.doi.org/10.1016/j.jcis.2009.12.040] [PMID: 20074742]

[67]   Roy S, Das PK. Antibacterial hydrogels of amino acid-based cationic amphiphiles. Biotechnol Bioeng 2008; 100(4): 756-64.
[http://dx.doi.org/10.1002/bit.21803] [PMID: 18318444]

[68]   Xu L, Shen Q, Huang L, Xu X, He H. Charge-mediated co-assembly of amphiphilic peptide and antibiotics into supramolecular hydrogel with antibacterial activity. Front Bioeng Biotechnol 2020; 8: 629452.
[http://dx.doi.org/10.3389/fbioe.2020.629452] [PMID: 33425884]

[69]   Li S, Dong S, Xu W, *et al.* Antibacterial Hydrogels. Adv Sci (Weinh) 2018; 5(5): 1700527.

[http://dx.doi.org/10.1002/advs.201700527] [PMID: 29876202]

[70]   Dai T, Wang C, Wang Y, Xu W, Hu J, Cheng Y. A nanocomposite hydrogel with potent and broad-spectrum antibacterial activity. ACS Appl Mater Interfaces 2018; 10(17): 15163-73.
[http://dx.doi.org/10.1021/acsami.8b02527] [PMID: 29648438]

[71]   Nandi N, Gayen K, Ghosh S, *et al.* Amphiphilic peptide-based supramolecular, noncytotoxic, stimuli-responsive hydrogels with antibacterial activity. Biomacromolecules 2017; 18(11): 3621-9.
[http://dx.doi.org/10.1021/acs.biomac.7b01006] [PMID: 28953367]

[72]   Magallanes M, Dijkstra J, Fierer J. Liposome-incorporated ciprofloxacin in treatment of murine salmonellosis. Antimicrob Agents Chemother 1993; 37(11): 2293-7.
[http://dx.doi.org/10.1128/AAC.37.11.2293] [PMID: 8285608]

[73]   Falkinham JO III, Macri RV, Maisuria BB, *et al.* Antibacterial activities of dendritic amphiphiles against nontuberculous mycobacteria. Tuberculosis (Edinb) 2012; 92(2): 173-81.
[http://dx.doi.org/10.1016/j.tube.2011.12.002] [PMID: 22209468]

[74]   Tripodo G, Perteghella S, Grisoli P, Trapani A, Torre ML, Mandracchia D. Drug delivery of rifampicin by natural micelles based on inulin: Physicochemical properties, antibacterial activity and human macrophages uptake. Eur J Pharm Biopharm 2019; 136: 250-8.
[http://dx.doi.org/10.1016/j.ejpb.2019.01.022] [PMID: 30685506]

[75]   Zhou C, Wang F, Chen H, *et al.* Selective Antimicrobial activities and action mechanism of micelles self-assembled by cationic oligomeric surfactants. ACS Appl Mater Interfaces 2016; 8(6): 4242-9.
[http://dx.doi.org/10.1021/acsami.5b12688] [PMID: 26820390]

[76]   Bush K, Courvalin P, Dantas G, *et al.* Tackling antibiotic resistance. Nat Rev Microbiol 2011; 9(12): 894-6.
[http://dx.doi.org/10.1038/nrmicro2693] [PMID: 22048738]

[77]   Wenzel RP, Edmond MB. Managing antibiotic resistance. N Engl J Med 2000; 343(26): 1961-3.
[http://dx.doi.org/10.1056/NEJM200012283432610] [PMID: 11136269]

[78]   Andersson DI. Persistence of antibiotic resistant bacteria. Curr Opin Microbiol 2003; 6(5): 452-6.
[http://dx.doi.org/10.1016/j.mib.2003.09.001] [PMID: 14572536]

[79]   Andersson DI, Levin BR. The biological cost of antibiotic resistance. Curr Opin Microbiol 1999; 2(5): 489-93.
[http://dx.doi.org/10.1016/S1369-5274(99)00005-3] [PMID: 10508723]

[80]   Armstead AL, Li B. Nanomedicine as an emerging approach against intracellular pathogens. Int J Nanomedicine 2011; 6: 3281-93. [eng.].
[PMID: 22228996]

[81]   Salick DA, Kretsinger JK, Pochan DJ, Schneider JP. Inherent antibacterial activity of a peptide-based beta-hairpin hydrogel. J Am Chem Soc 2007; 129(47): 14793-9.
[http://dx.doi.org/10.1021/ja076300z] [PMID: 17985907]

CHAPTER 5

Biological Importance of Some Functionalized Schiff Base-Metal Complexes

Mintu Thakur[1] and **Kinkar Biswas**[1,2,*]

[1] *Department of Chemistry, Raiganj University, Raiganj, Uttar Dinajpur 733134, India*

[2] *Department of Chemistry, University of North Bengal, Darjeeling 734013, India*

Abstract: Schiff base ligands or compounds are useful in modern inorganic chemistry. Numerous transition metal-based catalysts have been synthesized with Schiff base scaffolds. The application of such Schiff bases is also found in biological studies. Herein, we have discussed the various synthetic procedures of diversified Schiff base compounds and their metal complexes. The biological activity of those complexes has also been delineated in this chapter with special emphasis. Various metal complexes [Co(II), Ni(II), Cu(II), Zn(II) and Fe(III)] with different Schiff base compounds displayed anti-fungal activity. Similarly, anti-viral activity was seen with Co(II) and Pd(II) metal complexes. Many Schiff base-metal complexes are found, which showed anti-cancer activity against various carcinoma cells like HpG2, MCF-7, A549, HCT116, Caco-2 and PC-3. Similarly, the transition metal complexes (generally 1st and 2nd row) of Schiff bases also exhibited good anti-bacterial activity against various bacterial strains. The ionic-liquid-tagged Schiff bases have also been found to be good anti-microbial agents.

Keywords: Anti-bacterial, Anti-microbial, Biological activity, Metal complex, Schiff Base.

INTRODUCTION

Schiff base compounds are very useful and play an important role in various fields. These compounds showed important biological activities, which are useful in many catalytic reactions when combined with metal ions [1]. The synthesis of new Schiff bases and their metal complexes played an important role in the development of co-ordination chemistry. The chemistry of Schiff base and its metal complexes attract immense attention in the field of inorganic and organometallic chemistry.

* **Corresponding author Kinkar Biswas:** Department of Chemistry, University of North Bengal, Darjeeling 734013, India; Email: kinkar.chem@gmail.com

Tilak Saha, Manab Deb Adhikari and Bipransh Kumar Tiwary (Eds.)

Schiff bases and their complexes have been synthesized by the condensation of an amino compound (aliphatic and aromatic) with carbonyl compounds under dehydrated conditions. The invention of the Schiff base by Hugo Schiff in 1864 opened a new dimension in the field of chemistry. The Schiff base complexes were widely used for industrial purposes and also showed a broad range of biological activities like anti-fungal [2], anti-bacterial [3], antimalarial, antiproliferative, anti-inflammatory, anti-viral [4] and antipyretic, and some of these also show excellent catalytic activity in various reactions [5].

ROLE OF SCHIFF BASE IN CO-ORDINATION CHEMISTRY

Co-ordination chemistry is an important part of chemistry that gives a good concept about the stability of the structure of different complexes. When dissolved in water or other solvents, co-ordination compounds showed such properties which are completely different from those of the constituents. The co-ordination complex compounds are formed by the association of one or more than one molecule or anions with a central atom or ion, usually metal cations. In the formation of a stable complex, a cation or metal to which one or more neutral molecules or ions are bonded with a central metal atom is called a ligand. Ligands can be monodentate or polydentate and have the ability to form chelate complexes. The key breakthrough occurred when Alfred Werner proposed Co(III) ion complex formation with octahedral geometry consisting of ligands in 1893.

In classical co-ordination chemistry, there is an association between the ligands to central metal ions via their lone pairs of electrons residing on the main group of atoms. In co-ordination chemistry, a structure was first described by the number of sigma bonds that formed between ligands and the central metal atom is known as co-ordination number of this complex. The number of bonds depends mainly on the size, charge, electronic configuration of the central metal atom and ligand molecules or ions and the extent of interactions between s and p orbitals of ligands and d orbitals of the central metal atom or ion. In co-ordination, complex metals with small sizes lead to high co-ordination numbers, *e.g.,* $[Mo(CN)_8]^{4-}$ and small sizes of central atoms and ligands with large sizes lead to low co-ordination numbers, *e.g.,* $Pt[PC(me)_3]_2$. Further, the stability of the co-ordination complexes was vastly described by the crystal field theory (CFT) IN 1929 by Hans Bethe and by the ligand field theory (LFT) IN 1935.

There are diverse applications of co-ordination compounds in different fields, such as industrial catalysts in controlling reactivity, and they play ans essential role in biochemistry. The specific color of different metals in different complexes plays an important role in medicinal chemistry. The ligands are used in the case of treatment of problems due to the presence of some metal in toxic proportion in plants and animals. So, excess metals like copper and iron are taken away by chelating ligands D-penicillamine and desferrioxamine via the formation of the co-ordination complexes. Some co-ordination complexes of platinum are used as an inhibitor of the growth of tumors.

WHAT IS SCHIFF BASE LIGAND?

Schiff bases are versatile compounds synthesized from the condensation of primary amino compounds with aldehydes or ketones (Scheme 1). The high thermal stability of many Schiff bases and their complexes were useful attributes for their application as catalysts in reactions involving high temperatures. This activity of Schiff bases was usually increased by complexation therefore to understand the properties of both ligands and metal ions can lead to the synthesis of highly reactive compounds. Schiff bases are some of the most widely used organic compounds that are used as pigments and dyes, catalysts, intermediates in organic synthesis, and polymer stabilizers. Schiff bases have also exhibited a broad range of pharmacological activities such as anti-fungal, anti-bacterial, antimalarial, antitubercular, antiproliferative, anti-inflammatory, anti-viral and antipyretic properties. Imine or azomethine groups are present in various natural, natural-derived, and non-natural compounds. The imine group present in such compounds has been shown to be critical to their biological activities, which can be altered depending upon the type of substituent present on the aromatic rings.

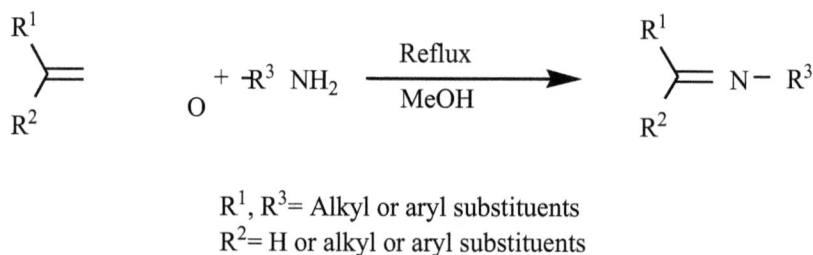

$$R^1, R^3 = \text{Alkyl or aryl substituents}$$
$$R^2 = \text{H or alkyl or aryl substituents}$$

Scheme 1. The general methods of preparation of Schiff base compounds.

This type of Schiff base is used for the preparation of different types of co-ordination complexes through co-ordination with several numbers of transition metals. The transition metals generally serve as a bridge or transition between the two sides of the periodic table. These have diverse applications in the different fields of chemistry, not only and also biological field. They have certain characteristic properties, which result from partially filled d shell. These are (i) the colour of the compounds due to d-d transition, (ii) many oxidation states are found, (iii) paramagnetism property has been found due to the presence of unpaired electrons. Actually, such types of characteristics were shown in Schiff base compounds due to the presence of the azomethine group, which increases the stability, flexibility and photochemical properties.

Chemistry of Azomethine Group

Schiff base compounds are prepared through the condensation reaction of primary amine and carbonyl compounds of aliphatic or aromatic with the general formula [-C=N-] that form a Schiff base, a stable imine. In this azomethine group, a pair of pi-electrons bonded between the carbon of carbonyl group and nitrogen of primary amine. A lone pair of electrons on nitrogen in the azomethine group adds distinct properties. The Azomethine group shows an intermediate property due to having an electronegative value of nitrogen in between the carbon and oxygen atom; electronegative values are C=2.5, N=3, and O=3.5. There also has a linear relationship between infra-red stretching frequency and inter-nuclear distance. From the relationship, bond length was observed C-N=1.47 A°, C=N = 1.29-1.31 A°.

Various Bond Energies in Schiff Base Ligands

The numerical values of bond energies can be calculated involving "C" atom from the concept of the heat of sublimation of carbon. The term, thermochemical bond energies are equal to the heat of atomization of molecules. Cottrell calculated the bond energy value of C-C, C=C, C=N. The typical bond energies are given as, E_{C-C} =347 kJmol^{-1}, $E_{C=C}$ = 611 kJmol^{-1}, $E_{C=N}$ = 615 kJmol^{-1}, E_{C-N} = 305 kJmol^{-1} as per previous data.

Mechanism of Formations of –C=N- Bond

Such bonds are mainly formed by the condensation reaction of primary amines with carbonyl compounds. The experimental condition of this reaction varied depending on the reactivity of choosing amine and carbonyl compounds of that reaction. In the addition of an amine to aldehyde or ketone, some by-products are also formed and removed by distillation or by using a suitable solvent to form an azeotropic mixture with water. This reaction mechanism mainly contains two steps. In the first step, amino group was added to the carbonyl group to form a carbinol amine, and in the second step, –C=N- bond was finally formed by the elimination of the water molecule.

BIOLOGICAL IMPORTANCE OF SCHIFF BASE COMPOUNDS

The Schiff base complexes were widely used for industrial purposes and also showed a broad range of biological activities. Now a day, researchers get attracted to the synthesis of diverse Schiff base compounds and studies of some new kinds of chemotherapeutic agents. Biochemists also get interested in the earlier work reported in the literature.

Anti-fungal Activities

Anti-fungal agents are used to preventing the further growth of fungi. But in medical science, antifungals are used as a medicine for the treatment of infections such as athlete's foot, ringworm, and thrush, which work by exploiting differences between mammalian and fungal cells. Normally, these anti-fungal substances kill off the fungal organism. Thiazole and benzothiazole-based Schiff base compounds have effective anti-fungal activity due to the presence of methoxy halogen and naphthyl group, which enhances fungicidal activities toward *Carvularia*. There are some other Schiff bases of quinazolinones that mainly showed anti-fungal activity against *Candida albicans*, *Trichophyton rubrum*, and *Aspergillus niger*. On the other hand, Schiff bases and their metal complexes also showed anti-fungal activities. For example, Schiff base-metal complexes which are formed between furan or furyl glyoxal with various amines, showed anti-fungal activity against *Helminthosporium*, *Graminium*, *Syncephalosturum racemosus* and *C. capsic*. *Alternaria brassicae* and *Alternaria brassicicolare* phytopathogenic fungi severely affect the production of most cruciferous crops like broccoli, cauliflower, mustard, turnip, cabbage, rape and radish. N-(salicylidene)-2-hydroxyaniline at the concentration of 500 ppm inhibited the growth of these fungi by 67-68% [6].

Joseyphus and Nair synthesised L1 and L2 Schiff base compounds and their metal complexes with Zn, which were tested against some bacterial species like *Staphylococcus aureus*, *Klebsiella pneumaniae*, *Proteus vulgaris*, *Escherichia coli* and *Pseudomonas aeruginosa* and fungal species as *Rhizoctonia, Aspergillus niger, Aspergillus flavus, Rhizopus stolonifer* and *Candida albicans* by a disc diffusion process. The metal complexes of these Schiff base ligands showed higher anti-microbial activity than free ligands. Generally, Zn ions are introduced as essential for the growth inhibitor effect [7]. El-Ajaily *et al.* synthesized the Schiff bases (HL1), namely; [(S,Z)-2-((2-hydroxy-1-phenylethylidene)am-no)-3-(4-hydroxyphenyl) propanoic acid and (HL2), namely; (E)-4-(2-(-,4-dinitrophenyl) hydrazono) methyl)-N,N-dimethylaniline] and also prepared their complexes with Co(II), Ni(II), Cu(II), Zn(II) and Fe(III) ions. These complexes were tested for anti-fungal activities against *Altarnaria*, *Aspergillus*, *Rhizopus stolonifer* and *Aspergillus niger* [8]. Thangagiri and his group synthesized Schiff base (L) and its metal complexes with nickel, copper, zinc and cobalt. These complexes were exposed to various fungi to establish their bioactivities which showed positive results. The designed complexes were found to be active against fungi like *Rhizoctonia bataicola*, *Aspergillus niger*, *Rhizopus stolonifer, Trichoderma harizanum, Candida albicans* and *Aspergillus flavus*. But it was noticed that metal complexes were found to be more active than ligands [9]. Govindaraj *et al.* designed Schiff base compounds L1(2-((2-hydrox--3-methoxybenzylidene)amino)benzoic acid) and L2 (2-((2-hydroxybenzylidene) amino)benzoic) and their metal complexes with Ni(II) which were tested for anti-fungal and anti-bacterial species by disc diffusion method against *Staphylococcus aureus* (Gram-positive), *Aspergillus niger* and *Klebsiella aerogenes* (Gram-negative) [10]. The structure of the Schiff base metal complexes is given in Table **1**.

Anti-viral Activity

The role of the vaccines may lead to the eradication of viral pathogens, such as smallpox, polio and rubella. In this way, virus-related hepatitis human immunodeficiency diseases have been the limitation of vaccine approaches. Normally viral diseases are one type of life-threatening for immune-compromised patients, and urgent treatment will be required to overcome this problem. Although many therapeutic options for viral infections are known, now a day's, available anti-viral agents are not still fully effective; it may be due to the high rate of virus mutation. These anti-viral drugs are a class of medication used specifically for treating viral infections like antibiotics; specific anti-virals are used for specific viruses. These agents are relatively harmless to the host and, therefore, can be used for infections. Several Schiff base complexes of salicylaldehyde and 1-amino-3-hydroxyguanidine tosylate are a good working station for the design of new anti-viral agents, which normally inhibit its growth by 50% when engaged at low concentration as 3.2 µM. Silver complexes in oxidation state (I) showed inhibition against *Cucumbar mosaic* viruses; glycine salicylaldehyde Schiff base Ag (I) also showed effective results up to 74.7% towards *C. mosaic virus* and on the other part Pd (II) complexes with benzyl bis(thiosemicarbazone) and 3,5-diacyl-12,4-triazole bis(4-methylthiosemicarbazone) were invited against the replication of wild type herpes simplex virus (HSV-1) and (HSV-2) strains, respectively. A few examples of Schiff base metal complexes have been displayed in Table **1** below.

Table 1. Various anti-microbial activities of diversified Schiff base-metal complex.

S. No.	Structure of IL-Schiff base	Metal complexes	Biological activity	Refs
1		Zn(II)	Shows anti-microbial activity against both bacteria (like *Staphylococcus aureus, Klebsiella pneumaniae, Proteus vulgaris, Escherichia coli* and *pseudomonas aeruginosa*) and fungi (as *Rhizoctonia, Aspergillus niger, Aspergillus flavus, Rhizopus stolonifer* and *Candida albicans*)	[7]
2		Co(II), Ni(II), Cu(II), Zn(II) and Fe(III)	Showed anti-fungal activities against *Alternaria, Aspergillus, Rhizopus stolonifer* and *Aspergillus niger*	[8]

(Table 1) cont.....

S. No.	Structure of IL-Schiff base	Metal complexes	Biological activity	Refs
3		Co(II), Ni(II), Cu(II) and Zn(II)	Showed anti-fungal activities against *Rhizoctonia bataicola, Aspergillus niger, Rhizopus stolonifer, Trichoderma harizanum, Candida albicans* and *Aspergillus flavus*	[9]
4		Ni(II)	Showed anti-fungal activities against *Aspergillus niger*	[10]
Anti-viral activity				
1		Co(II)	Anti-viral activity against HSV-1 and HIV-1.	[11]

(Table 1) cont.....

S. No.	Structure of IL-Schiff base	Metal complexes	Biological activity	Refs
2	MAp17	Not reported	Anti-viral activity against HIV and HIV-PIC	[12]
3	X=CH₂, O, CO, Y= H,Cl, Z=H, Et, W= H,Et, R₁=H,Bn, R₂= H,F	Not performed	Anti-viral activity against a panel of DNA and RNA viruses.	[13]

(Table 1) cont.....

S. No.	Structure of IL-Schiff base	Metal complexes	Biological activity	Refs
4	R^1 = C_6H_{11}, CH_2CH_2OH, C_6H_5, C_6H_4OH, $C_6H_4NO_2$	Not performed	Anti-viral activity against a series of ortho-poxviruses including cowpox, monkepox, variolovaccine and variola	[14]
5		Pd(II)	Anti-viral activity against the replication of wild-type herpes simplex virus (HSV-1) and (HSV-2) strains.	[15]
6		Pd(II)	Anti-viral activity against HSV-1 and HSV-2	[16]
Anti-cancer activity				
1		Ni(II), Co(II) and Zn(II)	Showed anti-cancer activities against such as liver cancer HpG2, breast cancer MCF-7, Lung carcinoma A549 and colorectal cancer HCT116	[18]

S. No.	Structure of IL-Schiff base	Metal complexes	Biological activity	Refs
2	**L1** **L2** **L3**	Pd(II) and Pt(II)	Showed anti-cancer activity against various human cancerous named Caco-2, HeLa, HepG2, MCF-7, and PC-3 and noncancerous named MCF-12A cell lines	[19]
3		Cu(II)	Showed anti-cancer activities against the tumor cells like colon cancer cells (HCT-116) and breast cancer cells (MDA-MB-231)	[20]
4		Cu(II) and Ni(II)	Showed anti-cancer activities against human breast cancer cells (MCF-7)	[21]

(Table 1) cont.....

S. No.	Structure of IL-Schiff base	Metal complexes	Biological activity	Refs
5		Co(II), Zn(II), Cu(II), Cr(III), Mn(II) and Ni(II)	Showed anti-cancer activities against breast and colon cell lines	[22]
6		Cu(II)	Showed anti-cancer activities against human hepatic carcinoma cell lines like Hep-G2	[23]
7		Co(II)	Showed anti-cancer activities against human skin cancer cell lines named A-431, HT-144 and SK-ME--30	[24]
8		Ru(III)	Showed anti-cancer activities against renal cancer cells (TK-10), melanoma cancer cells (UACC-62), and breast cancer cells (MCF-7)	[25]
Anti-bacterial activity				
1		Cu(II), Ni(II), Co(II) and Zn(II)	Showed anti-bacterial activities against *Bacillus subtilis*, *E. coli*, *Staphylococcus aureus* and *Salmonella typhi*	[27]

(Table 1) cont.....

S. No.	Structure of IL-Schiff base	Metal complexes	Biological activity	Refs
2		Cu(II), Ni(II), Co(II) and Zn(II)	Showed anti-bacterial activities against some Gram-positive bacteria like *Staphylococcus aureus* and Gram-negative bacteria like *Salmonella* Typhimurium, *Escherichia coli* and *Klebsiella pneumonia*	[28]
3		Pd(II), Ni(II) and Cu(II)	Showed anti-bacterial activities against Gram-positive bacteria like *Bacillus subtilis* and *Staphylococcus aureus* and Gram-negative bacteria like *Escherichia coli* and *Pseudomonas aeruginosa*	[29]
4	 $R= C_6H_5Cl , C_6H_5NO_2$	Ni(II), Cu(II)	Showed anti-bacterial activities against both types of Gram-positive bacteria such as *Staphylococcus aureus* and Gram-negative bacteria *Escherichia coli*.	[30]

(Table 1) cont.....

S. No.	Structure of IL-Schiff base	Metal complexes	Biological activity	Refs
5	R=H, Cl	Co(II), Cu(II), Ni(II)	Showed anti-bacterial activities against *Pseudomonas aeruginosa, Bacillus subtilis, Staphylococcus aureus* and *Escherichia coli.*	[31]
6		Zn(II), Cu(II)	Showed anti-bacterial activities against *Salmonella typhi, Bacillus subtilis, Staphylococcus* and *Klebsiella*	[32]
7		Cu(II), Ni(II), Co(II), Zn(II) and Mn(II)	Showed anti-bacterial activities against *S. aureus,* Enterococcus and Gram-negative bacteria like *K. pneumonia, E. coli.*	[33]

Knight *et al.* designed the Schiff base and its metal complex with Co (II), which was performed in anti-viral activity against HSV-1 and HIV-1 [11]. Al-Abed and his team prepared a class of arylene bis(methylketone) based Schiff base with contiguous lysines. These synthesized Schiff-based compounds were tested for an anti-viral activity for anti-HIV activity and inhibition against the nuclear import of the HIV-1PIC [12]. Jarrahpour and his research group synthesized Schiff bases of isatin, benzyllisatin and 5-fluoroisatin and these were tested for biological activities. The compounds were tested for anti-viral activity against a panel of DNA and RNA viruses [13]. Balakhnin and their team developed several compounds of pyrido[1,2-a]benzimidazoles which were screened for anti-viral activity against a series of ortho-poxviruses including cowpox, monkeypox, variolovaccine and variola [14]. Pd(II) complexes with benzyl bis(thiosemicarbazone) and 3,5-diacyl-1,2,4-triazole bis(4-methyl thio semicarbazone) were screened against the replication of wild-type *Herpes simplex virus* (HSV-1) and (HSV-2) strains [15]. Matesanz and his team synthesized bis(thiosemicarbazone) compounds and their Pd(II) complexes, which were tested for anti-viral activity against HSV-1 and HSV-2 strains, respectively [16].

Anti-cancer Activity

It has already been known that chelation is the reason and cure for many diseases, including cancer. Angiogenesis-dependent diseases, which are maintained by chemotherapy, immunotherapy and radiation therapy, may prevent the stimulating factors. This is one of the types of complex methods encompassing endothelial cell migration, proliferation and tube formation, which are normally good-regulated processes involving a number of stimulators.

Five Schiff bases 2-((3-chlorophenylimino) methyl)-5-(diethylamino) phenol, 2-((2, 4-dichlorophenylimino)methyl)-5-(diethylamino)phenol, 5-(diethylamino) methyl)phenol, 2-((2-chloro-4-methylphenylimino)methyl)-(diethylamino)phenol, and 5-(diethyl amino)-2-((2,6-diethylimino)methyl)phenol were synthesized and characterized by elemental analysis, FTIR, ^{1}H- and ^{13}C-NMR spectroscopy, single crystal X-ray diffraction and drug-DNA interaction studies result from UV-Vis spectroscopy and also exhibited electrochemistry complement that the compounds bind to DNA through electrostatic interactions. The cytotoxicity of these compounds has been studied against cancer cell lines (HeLa and MCF-7) and normal cell lines (BHK-21). So, these compounds played a vital role as anti-cancer agents. Complexes of Co(II), Ni(II) and Cu (II) with potential biologically active Schiff base ligand, bis(3-acetylcoumarine)thiocarbohydrazone showed cytotoxic activity [17]. Abd-Elzahe and his group synthesized a new Schiff base, (E)-2-(((5-methyl-4-phenylthiazol-2-yl)imino)methyl)phenol and its metal complexes with Ni(II), Co(II) and Zn(II). These complexes were studied against some different types of human tumor cell lines, such as liver cancer HpG2, breast cancer MCF-7, Lung carcinoma A549 and colorectal cancer HCT116 in comparison with the activity of the doxorubicin as a drug. This result showed that Zn metal complex exhibited potent inhibition against human TRK in four cell lines which are HepG2, MCF7, A549, and HCT116, by the ratio 80, 70, 61 and 64%, respectively, in comparison with the inhibition in untreated cells [18]. Onani and his team designed some Schiff bases as R-(phenyl)methanamine (L1), R-(pyridin-2-yl)methanamine (L2), and R-(furan-2-yl)methanamine (L3) (R-(E)-N-((1H-pyrrol-2-yl) methylene) and their metal complexes with Pd(II) and Pt(II) and which were studied for anti-cancer activity against various human cancerous named Caco-2, HeLa, HepG2, MCF-7, and PC-3 and noncancerous named MCF-12A cell lines. The result showed that instead of C5 complex, other complexes exhibited a reduction of the viability of the five cancerous cell lines by more than 60% [19]. Rama and Selvameena synthesized a new bidentate Schiff base ligand (HL1) and its metal complex with copper, which has been investigated as anti-cancer, anti-bacterial and anti-fungal agent. The reported Schiff base copper complex in this paper was tested for cytotoxicity against the tumor cells like colon cancer cells (HCT-116) and breast cancer cells (MDA-MB-231). Generally, the result showed that the complex is specific in action toward the breast cancer cell line and less toxic towards the colon cancer cell line [20].

Rangappan *et al.* synthesized Schiff base ligand from 4-chloro-*o*-phenylenediamine and 3,5-dichloro-2-hydroxyacetophenone and derived its metal complex with Cu(II) and Ni(II), which are homo and hetero binuclear oxygen bridged compounds. The anti-cancer activity of the derived metal complexes was studied by treating them with human breast cancer cells (MCF-7). The study showed that the complexes inhibited the growth of MCF-7 cells in a dose-dependent manner. In comparison, the highest level of cytotoxic activity was investigated in the homo-binuclear Cu (II) complex (IC_{50}= 7.3789). The results indicate that these complexes have important beneficial features for potential anti-cancer agents [21]. El-Halim *et al.* synthesized a new Schiff base (HL) from quinoline-2-carboxaldhyde with 2-aminophenol in a molar ratio of 1:1 and its metal complexes with Co(II), Zn(II), Cu(II), Cr(III), Mn(II) and Ni(II). The anti-cancer activity of Schiff base (HL) and its metal complexes were also tested against breast and colon cell lines. The metal complexes of synthesized Schiff base (HL) showed IC50 higher than that of HL, especially the Cu(II) complex, which resulted in the highest IC_{50} against the breast cell line [22]. Creaven and his research team derived some new Schiff bases from substituted aromatic aldehydes with 7-amino-4-methyl-quinolin-2(1H)-one and also prepared its metal complexes with Cu(II). These compounds were screened for their *in vitro* anti-cancer activity using human hepatic carcinoma cell lines like Hep-G2. Thus, the results reported and suggested that the copper(II) complexes of Schiff base ligands in the future have good potential as therapeutic agents [23]. Xiao *et al.* derived two new Schiff base compounds and their metal complexes with Co(II). These Schiff bases and metal complexes were studied *in vitro* antitumor activity and evaluated, in which human skin cancer cell lines named A-431, HT-144 and SK-MEL-30 were used in the screening tests. From the results, it was noticed that Schiff bases metal complexes showed significant growth inhibition activity on the three tumor cell lines (IC_{50}=11.3, B19.8 m/M), compared to their corresponding organic ligands Py, L1 and L2 (IC_{50}=90.8, B120.5 m/M) [24].

Ejidike and Ajibade derived Mononuclear Ru(III) complexes of the type [Ru(LL)Cl$_2$(H$_2$O)] where LL =monobasic tridentate Schiff base anion: (1Z)-N-(2-{(E)-[1- (2,4- dihydroxyphenyl) ethyllidene]amino} ethyl-N- phenylethani-midamide [DAE], 4- [(1E)-N-{2-[(Z)-(4-hydroxy-3 methyoxybenzylidene)amino] ethyl}ethanimidoyl] benzene-1-3-diol [HME], 4-[(1E)-N-{2-[(Z)-(4-hydroxy-3 methoxybenzylidene)amino] ethyl}ethanimidoyl]benzene-1,3-diol [HME), 4-

[(1*E*)-*N*-{2-[(*Z*) (3,4dimethoxybenzylidene)amino]ethyl}ethanimidoyl]benzene-1,3-diol [MBE], and *N*-(2-{(*E*)-[1-(2,4-dihydroxyphenyl)ethylidene]amino} ethyl) benzenecarboximidoyl chloride [DEE]). *In vitro* anti-cancer studies of the derived complexes against renal cancer cells (TK-10), melanoma cancer cells (UACC-62), and breast cancer cells (MCF-7) were investigated using the Sulforhodamine B assay, [Ru(DAE)Cl$_2$(H$_2$O)] resulted in the highest activity with IC$_{50}$ valves of 3.57 ± 1.09, 6.44 ± 0.38, and 9.06 ± 1.18 □M against MCF-7, UACC-62, and TK-10, respectively, order of activity being TK-10 < UACC-62 < MCF-7 [25].

Anti-bacterial Activity

An anti-microbial agent is one type of compound that kills microorganism or prevent their growth. The several numbers of infectious diseases caused by Gram-positive and Gram-negative pathogenic bacteria might be a serious threat to us. Nowadays, the chemistry of biological science has produced several compounds used as anti-bacterial agents. Generally, such types of compounds can provide great promise in this field. Schiff base and its supported metal complexes as an example [(Cu)$_2$(OH)(SAL)$_2$(OPD)$_2$ (AA)$_2$(N$_3$)$_2$] acted as a good anti-bacterial or anti-microbial agent against Gram-positive and Gram-negative bacteria (Table **1**) [26].

Imran *et al.* derived some new transition metal complexes of ciprofloxacin which derived from ciprofloxacin and *p*-substituted anilines. These were tested for anti-bacterial activity against several numbers of bacterial strains like *Bacillus subtilus*, *E. coli*, *Staphylococcus aueus* and *Sulmonella typhae* (Table **1**) [27]. Polo-Ceron and his group synthesized some new Schiff metal complexes with Cu(II), Ni(II), Co(II) and Zn(II), which were tested for biological activities against some Gram-positive bacteria like *Staphylococcus aureus* and Gram-negative bacteria like *Salmonella typhimurium*, *Escheria coli* and *Klebsiella pneumonia* [28]. Ko and his group derived a new ligand 4-((2-Hydroxy1-naphthyl)methylene amino)-1,5-dimethyl-2-phenyl-1H-pyrazol-3(2H)-one (HL) and their metal complexes Pd(II), Ni(II), Cu(II). These metal complexes were tested for their biological against Gram-positive bacteria like *Bacillus subtilis* and *Staphylococcus aureus* and Gram-negative bacteria like *Escherichia coli* and *Pseudomonas aeruginosa* [29]. Mohammed Hamad *et al.* synthesized some Schiff bases and their metal complexes of Cu(II) and Ni(II). These compounds showed an important effect against both types of Gram-positive bacteria, such as *Staphylococcus aurous* and Gram-negative bacteria *Escherichia coli* [30]. Parisi and his team derived Schiff base, 4-((2-hydroxybenzylidene)amino)-N-(thia-ol-2-yl) benzenesulfonamide (HL) and its metal complexes with Co(II), Ni(II) and Cu(II). This Schiff base and its metal complexes were tested against some bacterial strains such as *Pseudomonas aeruginosa*, *Bacillus subtilis*,

Staphylococcus aureus and *Escherichia coli*. This experimental result showed that the anti-bacterial activity of all metal complexes was better than that of the Schiff base [31]. Mohan and his group synthesized some Schiff bases and their metal complexes with Cu(II) and Zn(II), which are used in the field of pharmacology owing to their anti-microbial activities against some bacteria such as *Salmonella typhi, Bacillus subtilis, Staphylococcus* and *Klebsiella*. Generally, it was reported that metal complexes showed moderate activity compared to the free ligand [32]. Gupta *et al.* prepared Schiff base and its metal complexes with Cu(II), Ni(II), Co(II), Zn(II) and Mn(II) and which were tested for anti-bacterial activity against Gram-positive bacteria such as *S. aureus, Enterococcus* and Gram-negative bacteria like *K. pneumonia, E. coli* [33]. Yousif and their team developed Schiff base metal complexes with the metal ions Vo(II), Co(II), Pd(II), Rh(III) and Au(III). These Schiff base and metal complexes were screened for their *in vitro* anti-bacterial activity against *Escherichia coli, Salmonella typhi* and *Staphylococcus aureus* bacterial strain by the inhibition zone method using agar diffusion method [34]. Morgan *et al.* synthesized a new Schiff base ligand named (E)-2-(((3-aminophenyl) imino)methyl)phenol(HL) and several numbers of its metal complexes with Cr(III), Fe(III), Co(II), Ni(II), Zn(II), Cu(II), Cd(II) and Mn(II). These designed Schiff base and its metal complexes were screened for anti-microbial activities against some bacterial activities such as Gram-positive bacteria named *Bacillus subtilis* and *Staphylococcus* and Gram-negative bacteria named *Salmonella spp., Pseudomonas aeruginosa* and *Escherichia coli* and also screened against fungi like *Aspergillus fumigatus* and *Candida albicans* [35].

BIOLOGICAL ACTIVITY OF IONIC LIQUID TAGGED SCHIFF BASE METAL COMPLEXES

Rozwadowski and his group reported a new amino acid ionic liquid-supported Schiff bases, derivatives of salicyldehyde and several number of amino acids like L-threonine, L-valine, L-leucine, L-isoleucine and L-histidine, which have been investigated by various spectroscopic techniques such as NMR, UV-Vis, IR, MS *etc.* and also by deuterium isotope effects on ^{13}C-NMR chemical shifts. Their experimental observation showed that in all studied amino acid ionic liquid-supported Schiff bases instead of L-histidine derivative, a proton transfer equilibrium existed and the presence of the COO⁻ group stabilized the proton transferred NH-form. Normally it is clear that the stability of their synthesized compounds tetrabutylammonium salts of amino acid Schiff base derivatives of salicylaldehyde which exists in tautomeric equilibrium state supported by a bifurcated intermolecular hydrogen bond. The presence of the imidazole ring weakens the interactions between the COO⁻ and NH⁻ groups that stabilize the proton transferred form (Scheme **2**).

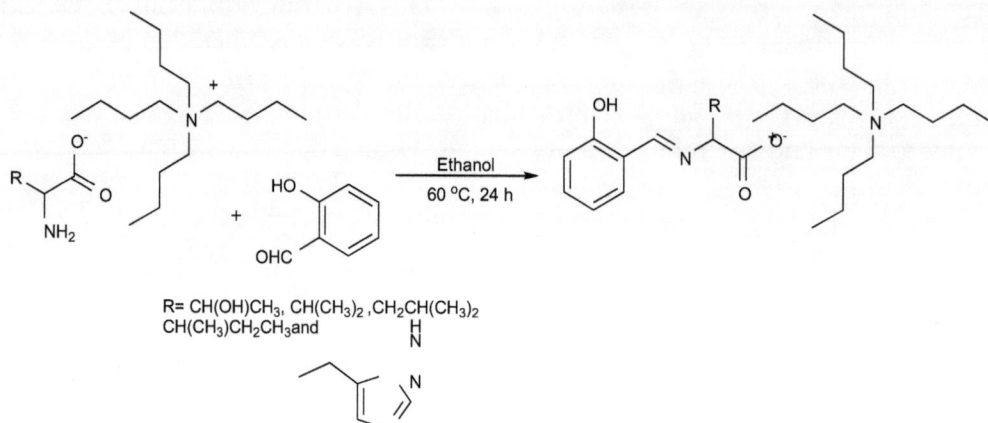

R= CH(OH)CH$_3$, CH(CH$_3$)$_2$, CH$_2$CH(CH$_3$)$_2$
CH(CH$_3$)CH$_2$CH$_3$ and

Scheme 2. Synthesis of amino acid ionic liquid-supported Schiff base and its application.

Khungar and his group synthesized an imidazolium ionic liquid tagged Schiff base through three steps from 2,4-dihydroxybenzaldehyde by selective alkylation with 1,3-dibromopropane, followed by reaction with 1-methylimidazole and Schiff base formation with aromatic amines (Scheme 3). This reported compound was evaluated for anti-bacterial and anti-fungal activities, and it showed the inhibition of both Gram-positive and Gram-negative bacteria [36].

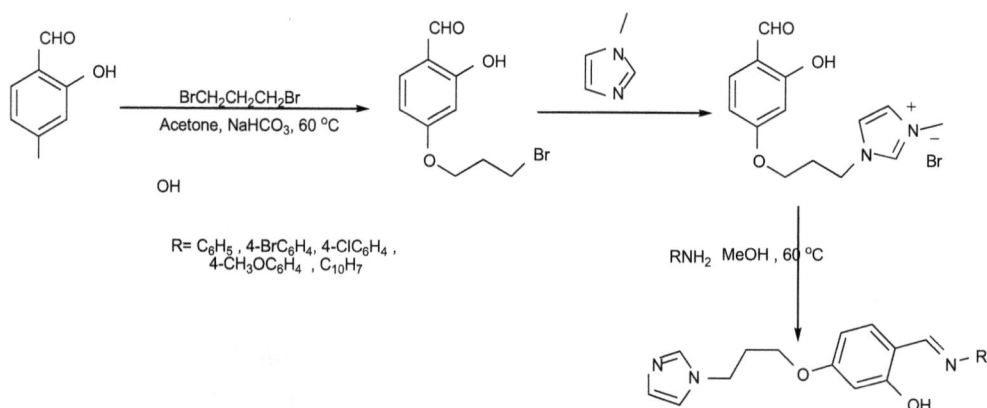

R= C$_6$H$_5$, 4-BrC$_6$H$_4$, 4-ClC$_6$H$_4$,
4-CH$_3$OC$_6$H$_4$, C$_{10}$H$_7$

Scheme 3. Synthesis and anti-microbial activities of ionic liquid-supported Schiff base.

Sinha and his group synthesized two ionic liquid-supported Schiff base Cu(II) complexes of 1-{2-[(2-hydroxybenzylidene)amino]ethyl}-3-methylimidazolium hexafluorophosphate and characterized by different analytical and spectroscopic methods such as elemental analysis, magnetic susceptibility, UV–Vis, IR and NMR spectroscopy, and mass spectrometry. The Schiff base ligand was found to act as a potential bidentate chelating ligand with N, O donor sites and which formed 1:2 metal chelates with Cu(II) salts. The synthesized Cu(II) complexes were tested for biological activity and did not show anti-bacterial activity against two commonly known bacteria, *viz.*, *B. subtilis* and *E. coli* [37].

CONCLUSION

The importance of novel Schiff base ligands and their complexes with transition metals is a demanding area in biochemistry. The synthesis of new Schiff bases attracts immense attention in the field of chemical research. In this chapter, we have focused on a few structurally diversified Schiff bases and their metal complexes which are biologically active in different respect (anti-fungal, anti-bacterial, anti-cancer and anti-viral). A few Schiff bases tagged with ionic liquid (green solvent and catalyst) having biological activity have also been discussed in this article. We hope there is ample scope for this particular research in the near future.

CONSENT FOR PUBLICATION

Not applicable.

CONFLICT OF INTEREST

The authors declare no conflict of interest, financial or otherwise.

ACKNOWLEDGEMENT

Declared none.

REFERENCES

[1]   Tulu MM, Yimer AM. Catalytic studies on Schiff base complexes of Co (II) and Ni (II) using benzoylation of phenol. Mod Chem Appl 2018; 6: 260.
        [http://dx.doi.org/10.4172/2329-6798.1000260]

[2]   Wei L, Tan W, Zhang J, *et al.* Synthesis, characterization, and anti-fungal activity of schiff bases of inulin bearing pyridine ring. Polymers (Basel) 2019; 11(2): 371.
        [http://dx.doi.org/10.3390/polym11020371] [PMID: 30960355]

[3]   da Silva CM, da Silva DL, Modolo LV, *et al.* Schiff bases: A short review of their antimicrobial activities. J Adv Res 2011; 2(1): 1-8.
        [http://dx.doi.org/10.1016/j.jare.2010.05.004]

[4]   Rana KA, Pandurangan AN, Singh NA, Tiwari AK. A systemic review of schiff bases as an analgesic, anti-inflammatory. Int J Curr Pharm Res 2012; 4(2): 5-11.

[5]   Shah SS, Shah D, Khan I, Ahmad S, Ali U. Synthesis and antioxidant activities of Schiff bases and their complexes: An updated review. 2020; 6936-63.

[6]   Rehman W, Baloch MK, Muhammad B, Badshah A, Khan KM. Characteristic spectral studies and *in vitro* antifungal activity of some Schiff bases and their organotin (?) complexes. Chin Sci Bull 2004;

49(2): 119-22.
[http://dx.doi.org/10.1360/03wb0174]

[7]   Joseyphus RS, Nair MS. Antibacterial and anti-fungal studies on some schiff base complexes of zinc (II). Mycobiology 2008; 36(2): 93-8.
[http://dx.doi.org/10.4489/MYCO.2008.36.2.093] [PMID: 23990740]

[8]   Miloud MM, El-ajaily MM, Al-noor TH, Al-barki NS. Anti-fungal activity of some mixed ligand complexes incorporating Schiff bases. J Bacteriol Mycol (Monroe Township) 2020; 7(1): 1122.

[9]   Munjal M. Synthesis, characterization and anti-fungal activity of transition metal (II) complexes of schiff base derived from p-aminoacetanilide and salicylaldehyde. J Pharmacogn Phytochem 2019; 7(6): 864-6.

[10]   Govindaraj V, Ramanathan S, Murgased S. Synthesis, Characterization, Antibacterial, Antifungal Screening and Cytotoxic Activity Of Schiff Base Nickel (II) Complexes with Substituted Benzylidine Aminobenzoic Acid. Der Chemica Sinica 2018; 9(3): 736-45.

[11]   Chang EL, Simmers C, Knight DA. Cobalt complexes as anti-viral and anti-bacterial agents. Pharmaceuticals (Basel) 2010; 3(6): 1711-28.
[http://dx.doi.org/10.3390/ph3061711] [PMID: 27713325]

[12]   Al-Abed Y, Dubrovsky L, Ruzsicska B, Seepersaud M, Bukrinsky M. Inhibition of HIV-1 nuclear import *via* schiff base formation with arylene bis(methylketone) compounds. Bioorg Med Chem Lett 2002; 12(21): 3117-9.
[http://dx.doi.org/10.1016/S0960-894X(02)00642-X] [PMID: 12372514]

[13]   Jarrahpour A, Khalili D, De Clercq E, Salmi C, Brunel J. Synthesis, antibacterial, antifungal and antiviral activity evaluation of some new bis-Schiff bases of isatin and their derivatives. Molecules 2007; 12(8): 1720-30.
[http://dx.doi.org/10.3390/12081720] [PMID: 17960083]

[14]   Kotovskaya SK, Baskakova ZM, Charushin VN, *et al.* Synthesis and antiviral activity of fluorinated pyrido[1,2-a]benzimidazoles. Pharm Chem J 2005; 39(11): 574-8.
[http://dx.doi.org/10.1007/s11094-006-0023-9]

[15]   Genova P, Varadinova T, Matesanz AI, Marinova D, Souza P. Toxic effects of bis(thiosemicarbazone) compounds and its palladium(II) complexes on herpes simplex virus growth. Toxicol Appl Pharmacol 2004; 197(2): 107-12.
[http://dx.doi.org/10.1016/j.taap.2004.02.006] [PMID: 15163546]

[16]   Matesanz AI, Pérez JM, Navarro P, Moreno JM, Colacio E, Souza P. Synthesis and characterization of novel palladium(II) complexes of bis(thiosemicarbazone). Structure, cytotoxic activity and DNA binding of Pd(II)-benzyl bis(thiosemicarbazonate). J Inorg Biochem 1999; 76(1): 29-37.
[http://dx.doi.org/10.1016/S0162-0134(99)00105-1] [PMID: 10530004]

[17]   Uddin N, Rashid F, Ali S, *et al.* Synthesis, characterization, and anticancer activity of Schiff bases. J Biomol Struct Dyn 2020; 38(11): 3246-59.
[http://dx.doi.org/10.1080/07391102.2019.1654924] [PMID: 31411114]

[18]   Abd-Elzaher MM, Labib AA, Mousa HA, Moustafa SA, Ali MM, El-Rashedy AA. Synthesis, anti-cancer activity and molecular docking study of Schiff base complexes containing thiazole moiety. Beni-Suef University Journal of Basic and Applied Sciences 2016; 5(1): 85-96.

[19]   Mbugua SN, Sibuyi NRS, Njenga LW, *et al.* New Palladium(II) and Platinum(II) Complexes Based on Pyrrole Schiff Bases: Synthesis, Characterization, X-ray Structure, and Anticancer Activity. ACS Omega 2020; 5(25): 14942-54.
[http://dx.doi.org/10.1021/acsomega.0c00360] [PMID: 32637768]

[20]   Rama I, Selvameena R. Synthesis, structure analysis, anti-bacterial and *in vitro* anti-cancer activity of new Schiff base and its copper complex derived from sulfamethoxazole. J Chem Sci 2015; 127(4): 671-8.

[http://dx.doi.org/10.1007/s12039-015-0824-z]

[21]   Parasuraman B, Rajendran J, Rangappan R. An Insight into Antibacterial and Anticancer Activity of Homo and Hetero Binuclear Schiff Base Complexes. Orient J Chem 2017; 33(3): 1223-34.
[http://dx.doi.org/10.13005/ojc/330321]

[22]   Abd El-Halim HF, Mohamed GG, Anwar MN. Antimicrobial and anticancer activities of Schiff base ligand and its transition metal mixed ligand complexes with heterocyclic base. Appl Organomet Chem 2018; 32(1): e3899.
[http://dx.doi.org/10.1002/aoc.3899]

[23]   Creaven BS, Duff B, Egan DA, *et al.* Anticancer and antifungal activity of copper(II) complexes of quinolin-2(1H)-one-derived Schiff bases. Inorg Chim Acta 2010; 363(14): 4048-58.
[http://dx.doi.org/10.1016/j.ica.2010.08.009]

[24]   Xiao YJ, Diao QC, Liang YH, Zeng K. Two novel Co(II) complexes with two different Schiff bases: inhibiting growth of human skin cancer cells. Braz J Med Biol Res 2017; 50(7): e6390.
[http://dx.doi.org/10.1590/1414-431x20176390] [PMID: 28678922]

[25]   Ejidike IP, Ajibade PA. Synthesis, characterization, anti-cancer, and antioxidant studies of Ru (III) complexes of monobasic tridentate Schiff bases. Bioinorg Chem Appl 2016; 2016: 1-11.
[http://dx.doi.org/10.1155/2016/9672451] [PMID: 27597814]

[26]   Prasad MS, Neeraja G, Anil Kumar B, Mohana K. Copper Schiff Base Complex and its Antimicrobial and Anticancer Activities. J Chem Pharm Sci 2017; 10: 1406-9. [JCPS].

[27]   Imran M, Iqbal J, Iqbal S, Ijaz N. *In vitro* anti-bacterial studies of ciprofloxacin-imines and their complexes with Cu (II), Ni (II), Co (II), and Zn (II). Turk J Biol 2007; 31(2): 67-72.

[28]   Londoño-Mosquera JD, Aragón-Muriel A, Polo Cerón D. Synthesis, antibacterial activity and DNA interactions of lanthanide(III) complexes of N(4)-substituted thiosemicarbazones. Univ Sci (Bogota) 2018; 23(2): 141-69.
[http://dx.doi.org/10.11144/Javeriana.SC23-2.saaa]

[29]   Al Zoubi W, Al-Hamdani AAS, Putu Widiantara I, Hamoodah RG, Ko YG. Theoretical studies and antibacterial activity for Schiff base complexes. J Phys Org Chem 2017; 30(12): e3707.
[http://dx.doi.org/10.1002/poc.3707]

[30]   Mohammed Hamad MN, Hussein MB. Synthesis and antibacterial activity of 3-nitrobenzaldehyde semicarbazone ligand and its Ni (II) and Cu (II) Complexes. J Microbiol Exp 2020; 8(5): 163-5.
[http://dx.doi.org/10.15406/jmen.2020.08.00302]

[31]   Reiss A, Cioaterǎ N, Dobriţescu A, *et al.* Bioactive Co(II), Ni(II), and Cu(II) Complexes Containing a Tridentate Sulfathiazole-Based (*ONN*) Schiff Base. Molecules 2021; 26(10): 3062.
[http://dx.doi.org/10.3390/molecules26103062] [PMID: 34065538]

[32]   Omanakuttan A, Priyanka G, Mohan RD. Synthesis Characterisation and anti-microbial properties of two Salicylaldimine Schiff base complexes of transition metals. IOP Conference Series: Materials Science and Engineering. 561(1): 012050.
[http://dx.doi.org/10.1088/1757-899X/561/1/012050]

[33]   Saini RP, Kumar V, Gupta AK, Gupta GK. Synthesis, characterization, and antibacterial activity of a novel heterocyclic Schiff's base and its metal complexes of first transition series. Med Chem Res 2014; 23(2): 690-8.
[http://dx.doi.org/10.1007/s00044-013-0657-6]

[34]   Yousif E, Majeed A, Al-Sammarrae K, Salih N, Salimon J, Abdullah B. Metal complexes of Schiff base: Preparation, characterization and antibacterial activity. Arab J Chem 2017; 10: S1639-44.
[http://dx.doi.org/10.1016/j.arabjc.2013.06.006]

[35]   Nozha SG, Morgan SM, Ahmed SEA, *et al.* Polymer complexes. LXXIV. Synthesis, characterization and antimicrobial activity studies of polymer complexes of some transition metals with bis-bidentate Schiff base. J Mol Struct 2021; 1227: 129525.

[http://dx.doi.org/10.1016/j.molstruc.2020.129525]

[36]   Khungar B, Rao MS, Pericherla K, *et al.* Synthesis, characterization and microbiocidal studies of novel ionic liquid tagged Schiff bases. C R Chim 2012; 15(8): 669-74.
[http://dx.doi.org/10.1016/j.crci.2012.05.023]

[37]   Saha S, Brahman D, Sinha B. Cu (II) complexes of an ionic liquid-based Schiff base [1-{2-(2-hydroxy benzylidene amino) ethyl}-3-methylimidazolium] Pf6: Synthesis, characterization and biological activities. J Serb Chem Soc 2015; 80(1): 35-43.
[http://dx.doi.org/10.2298/JSC140201078S]

CHAPTER 6

Metal-Organic Frameworks (MOFs) for the Antimicrobial Applications

Nazeer Abdul Azeez[1], Sapna Pahil[2], Surendra H. Mahadevegowda[3] and **Sudarshana Deepa Vijaykumar[4,*]**

[1] Department of Biotechnology, Bannari Amman Institute of Technology, Sathyamangalam, Erode, Tamil Nadu-638 401, India

[2] Department of Microbiology and Immunology, University of Michigan Medical School, Ann Arbor, Michigan, USA

[3] Department of Chemistry, School of Sciences, National Institute of Technology, Tadepalligudem, Andhra Pradesh – 534 101, India

[4] Department of Biotechnology, National Institute of Technology, Tadepalligudem, Andhra Pradesh –534 101, India

Abstract: Metal-Organic Frameworks (MOFs) are a class of porous crystalline materials made-up of transition-metal cations linked with multidentate organic ligands by the coordination bonds. The strong, flexible frameworks and the porous structure of the MOFs establish them as an effective carriers of various functional compounds, such as gases, drugs, and anti-microbial agents. The MOFs render high loading capacity and sustained release, which is the desired property in anti-microbial applications. Similar porous material for the anti-microbial application is Zeolite, however, it is more complex to synthesize than MOFs. Currently, MOFs are used mainly in catalysis, gas separation and storage, and water purification applications. In the applications as anti-microbial agents, MOFs are just emerging into the field application from the laboratory scale. Hence, this chapter discusses the properties, synthetic procedures, anti-bacterial mechanisms and various forms of MOFs for anti-microbial applications. The MOFs are often doped with metal nanoparticles, polymers, and metal-polymer complexes. Each category of MOFs has a different mechanistic approach to inhibiting microbial colony growth. In this regard, this chapter will provide sufficient information on the MOFs, which will help to understand their significance in anti-microbial applications and their scope.

Keywords: Anti-microbial, Metal-Organic Frameworks, Nanoparticles, Polymers, Porous materials.

---
* **Corresponding author Sudarshana Deepa Vijaykumar:** Department of Biotechnology, National Institute of Technology, Tadepalligudem, Andhra Pradesh – 534 101, India; E-mail: sudarshanadeepa@nitandhra.ac.in

INTRODUCTION

Microbial contamination negatively impacts various industries such as food, livestock, medical, and environmental management, threatening millions of lives globally [1]. Measures to eradicate these microbial contaminations through antibiotics have further raised the antibiotic resistance of the microbes targeted. World Health Organization (WHO) has urged that antibiotic abuse leading to drug resistance of the microbes has led to a greater challenge to public health around the world [2]. In 2017, WHO announced the priority list of bacteria for which new antibiotic development is urgently required [3]. According to WHO, in 2016, a lower respiratory infection caused 3 million deaths, diarrhea caused 1.4 million, and tuberculosis caused 1.3 million deaths globally [4]. Antibiotic resistance, on the other hand, is usually a natural process, but the misuse of antibiotics on humans and animals further worsens the situation. This makes it hard to treat the growing infection cases such as tuberculosis, gonorrhea and salmonellosis. Economically, antibiotic resistance casts a burden by an extended hospital stay for treatment, prolonged medication, and increased mortality states WHO [5]. In the current situation, efficient new methods or systems are urgently require combating bacterial infections due to the existence of multi-drug resistance bacteria [6]. The use of antibiotics to cure infections caused by many pathogens is becoming inefficient, and fewer antibiotics have been marketed in recent years [7]. Hence, it is an exigent need to develop novel anti-microbial agents or systems to restrict the growing issues of microbial contamination and antibiotic resistance.

Due to the rise in the number of antibiotic-resistant pathogens, the attention of various research communities was drawn to develop an efficient anti-microbial agent alternative to antibiotics. Developing new anti-microbial agents came up in three eras. The first era discovered anti-microbial molecules, such as penicillin, the second era marked the development of metallic carriers for antibiotics, and the third era developed the nontoxic metal-organic carriers for anti-microbial agents [8]. A breakthrough in the third-era discovery is the formulation of metal-organic frameworks (MOFs). The MOFs are polymeric porous crystalline metal-organic frameworks with large specific surface areas, high porosity, well-dispersed active centers and tunable functional groups with appropriate metal ions and organic linkers. A general schematic representation of MOF is elucidated in Fig. (**1**). Structural building blocks of MOFs have two major components;

i) Metal ions or metal ion clusters forming a continuous framework through coordination bonds or covalent bonds using

ii) Linkers made up of organic molecules [9, 10].

The metal ions and ligands arrange geometrically, rendering corresponding physicochemical and functional properties to MOFs, which can be altered by varying the arrangement of the framework [8]. By selecting the appropriate geometry of the framework, the composition and size of the pores can be controlled, thereby determining the porosity and specific surface area of the skeleton corresponding to various applications [11, 12].

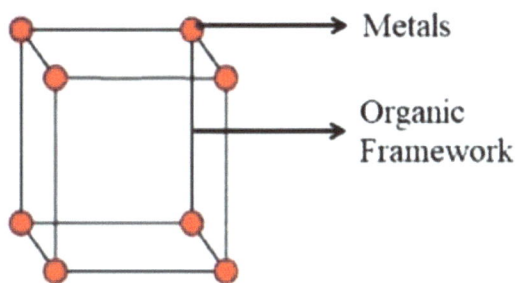

Fig. (1). General schematic representation of a MOF.

MOFs are either used to carry the anti-microbial agents with the desired porosity and crystal structure for the optimal loading and release, or they are used as antibiotics themselves, having metal ions that are lethal to microbial cells [13, 14]. This chapter discusses the synthesis methods, anti-microbial activities, and the prospects of the anti-microbial MOFs.

SYNTHESIS OF MOFs

Synthesis of MOFs is easier than other porous materials, such as zeolites, as the nucleation, crystal growth and template removal for MOFs require mild conditions [14]. The synthesis of zeolite is complex and expensive [15]. Whereas, MOFs are simple to synthesize and effective in functionality. MOFs are formed by the coordination of polymers with an open framework containing potential pores [16]. MOFs are formed by the self-assembly of polymeric linking organic ligands and metallic ions. The metals used to synthesize MOFs are copper [Cu], zinc [Zn], aluminium [Al], iron [Fe], chromium [Cr] and gadolinium [Gd]. The organic linkers used are 2,5- dihydroxyterephthalate (DOT), 2,20- bipyridine-5,50- dicarboxylate (BPYDC), N,N-diethylformamide (DEF), p-terphenyl-4,-0-dicarboxylate (TPDC), biphenyl-4,40-dicarboxylate (BPDC), 2-methylimidazolate (MIM), 2-formylimidazolate (FIM), pyrazine-2,--dicarboxylate (PZDC), 4,4'-bipyridine, 1,10-phenanthroline (phen), 4,4-[hexafluoroisopropylidene] dibenzoate (HFBBA), m-benzenedicarboxylate (m-BDC) [17]. The parameters that control the formation of MOFs during the synthesis process are temperature, reaction time, pressure, pH, and solvent system. Various synthetic approaches are diffusion method, hydrothermal method, microwave assisted, electrochemical, mechanochemical, and ultrasonic method [18 - 23]. The synthetic procedures of various MOFs are tabulated in Table **1**. The basic synthetic procedure of MOFs is roughly deliberated in Fig. (**2**).

Table 1. The synthetic procedures of various MOFs.

S. No.	MOF	Formula	Synthetic Method	References
1	MOF-74	Zn_2DOT	Microwave assisted; Solvothermal	[24, 25]
2	MOF-101	$Cu_2(BDC-Br)_2(H_2O)_2$	Solvothermal	[26]
3	MOF-177	$Zn_4O(BTB)_2$	Solvothermal	[27]
4	MOF-235	$(Fe_3O(BDC)_3(DMF)_3)(FeCl_4).(DMF)_3$	Solvothermal	[28]
5	MOF-253	$Al(OH)(BPYDC)$	Hydrothermal	[29]
6	MIL-53	$Al(OH)(BDC)$	Hydrothermal; Ultrasonicated	[30, 31]
7	MIL-53[Al]-NH2	$Al(OH)(BDC-NH_2)$	Hydrothermal	[32]
8	MIL-88A	$Fe_3O(MeOH)_3(O_2CCH)CHCO_2)_3.MeCO_2. nH_2O$	Hydrothermal; Ultrasonicated	[33, 34]
9	MIL-88-Fe	$Fe_3O(MeOH)_3(O_2C(CH_2)_2CO_2)_3.AcO.(MeOH)_{4.5}$	Microwave assisted; Solvothermal	[33, 34]
10	MIL-100-Fe	$Fe^{III}_3 O(H_2O)_2F.(BTC)_2. nH2O$	Hydrothermal; Microwave assisted	[31, 35]
11	MIL-101	$Cr_3O(H2O)_2F.(BDC)_3. nH_2O$	Hydrothermal; Solvothermal	[36, 37]
12	IRMOF-1[MOF-5]	$Zn_4O(BDC)_3. 7DEF.3H_2O$	Hydrothermal; Solvothermal	[38, 39]
13	IRMOF-16	$Zn_4O(TPDC)_3. 17DEF.2H_2O$	Solvothermal	[40]
14	UiO-66	$Zr_6O_6(BDC)_6$	Solvothermal	[41]
15	UiO-67	$Zr_6O_6(BPDC)_6$	Mechanochemical; Electrochemical	[42, 43]
16	UiO-68	$Zr_6O_6(TPDC)_6$	Solvothermal	[44]
17	HKUST-1 [MOF-199]	$Cu_3(BTC)_2$	Solvothermal	[45]
18	ZIF-8	$Zn(MIM)_2$	Diffusion	[46]
19	ZIF-90	$Zn(FIM)_2$	Chemical reaction	[47]
20	CPL-2	$Cu_2(PZDC)_2(4,4'-BPY)$	Hydrothermal	[48]
21	F-MOF-1	$[Cu(HFBBA)(phen)_2](H_2 HFBBA)_2(H_2O)(HCO_2)$	Solvothermal	[49]
22	MOP-1	$Cu_{24}(m-BDC)_{24}(DMF)_{14}(H_2O)_{10}$	Solvothermal	[50]

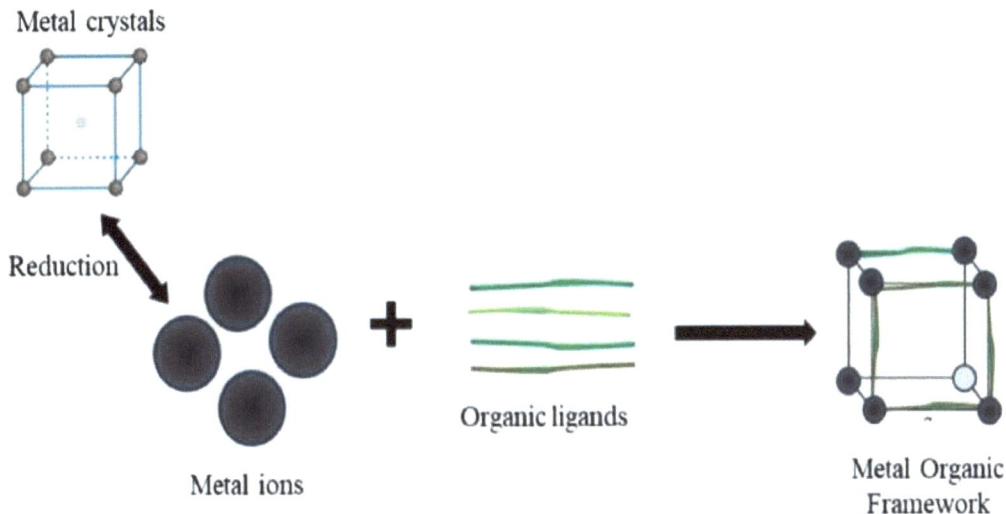

Fig. (2). Principle synthetic scheme of MOFs.

Diffusion Method

This method is preferred for poorly soluble and single crystal yield appropriate for X-ray diffraction analysis as a substitute for amorphous or poly-crystalline. The synthetic procedure is diagrammatically depicted in Fig. (3). It involves the gradual introduction of each building component into a system to interact. Two major approaches of the diffusion method are:

[i] the one where two layers with different densities are formed. One phase is the precipitant solvent, and the other one encompasses the product in a solvent. The layers are divided by a thin layer of interfacial solvent. The precipitant gradually diffuses from its layer through the interfacial solvent and enters the other layer, and interacts with the product. Crystal growth occurs at the interface.

[ii] in the second method, the reactants gradually diffuse across the physical barriers, such as gels, through a concentration gradient. The physical barriers reduce the diffusion rate and forbid the precipitation of bulk particles [51].

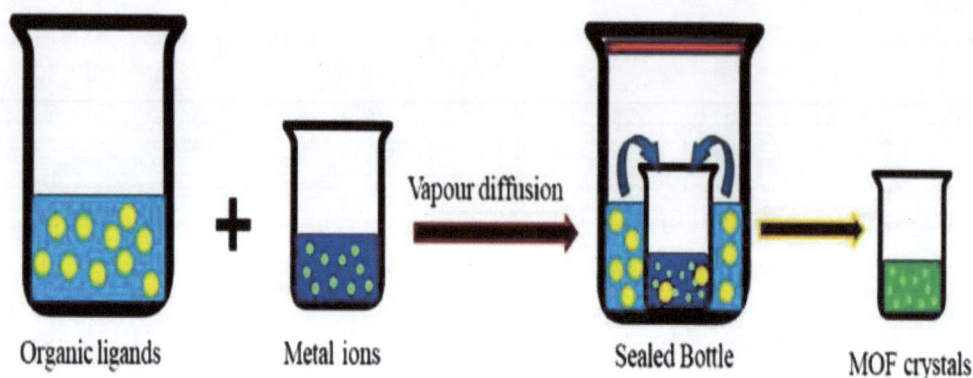

Fig. (3). Diffusion mode of MOF synthesis where the organic ligand solution is diffused into metal ion solution through either vapour or membrane transfer.

Hydrothermal Method

This is the commonly used method for the preparation of amorphous zeolite. The basic principle is the self-assembly of products from soluble precursors under increased pressure. The reactants are taken in a tightly closed autoclave and heated from 80 to 260 °C for a long period (in some cases, the heating goes on for many days) which increases the pressure of the closed system. The increased pressure accelerated the reaction of the components added to the system. Post reaction, the system is rapidly cooled to obtain the crystals [52].

Microwave-assisted Method

The increase in temperature for a long period and longer reaction time of the hydrothermal method are eliminated in this process. Microwaves help initiate the reaction and are catalyzed by the increasing pressure inside the closed reaction system. This method is used to produce finer metal and metallic oxide particles. The varying parameters enable the optimization of the shape and size of MOFs produced. Microwaves produce uniform seeding conditions even though the microwave technique is generally incompetent to produce crystals. Thus, the reaction time can be reduced, and the crystal shape and size can be controlled in this method. Post reaction, the saturation point and the temperature reduction rate play vital roles in the formation of the crystals [53].

Electrochemical Method

The electrochemical method is used for large-scale industrial production of MOF powders. This rapid synthesis method does not require elevated temperature and anions, such as nitrates from metal salts. However, in bulk production (larger MOFs), there is a limitation in crystallization due to the production of the metal ions *in situ* near the support surface. This decreases the undesirable accumulation of crystals during the synthesis of the membrane. While cooling, there is a probability for thermal-induced cracking; thus, the slighter temperatures are employed. An incompatibility in thermal growth coefficients between the different support structures and the MOF leads to this cracking. In this respect, MOF shows a negative thermal expansion coefficient. This renders the electrochemical method the advantage of tunability in various parameters for optimization of the yield [17].

Mechanochemical Method

Mechanochemical synthesis is a rapid technique that demands the synthesis procedure without the use of organic solvents. MOFs to entrap smaller particles are synthesized using this technique. Briefly, mechanical forces by milling or grinding are applied to initiate the chemical reaction by breaking the intra-molecular bonds. Mechanochemistry is an established concept in synthetic chemistry, which is currently used in multicomponent (ternary and higher) reactions to establish co-crystals that are pharmaceutically active, and in other fields such as in inorganic solid-state chemistry, polymer science, and organic synthesis. There are various advantages to choosing mechanochemical synthesis to produce MOF synthesis, such as no residual organic solvent in the products, ambient operational temperature, lesser reaction time [10-20 min] and better yield. The organic solvent could be easily evaporated in the preparation of MOF. When the seeding materials are metal oxides, the by-product is water. Further, metal oxide seeds cannot be effectively used in other methods as they are not soluble in most solvents, which makes this process ideal for the synthesis of metallic oxide MOFs. The organic linkers with low melting points and hydrated metal salts, specifically those that encompass essential anions, are used as precursors in this method. Metal acetates or carbonates are used to improve the crystalline property. The acetic acid deposited in the pores as by-products while using metal acetates is removed by thermal activation [54].

Ultrasonication Method

Ultrasonication is yet another mechanical activation method where ultrasonic sounds of more than 10 MHz are used to precede the chemical reaction in the reactor system. The ultrasonic waves create cavitation and microjets on the solid surface that are in the sonic range. When the reacting solution is homogeneous chemical reactions occur in the cavity (severe situations), at the interface (intermediate pressures and temperatures) or in the bulk media, where the shear forces are intense. The creation of radicals with an additional reaction, breakage of the bond, and establishment of molecules in an excited mode can happen by severe conditions and strong shear forces. Ultrasonic waves enhance the dissolution rate of the seed compounds. For the synthesis of MOF, the main objective of the ultrasonication technique is a quick, environmentally friendly, energy-efficient, a facile process that can be operated at ambient temperatures. The future of MOF synthesis is majorly relying on this method due to its efficient timing [55].

MOFs IN ANTIMICROBIAL APPLICATION

MOFs are the recently preferred anti-microbial materials due to their porous structure to carry the anti-microbial load and release the biologically active metal ions. The MOFs, among the other porous materials such as zeolites, mesoporous silica, and activated carbon, have several advantages, including the larger pore size, tuneable geometry, biocompatibility, and their inherent bactericidal nature [56].

MOFs can act as both anti-microbial agents themselves as well as can carry the anti-microbial agents such as nitric oxide, hydrogen sulfide, carbon monoxide, and anti-bacterial pharmaceuticals as penicillin, glycopeptides, aminoglycosides, macrolides, cephalosporin, and metal ions.

Anti-bacterial Mechanism of MOFs

The ionic surface of metal co-ordinates and the nanometric dimension of MOFs render bactericidal ability through increased membrane permeability, reducing the proton motive force, de-energization of cells, efflux of phosphate, leakage of cellular content, and disruption of DNA replication [57]. Hence, the anti-bacterial efficiency of the MOFs depends on the release kinetics of metal ions from the framework. This hypothesis was postulated and studied by Berchel *et al.* 2011 [58]. Adapting the Pearson hard and soft (Lewis) acids and bases (HSAB) theory to evaluate the hypothesis, the ease of cation release was reckoned to depend on the relative hardness of cations (Lewis acids) and organic linkers (Lewis bases) of the MOF. A soft acid linked by a hard base provides only moderate stability to the overall structure. Hence, they are more susceptible to degradation and release cations more easily. Whereas the soft acids-soft bases and hard acids-hard bases pairs provide better stability to MOFs [59]. Hence, choosing the metal ions and respective organic ligands is vital for effective anti-bacterial efficacy.

MOFs as the Antibiotic Carrier

The MOFs have high drug loading efficiency due to high surface area and porosity and deliver the antibiotics by sustained release kinetics through matrix degradation. Further the MOFs are optically active and thus detected by imaging techniques *in vivo* and release by photostimulation [60]. The three strategies for loading the antibiotics into MOFs are direct assembly, encapsulation and post-synthesis loading techniques. The loading techniques are selected based on the location of the antibiotics and the effective interactions between bioactive compounds and the MOF components [61].

Direct Assembly Technique

The direct assembly technique is employed when the organic linker in the MOF itself is the antibiotic or its pro-drug molecule. Since the bioactive element itself is the structural moiety, the loading efficiency is high, and the load is distributed uniformly. The release is time-dependent steady matrix degradation controlled; hence, the programmed release can render the desired anti-microbial result. However, the critical issue in this strategy is that there is the probability that the antibiotics may lose their functional properties during the extreme treatments in the synthesis procedures. This technique is preferably used by various researchers to load and site specifically release photo dynamic therapy (PDT) or photo thermal therapy (PTT) agents [62]. PDT and PTT are successful treatment methods against the drug-resistant microbes. Technical progress has been made in this technique for antibiotic loading, and some formulations, such as Porphyrin for PDT, have acquired US FDA approval [63].

Encapsulation Technique

Anti-bacterial activity of MOF themselves is not much reported in the literature, however, a combination of MOF with other bactericidal agents like antibiotics can provide synergistic and alternative anti-microbial therapy. In the encapsulation technique, the antibiotics are loaded into the cavities or voids present in MOFs and interact with ions or linkers either through covalent or non-covalent interactions. The encapsulation technique is ideal to carry the nucleic acids, enzymes, and proteins. Size of the antibiotic is the deciding parameter to choose the type of MOF required as the cavities in the MOFs should be able to accommodate the load. Hence, the micro-porous MOFs cannot be used in this technique as the antibiotic sizes are usually larger to get accumulated. Zeolitic Imidazolate Frameworks (ZIF-8) family of MOFs have become the better choice for encapsulation due to their large cavity size and stability at neutral pH, thereby keeping the loaded antibiotics intact within the matrix and quickly released only in acidic conditions. However, the larger cavity size would lead to burst release decreasing the linearity of the delivery profile [64]. Further, Nabipour and coworkers evaluated the anti-bacterial efficiency of gentamicin-loaded $Zn_2[bdc]_2[dabco]$ MOF for both Gram-positive and negative bacteria, and this study describes the potential utility of pH-controlled release of gentamicin from MOF [65].

Zinc-based three different nano metal–organic frameworks (nMOF's) show an excellent anti-bacterial synergetic effect with ampicillin and kanamycin. The developed synergetic platform through nMOF/antibiotic formulations shows an effective synergetic effect against both Gram-positive and negative bacterial strains [66]. In general, the mechanistic study of the anti-bacterial effect associated with the damage of bacterial cells shows an effective cell disruption when nMOF mixed with ampicillin or kanamycin. Even though, zinc-based nMOF may not have a significant bactericidal effect, the frameworks can act as a reservoir for Zn^{+2} ions and can undergo interaction with bacterial cell walls to enhance the synergized bactericidal effect with an antibiotic. Overall, the study of Zn-based nMOF with antibiotics indicates that Zn-nMOF/antibiotic formulations show higher anti-bacterial effects than antibiotics alone.

Post-synthesis Loading Technique

Due to the adaptive nature of MOF for post-synthesis modification, various MOF-based antibiotic loading techniques were developed for anti-microbial applications. In this technique, the MOF is prepared first, and then the anti-microbial agent is loaded into the framework, where they interact with either the organic linkers by covalent interactions or with the metal nodes by coordination bonds [67]. One of the stimulating examples is work reported by Wang *et al.*, here authors used zirconium-based nano-sized MOF post functionalization with the carboxyl-functionalized diiodo-substituted BODIPY, and the predesigned ligand was incorporated into MOF by solvent-assisted ligands exchange to enhance the singlet oxygen production efficiency for photodynamic therapy application.

However, the post-synthesis loading technique is mainly used for the loading of nucleic acids that may be disintegrated by the synthetic conditions. Usually, the loaded antibiotics remain in the outer vicinity of MOFs in this technique. The open metal sites of the MOFs provide the perfect place for antibiotics to interact without any change in their functional efficacy and get better among many synthetic provocations [68]. Zhu *et al.* showed that zirconium-based uniform-sized nanoparticles could be utilized as a carrier to deliver alendronate molecules to cancer cells. This study indicates the alendronate molecules immobilized MOF nanoparticles can kill the cancer cells efficiently with higher biocompatibility compared to free alendronate molecules. These studies indicate for enhanced bioactive-MOF interaction, modulators could be used while synthesis that provides functional groups for binding.

Gas Storage and Delivery Technique

In this method, gas molecules are exogenously delivered from outside of the body, at present scenario the development of better targeted and specific approaches is the growing interest in this arena. The gaseous molecules like nitric oxide [NO], carbon monoxide [CO] and hydrogen sulfide [H_2S] are mainly considered for gas storage by MOF despite their bad public image. Among these, due to the vast biological role of NO [69], the development of NO-releasing vehicles is an emerging area in the development of new material-based anti-bacterial therapies. This diatomic gaseous molecule is referred to as a gasotransmitter due to its involvement in signaling process [70]. Due to the high porous and large surface area of MOF, it serves as a storage and delivery agent for gasotransmitters and is useful in anti-bacterial therapy. Nitric oxide is a short-lived free radical and diatomic gaseous molecule. Due to the lipophilic nature of NO molecules, they can pass through biological membranes [71]. Due to its high reactivity, NO could be considered one of the fast-acting gaseous bactericidal agents compared to other existing gaseous anti-bacterial molecules. The earlier studies related to the NO-releasing strategy indicate that macromolecular-based NO donor systems are efficient anti-bacterial compared to small molecules-based NO donor [72, 73]. Further, the hitherto studies on MOF-based NO delivery shows that the efficacy of the uptake of bioactive gaseous molecule by scaffold depends on the availability of the active sites like amine and thiol functional groups, and the metal sites, which are capable of further coordination with a guest gaseous molecule. A recent study showed that titanium metal-organic framework obtained by the high valence metathesis and oxidation (HVMO) method have high binding scavenger property towards NO over water and CO_2 [74, 75].

Xu and coworkers have demonstrated that the 4-[methylamino] pyridine-modified copper-based MOF serves as an efficient NO-loading vehicle [76]. The Fourier transform infrared (FTIR) and Raman study was carried out to confirm the loading of NO in the modified MOF. The concentration of NO-releasing from the scaffold is evaluated using the Griess assay. A study of NO-releasing from NO-loaded MOF shows that NO was released drastically on the first day with a concentration of 0.197 µmol L^{-1}. Thereafter, the release amount of NO was decreased to 0.033 µmol L^{-1} on the 3rd day and further reduced to a very small amount on the 5th day. The controlled release of NO was achieved with an average release rate of 1.74 nmol L^{-1} h^{-1} for more than 14 days by incorporating NO-loaded MOF particles into the core layer of the coaxial nanofiber by electrospinning method using gelatin and poly-caprolactone. The *in vivo* diabetic wound healing study reveals the prepared NO-loaded composite scaffolds could serve as an efficient therapeutic agent for diabetic wound healing.

Release Characteristics from MOFs

MOFs are highly stimuli-responsive materials used for the sustainable programmed release of drugs. They are responsive to various stimuli, including pH, temperature, pressure, optical waves, magnetic field, and ionic charges.

The pH-dependent release takes place by the principle that an acidic environment breaks the coordination bonds and releases the antibiotics. It is the most preferred strategy among the other stimuli. Photosensitive release is a non-invasive, high spatial and temporal control release approach and is thus preferred as an attractive option. Photo-controlled release by visible, UV or NIR light has been reported to attenuate the systemic toxicity of metal ions. With recent advancements in nano-biotechnology, a myriad of nanomaterials and MOFs have been fabricated as light-responsive drug delivery systems [77]. The PDT and PTT MOFs are employed in this strategy. Pressure-dependent release of antibiotics utilizes the porosity collapse by increasing the pressure [78].

Anti-bio Fouling Potency

The MOFs have been reported to have anti-bio fouling potency by various research groups. The bactericidal properties of metal ions prevent bacterial colonization. The organic linkers with charged functional groups can potentially repel the like-charged bio-film materials. These properties render the self-cleaning functionality to MOFs [79]. Biofilms are formed by organized bacterial colonies and a 3D matrix of extracellular polymeric bio substances, and extracellular DNA (eDNA) [80]. These bio-substances reinforce the bacterial colonies and strengthen the biofilm [81]. MOFs render peroxidase-like catalytic activity to liberate –OH radicals that kill the bacteria [82]. These properties of MOFs qualify them to be an excellent anti-bio fouling agent.

CONCLUSION

In conclusion, this chapter summarized the various synthetic methods to access the MOFs and discussed their anti-microbial applications through different methods. Due to the larger surface area and highly porous nature of MOF, these assembled frameworks attracted the attention of a greater number of research communities for employing them in anti-microbial applications. MOF allows for the loading of anti-bacterial agents; in addition to their stimuli-responsiveness and biocompatibility, MOFs could be considered as the promising candidates in the development of efficient anti-bacterial technique in the arena of emerging research on discoveries of new materials for biomedical applications.

CONSENT FOR PUBLICATION

Not applicable.

CONFLICT OF INTEREST

The authors declare no conflict of interest, financial or otherwise.

ACKNOWLEDGEMENT

Declared none.

REFERENCES

[1]   Shen M, Duan N, Wu S, Zou Y, Wang Z. Polydimethylsiloxane Gold Nanoparticle Composite Film as Structure for Aptamer-Based Detection of Vibrio parahaemolyticus by Surface-Enhanced Raman Spectroscopy. Food Anal Methods 2019; 12(2): 595-603.
[http://dx.doi.org/10.1007/s12161-018-1389-5]

[2]   Liao X, Ma Y, Daliri EBM, *et al.* Interplay of antibiotic resistance and food-associated stress tolerance in foodborne pathogens. Trends Food Sci Technol 2020; 95: 97-106.
[http://dx.doi.org/10.1016/j.tifs.2019.11.006]

[3]   World Health Organization. Antibiotic resistance. 2020. Available from: https://www.who.int/en/news-room/fact-sheets/detail/antibiotic-resistance

[4]   World Health Organization. The top 10 causes of death. 2016. Available from: https://www.who.int/news-room/fact-sheets/detail/the-top-10-causes-of-death

[5]   Organization WH. The top 10 causes of death. 2020. Available from: https://www.who.int/news-room/fact-sheets/detail/antibiotic-resistance

[6]   Taylor PW, Stapleton PD, Luzio JP. New ways to treat bacterial infections. Drug Discov Today 2002; 7> (21): 1086-91.
[http://dx.doi.org/10.1016/S1359-6446(02)02498-4]

[7]   Coates ARM, Halls G, Hu Y. Novel classes of antibiotics or more of the same? Br J Pharmacol 2011; 163(1): 184-94.
[http://dx.doi.org/10.1111/j.1476-5381.2011.01250.x] [PMID: 21323894]

[8]   Shen M, Forghani F, Kong X, *et al.* Antibacterial applications of metal–organic frameworks and their composites. Compr Rev Food Sci Food Saf 2020; 19(4): 1397-419.
[http://dx.doi.org/10.1111/1541-4337.12515] [PMID: 33337086]

[9]   Hosny NM. Anticancer and antimicrobial MOFs and their derived materials. Appl Met Fram Their Deriv Mater 2020; 263-85.

[10]   Li P, Li J, Feng X, Li J, *et al.* Metal-organic frameworks with photocatalytic bactericidal activity for integrated air cleaning. Nat Commun. Nature Publishing Group 2019; 10(1): 1-10.

[11]   Pettinari C, Marchetti F, Mosca N, Tosi G, Drozdov A. Application of metal− organic frameworks. Polym Int Wiley Online Library 2017; 66(6): 731-44.

[12]   Beg S, Rahman M, Jain A, *et al.* Nanoporous metal organic frameworks as hybrid polymer–metal

composites for drug delivery and biomedical applications. Drug Discov Today 2017; 22(4): 625-37.
[http://dx.doi.org/10.1016/j.drudis.2016.10.001]

[13]   Chen M, Long Z, Dong R, *et al.* Titanium Incorporation into Zr☐Porphyrinic Metal–Organic Frameworks with Enhanced Anti-bacterial Activity against Multidrug☐Resistant Pathogens. Small. Wiley Online Library 2020; 16(7): 1906240.

[14]   Wu Y, Luo Y, Zhou B, Mei L, Wang Q, Zhang B. Porous metal-organic framework (MOF) Carrier for incorporation of volatile antimicrobial essential oil. Food Control 2019; 98: 174-8.
[http://dx.doi.org/10.1016/j.foodcont.2018.11.011]

[15]   Azeez NA, Dash SS, Gummadi SN, Deepa VS. Nano-remediation of toxic heavy metal contamination: Hexavalent chromium [Cr [VI]]. Chemosphere 2020; 129204.

[16]   Batten SR, Champness NR, Chen X-M, Garcia-Martinez J, Kitagawa S, Öhrström L, *et al.* Coordination polymers, metal–organic frameworks and the need for terminology guidelines. CrystEngComm. Royal Society of Chemistry 2012; 14(9): 3001-4.

[17]   Safaei M, Foroughi MM, Ebrahimpoor N, Jahani S, Omidi A, Khatami M. A review on metal-organic frameworks: synthesis and applications. TrAC Trends Anal Chem 2019; 118: 401-25.
[http://dx.doi.org/10.1016/j.trac.2019.06.007]

[18]   de Lima Neto OJ, de Oliveira Frós AC, Barros BS, de Farias Monteiro AF, Kulesza J. Rapid and efficient electrochemical synthesis of a zinc-based nano-MOF for Ibuprofen adsorption. New J Chem. Royal Society of Chemistry 2019; 43(14): 5518-24.

[19]   Yu B, Ye G, Chen J, Ma S. Membrane-supported 1D MOF hollow superstructure array prepared by polydopamine-regulated contra-diffusion synthesis for uranium entrapment. Environ Pollut 2019; 253: 39-48.
[http://dx.doi.org/10.1016/j.envpol.2019.06.114] [PMID: 31302401]

[20]   Bhattacharyya S, Rambabu D, Maji TK. Mechanochemical synthesis of a processable halide perovskite quantum dot–MOF composite by post-synthetic metalation. J Mater Chem A. Royal Society of Chemistry 2019; 7(37): 21106-11.

[21]   Sun S, Huang M, Wang P, Lu M. Controllable hydrothermal synthesis of Ni/Co MOF as hybrid advanced electrode materials for supercapacitor. J Electrochem Soc. IOP Publishing 2019; 166(10): A1799.

[22]   Arul P, Gowthaman NSK, John SA, Lim HN. Ultrasonic assisted synthesis of size-controlled cu-metal–organic framework decorated graphene oxide composite: sustainable electrocatalyst for the trace-level determination of nitrite in environmental water samples. ACS Omega 2020.

[23]   Wang Y, Ge S, Cheng W, *et al.* Microwave Hydrothermally Synthesized Metal–Organic Framework-5 Derived C-doped ZnO with Enhanced Photocatalytic Degradation of Rhodamine B. Langmuir 2020; 36(33): 9658-67.
[http://dx.doi.org/10.1021/acs.langmuir.0c00395] [PMID: 32787068]

[24]   Chen C, Feng X, Zhu Q, *et al.* Microwave-assisted rapid synthesis of well-shaped MOF-74 [Ni] for CO2 efficient capture. Inorg Chem 2019; 58(4): 2717-28.
[http://dx.doi.org/10.1021/acs.inorgchem.8b03271] [PMID: 30720271]

[25]   Hu J, Chen Y, Zhang H, *et al.* TEA-assistant synthesis of MOF-74 nanorods for drug delivery and *in-vitro* magnetic resonance imaging. Microporous Mesoporous Mater 2021; 315: 110900.
[http://dx.doi.org/10.1016/j.micromeso.2021.110900]

[26]   Guan X, Wang Y, Cai W. A composite metal-organic framework material with high selective adsorption for dibenzothiophene. Chin Chem Lett 2019; 30(6): 1310-4.
[http://dx.doi.org/10.1016/j.cclet.2019.02.029]

[27]   Sangeetha S, Krishnamurthy G. Fabrication of MOF-177 for electrochemical detection of toxic $\hbox{Pb}^{2+}$ Pb2+ and $\hbox{Cd}^{2+}$ Cd2+ ions. Bull Mater Sci 2020; 43(1): 1-8.

[28] Tran NT, Kim D, Yoo KS, Kim J. Synthesis of Cu☐doped MOF☐235 for the Degradation of Methylene Blue under Visible Light Irradiation. Bull Korean Chem Soc. Wiley Online Library 2019; 40(2): 112-7.

[29] Qin S-J, Yan B. A facile indicator box based on Eu3+ functionalized MOF hybrid for the determination of 1-naphthol, a biomarker for carbaryl in urine. Sensors Actuators B Chem 2018; 259: 125-32.

[30] Han L, Zhang J, Mao Y, Zhou W, Xu W, Sun Y. Facile and Green Synthesis of MIL-53 [Cr] and Its Excellent Adsorptive Desulfurization Performance. Ind Eng Chem Res. ACS Publications 2019; 58(34): 15489-96.

[31] Abdpour S, Kowsari E, Moghaddam MRA. Synthesis of MIL-100 [Fe]@ MIL-53 [Fe] as a novel hybrid photocatalyst and evaluation photocatalytic and photoelectrochemical performance under visible light irradiation. J Solid State Chem 2018; 262: 172-80.

[32] Guan Y, Xia M, Wang X, Cao W, Marchetti A. Water-based preparation of nano-sized NH2-MIL-53 [Al] frameworks for enhanced dye removal. Inorganica Chim Acta 2019; 484: 180-4.

[33] Liu N, Huang W, Zhang X, *et al.* Ultrathin graphene oxide encapsulated in uniform MIL-88A [Fe] for enhanced visible light-driven photodegradation of RhB. Appl Catal B Environ 2018; 221: 119-28.

[34] Amaro-Gahete J, Klee R, Esquivel D, Ruiz JR, Jimenez-Sanchidrian C, Romero-Salguero FJ. Fast ultrasound-assisted synthesis of highly crystalline MIL-88A particles and their application as ethylene adsorbents. Ultrason Sonochem 2019; 50: 59-66.

[35] Yuan B, Wang X, Zhou X, Xiao J, Li Z. Novel room-temperature synthesis of MIL-100(Fe) and its excellent adsorption performances for separation of light hydrocarbons. Chem Eng J 2019; 355: 679-86.
[http://dx.doi.org/10.1016/j.cej.2018.08.201]

[36] Vo TK, Kim J-H, Kwon HT, Kim J. Cost-effective and eco-friendly synthesis of MIL-101 [Cr] from waste hexavalent chromium and its application for carbon monoxide separation. J Ind Eng Chem 2019; 80: 345-51.

[37] Tan B, Luo Y, Liang X, *et al.* Mixed-solvothermal synthesis of MIL-101 [Cr] and its water adsorption/desorption performance. Ind Eng Chem Res. ACS Publications 2019; 58(8): 2983-90.

[38] Tzitzios V, Kostoglou N, Giannouri M, Basina G, Tampaxis C, Charalambopoulou G, *et al.* Solvothermal synthesis, nanostructural characterization and gas cryo-adsorption studies in a metal–organic framework [IRMOF-1] material. Int J Hydrogen Energy 2017; 42(37): 23899-907.

[39] Naimi Joubani M, Zanjanchi MA, Sohrabnezhad S. A novel Ag/Ag3PO4-IRMOF-1 nanocomposite for anti-bacterial application in the dark and under visible light irradiation. Appl Organomet Chem. Wiley Online Library 2020; 34(5): e5575.

[40] Vignatti C, Luis-Barrera J, Guillerm V, *et al.* Squaramide-IRMOF-16 analogue for catalysis of solvent-free, epoxide ring-opening tandem and multicomponent reactions. ChemCatChem 2018; 10(18): 3995-8.
[http://dx.doi.org/10.1002/cctc.201801127]

[41] Chowdhuri AR, Laha D, Chandra S, Karmakar P, Sahu SK. Synthesis of multifunctional upconversion NMOFs for targeted antitumor drug delivery and imaging in triple negative breast cancer cells. Chem Eng J 2017; 319: 200-11.
[http://dx.doi.org/10.1016/j.cej.2017.03.008]

[42] Zhang T, Wei JZ, Sun XJ, *et al.* Continuous and Rapid Synthesis of UiO-67 by Electrochemical Methods for the Electrochemical Detection of Hydroquinone. Inorg Chem 2020; 59(13): 8827-35.
[http://dx.doi.org/10.1021/acs.inorgchem.0c00580] [PMID: 32623890]

[43]   Ali-Moussa H, Navarro Amador R, Martinez J, Lamaty F, Carboni M, Bantreil X. Synthesis and post-synthetic modification of UiO-67 type metal-organic frameworks by mechanochemistry. Mater Lett 2017; 197: 171-4.
[http://dx.doi.org/10.1016/j.matlet.2017.03.140]

[44]   Tan C, Han X, Li Z, Liu Y, Cui Y. Controlled exchange of achiral linkers with chiral linkers in Zr-based UiO-68 metal–organic framework. J Am Chem Soc. J Am Chem Soc 2018; 140(47): 16229-36.
[http://dx.doi.org/10.1021/jacs.8b09606] [PMID: 30392361]

[45]   Zhang L-H, Zhu Y, Lei B-R, Li Y, Zhu W, Li Q. Trichromatic dyes sensitized HKUST-1 [MOF-199] as scavenger towards reactive blue 13 *via* visible-light photodegradation. Inorg Chem Commun 2018; 94: 27-33.

[46]   Karimi A, Vatanpour V, Khataee A, Safarpour M. Contra-diffusion synthesis of ZIF-8 layer on polyvinylidene fluoride ultrafiltration membranes for improved water purification. J Ind Eng Chem 2019; 73: 95-105.

[47]   Jiang Z, Wang Y, Sun L, *et al.* Dual ATP and pH responsive ZIF-90 nanosystem with favorable biocompatibility and facile post-modification improves therapeutic outcomes of triple negative breast cancer *in vivo.* Biomaterials 2019; 197: 41-50.
[http://dx.doi.org/10.1016/j.biomaterials.2019.01.001] [PMID: 30640136]

[48]   Xiang H, Ameen A, Shang J, *et al.* Synthesis and modification of moisture-stable coordination pillared-layer metal-organic framework (CPL-MOF) CPL-2 for ethylene/ethane separation. Microporous Mesoporous Mater 2020; 293: 109784.
[http://dx.doi.org/10.1016/j.micromeso.2019.109784]

[49]   Prasetyo N, Pambudi FI. Toward hydrogen storage material in fluorinated zirconium metal-organic framework [MOF-801]: A periodic density functional theory [DFT] study of fluorination and adsorption. Int J Hydrogen Energy 2021; 46(5): 4222-8.

[50]   Sánchez-González E, López-Olvera A, Monroy O, Aguilar-Pliego J, Flores JG, Islas-Jácome A, *et al.* Synthesis of vanillin *via* a catalytically active Cu [II]-metal organic polyhedron. CrystEngComm. Royal Society of Chemistry 2017; 19(29): 4142-6.

[51]   Abazari R, Mahjoub AR, Shariati J. Synthesis of a nanostructured pillar MOF with high adsorption capacity towards antibiotics pollutants from aqueous solution. J Hazard Mater 2019; 366: 439-51.

[52]   Wang X, Li Q, Yang N, *et al.* Hydrothermal synthesis of NiCo-based bimetal-organic frameworks as electrode materials for supercapacitors. J Solid State Chem 2019; 270: 370-8.

[53]   Kang X, Fu G, Song Z, *et al.* Microwave-assisted hydrothermal synthesis of MOFs-derived bimetallic CuCo-N/C electrocatalyst for efficient oxygen reduction reaction. J Alloys Compd 2019; 795: 462-70.

[54]   Friščić T, Mottillo C, Titi HM. Mechanochemistry for synthesis. Angew Chemie Int Ed. Wiley Online Library 2020; 59(3): 1018-29.

[55]   Wang TJ, Liu X, Li Y, Li F, Deng Z, Chen Y. Ultrasonication-assisted and gram-scale synthesis of Co-LDH nanosheet aggregates for oxygen evolution reaction. Nano Res 2020; 13(1): 79-85.
[http://dx.doi.org/10.1007/s12274-019-2575-5]

[56]   Zhang Z, Wagner VE. Anti-microbial coatings and modifications on medical devices. Springer 2017; pp. 171-88.
[http://dx.doi.org/10.1007/978-3-319-57494-3]

[57]   Li Y, Yang L, Zhao Y, Li B, Sun L, Luo H. Preparation of AgBr@ SiO2 core@ shell hybrid nanoparticles and their bactericidal activity. Mater Sci Eng C 2013; 33(3): 1808-12.

[58]   Berchel M, Le Gall T, Denis C, *et al.* A silver-based metal–organic framework material as a 'reservoir' of bactericidal metal ions. New J Chem. Royal Society of Chemistry 2011; 35(5): 1000-3.

[59]   Wyszogrodzka G, Marszałek B, Gil B, Dorożyński P. Metal-organic frameworks: mechanisms of antibacterial action and potential applications. Drug Discov Today 2016; 21(6): 1009-18.

[60]   Wang Y, Yan J, Wen N, *et al.* Metal-organic frameworks for stimuli-responsive drug delivery. Biomaterials 2020; 230: 119619.
[http://dx.doi.org/10.1016/j.biomaterials.2019.119619] [PMID: 31757529]

[61]   Wang L, Zheng M, Xie Z. Nanoscale metal–organic frameworks for drug delivery: A conventional platform with new promise. J Mater Chem B Mater Biol Med 2018; 6(5): 707-17.
[http://dx.doi.org/10.1039/C7TB02970E] [PMID: 32254257]

[62]   Li B, Wang X, Chen L, *et al.* Ultrathin Cu-TCPP MOF nanosheets: A new theragnostic nanoplatform with magnetic resonance/near-infrared thermal imaging for synergistic phototherapy of cancers. Theranostics 2018; 8(15): 4086-96.
[http://dx.doi.org/10.7150/thno.25433] [PMID: 30128038]

[63]   Lu K, He C, Lin W. Nanoscale metal-organic framework for highly effective photodynamic therapy of resistant head and neck cancer. J Am Chem Soc 2014; 136(48): 16712-5.
[http://dx.doi.org/10.1021/ja508679h] [PMID: 25407895]

[64]   Chen L, Luque R, Li Y. Encapsulation of metal nanostructures into metal–organic frameworks. Dalton Trans 2018; 47(11): 3663-8.
[http://dx.doi.org/10.1039/C8DT00092A] [PMID: 29492493]

[65]   Nabipour H, Soltani B, Nasab NA. Gentamicin Loaded Zn 2 [bdc] 2 [dabco] Frameworks as Efficient Materials for Drug Delivery and Antibacterial Activity. J Inorg Organomet Polym Mater 2018; 28(3): 1206-13.

[66]   Bhardwaj N, Pandey SK, Mehta J, Bhardwaj SK, Kim K-H, Deep A. Bioactive nano-metal–organic frameworks as anti-microbials against Gram-positive and Gram-negative bacteria. Toxicol Res 2018; 7(5): 931-41.

[67]   Wang W, Wang L, Li Z, Xie Z. BODIPY-containing nanoscale metal–organic frameworks for photodynamic therapy. Chem Commun (Camb) 2016; 52(31): 5402-5.
[http://dx.doi.org/10.1039/C6CC01048B] [PMID: 27009757]

[68]   Zhu X, Gu J, Wang Y, *et al.* Inherent anchorages in UiO-66 nanoparticles for efficient capture of alendronate and its mediated release. Chem Commun (Camb) 2014; 50(63): 8779-82.
[http://dx.doi.org/10.1039/C4CC02570A] [PMID: 24967656]

[69]   Palmer RMJ, Ferrige AG, Moncada S. Nitric oxide release accounts for the biological activity of endothelium-derived relaxing factor. Nature 1987; 327(6122): 524-6.
[http://dx.doi.org/10.1038/327524a0] [PMID: 3495737]

[70]   Mustafa AK, Gadalla MM, Snyder SH. Signaling by Gasotransmitters. Sci Signal 2009; 2(68): re2-2.
[http://dx.doi.org/10.1126/scisignal.268re2] [PMID: 19401594]

[71]   Figueroa XF, Lillo MA, Gaete PS, Riquelme MA, Sáez JC. Diffusion of nitric oxide across cell membranes of the vascular wall requires specific connexin-based channels. Neuropharmacology 2013; 75: 471-8.
[http://dx.doi.org/10.1016/j.neuropharm.2013.02.022] [PMID: 23499665]

[72]   Hetrick EM, Shin JH, Stasko NA, Johnson CB, Wespe DA, Holmuhamedov E. Bactericidal efficacy of nitric oxide-releasing silica nanoparticles. ACS Nano 2008; 2(2): 235-46.

[73]   Ishima Y, Sawa T, Kragh-Hansen U, Miyamoto Y, Matsushita S, Akaike T, *et al.* S-Nitrosylation of human variant albumin Liprizzi [R410C] confers potent anti-bacterial and cytoprotective properties. J Pharmacol Exp Ther. ASPET 2007; 320(3): 969-77.
[PMID: 17135341]

[74]   Zou L, Feng D, Liu TF, *et al.* A versatile synthetic route for the preparation of titanium metal–organic frameworks. Chem Sci (Camb) 2016; 7(2): 1063-9.

[http://dx.doi.org/10.1039/C5SC03620H] [PMID: 29896371]

[75]   Jensen S, Tan K, Feng L, Li J, Zhou HC, Thonhauser T. Porous Ti-MOF-74 Framework as a Strong-Binding Nitric Oxide Scavenger. J Am Chem Soc 2020; 142(39): 16562-8.
[http://dx.doi.org/10.1021/jacs.0c02772] [PMID: 32876449]

[76]   Zhang P, Li Y, Tang Y, *et al.* Copper-based metal–organic framework as a controllable nitric oxide-releasing vehicle for enhanced diabetic wound healing. ACS Appl Mater Interfaces. ACS Appl Mater Interfaces 2020; 12(16): 18319-31.
[http://dx.doi.org/10.1021/acsami.0c01792] [PMID: 32216291]

[77]   Cai W, Wang J, Chu C, Chen W, Wu C, Liu G. Metal–organic framework□based stimuli□responsive systems for drug delivery. Adv Sci (Weinh) 2019; 6(1): 1801526.
[http://dx.doi.org/10.1002/advs.201801526] [PMID: 30643728]

[78]   Teplensky MH, Fantham M, Li P, Wang TC, Mehta JP, Young LJ, *et al.* Temperature treatment of highly porous zirconium-containing metal-organic frameworks extends drug delivery release. J Am Chem Soc 2017; 139(22): 7522-32.

[79]   Prince JA, Bhuvana S, Anbharasi V, Ayyanar N, Boodhoo KVK, Singh G. Self-cleaning Metal Organic Framework [MOF] based ultra filtration membranes-A solution to bio-fouling in membrane separation processes. Sci Rep. Nature Publishing Group 2014; 4(1): 1-9.

[80]   Berk V, Fong JCN, Dempsey GT, *et al.* Molecular architecture and assembly principles of Vibrio cholerae biofilms. Science 2012; 337(6091): 236-9.
[http://dx.doi.org/10.1126/science.1222981] [PMID: 22798614]

[81]   Davies D. Understanding biofilm resistance to anti-bacterial agents. Nat Rev Drug Discov. Nature Publishing Group 2003; 2(2): 114-22.
[PMID: 12563302]

[82]   Yin W, Yu J, Lv F, *et al.* Functionalized nano-MoS2 with peroxidase catalytic and near-infrared photothermal activities for safe and synergetic wound anti-bacterial applications. ACS Nano. ACS Nano 2016; 10(12): 11000-11.
[http://dx.doi.org/10.1021/acsnano.6b05810] [PMID: 28024334]

CHAPTER 7

Biogenic Metal Nanoparticles: A Sustainable Alternative to Combat Drug-Resistant Pathogens

Palas Samanta[1], **Sukhendu Dey**[2], **Sushobhon Sen**[3] and **Manab Deb Adhikari**[3,*]

[1] *Department of Environmental Science, Sukanta Mahavidyalaya, University of North Bengal, Dhupguri, West Bengal, India.*

[2] *Department of Environmental Science, University of Burdwan, West Bengal, India.*

[3] *Department of Biotechnology, University of North Bengal, Darjeeling, West Bengal, India.*

Abstract: The natural environment acts as the largest 'bio-laboratory" of yeast, algae, fungi, plants *etc.,* which are used as an abundant source of biomolecules. These different biomolecules play vital roles in the formation of different biogenic metals or metalloid nanoparticles. Recently, the overburden from the different microbial diseases has increased rapidly in different application sectors, viz., drug delivery, DNA analysis, cancer treatment, antimicrobial agents, water treatment and biosensor and catalysts, as a result of multipurpose work occurrence globally. The indiscriminate and arbitrary use of antibiotics in clinical practice has spurred the emergence of potentially life-threatening multidrug-resistant pathogens. In the quest for novel antimicrobial agents, the current interest is to develop potent antimicrobial agents which exhibit broad-spectrum bactericidal activity and possess a mechanism of action that does not readily favor the development of resistance. The use of nanoscale materials as bactericidal agents represents a novel paradigm in antibacterial therapeutics. Actually, eco-friendly, sustainable modern approaches, such as green syntheses of different biogenic metals or metalloid nanoparticles, are cost-effective and environment-friendly, and they are used as strong antimicrobial agents. This chapter focuses on synthesizing biogenic metal or metalloid nanoparticles with special emphasis on microbial synthesis, particularly from yeast, bacteria, algae, fungi, plants extract, etc. Finally, a detailed description of the biosynthesis mechanism using different green sources, along with their antimicrobial activity and mode of action, has been presented.

Keywords: Antibiotic, Antimicrobial agents, Biogenic metal nanoparticles, Biomolecules, Copper oxide nanoparticles, Green synthesis, Gold nanoparticles, Iron nanoparticles, Metalloid nanoparticles, Multidrug-resistant pathogens, Nanoparticles, Photo activation, Phytochemicals, Reactive oxygen species, Silver nanoparticles, Titanium dioxide nanoparticles, Zinc oxide nanoparticles.

* **Corresponding author Manab Deb Adhikari:** Department of Biotechnology, University of North Bengal, Darjeeling, West Bengal, India. E-mail: madhikari@nbu.ac.in

Tilak Saha, Manab Deb Adhikari and Bipransh Kumar Tiwary (Eds.)

INTRODUCTION

One serious global issue in modern biomedicine and healthcare regime is the emergence of multidrug-resistant bacterial strains. These pathogens have evolved mechanisms to evade the action of most commercially available antibiotics, underscoring the need for novel and potent antibacterial agents. The problem, especially the biogenic nanoparticles. The terminology 'biogenesis' was first coined by Henry Charlton Bastian, which means manufacturing a new life form from nonliving components/materials. Biogenic metal nanoparticles (NPs) technology is a hot topic in modern science, especially nanotechnology. This novel branch of nanoscience research includes materials science, physics, chemistry and biological science. Nanoparticles represent a particularly innovative regime, displaying unique properties with potentially wide-ranging therapeutic applications. Generally, nanoparticles (NPs), ≤ 100 nm in particle dimension, are primarily produced through bottom-up and top-down strategies. In the bottom-up approach, molecules and atoms are assembled to form molecular structures in the nanometer range, whereas, in the top-down approach, bulk raw materials are gradually broken down into nano-sized materials. Among these two methods, scientists world-wide used bottom-up-based chemical or biological approaches for nanoparticles (NPs) synthesis.

The synthesis of nanoparticles can control the shape, property and particle size of NPs. Crystalline property, among them, has been considered as prime in chemical science that could be used for vital applications viz., biosensor, bio-medical, catalyst for the bacterial biotoxin boycott and lower price electrode [1, 2]. The promising applications of NPs are nanowires, nanosheets and nanotubes, which have acquired more care in nanoscience [3, 4]. NPs function as a bridge between bulk materials and molecular or atomic structures. So, they are a very good choice for applications like electrochemistry, catalysis, biotechnology, trace materials and medical [5, 6].

Nature has provided lots of ways and insight for the synthesis of commercial nanomaterials. The natural environment act as the largest 'bio-laboratory" including yeast, bacteria, algae, fungi, plant extracts and waste materials, which are considered as eco-friendly materials for the synthesis of NPs with effective applications [7 - 10]. The biological pathway, which includes various types of microorganisms, has been used for synthesize various metallic NPs, which has benefits over chemical methods as the biological pathway is cost-effective, energy saving and greener. In addition to this, the coating of bio-molecules on the NPs surface makes them biocompatible in parallel to the NPs prepared by different chemical methods [11, 12]. The biocompatibility of bioinspired NPs offers very interesting use in biomedicine and allied fields [13]. The biogenic methods lead to designing NPs with varied sizes and interesting morphologies [14]. For example, sulfate-reducing bacteria are used to modify the shape and size of NPs by cell-soluble protein extract method [15]. These modified shaped-NPs *via* biogenic enzymatic methods were superior to the NPs synthesized through a chemical process as biogenic enzymatic methods used minimum expensive chemicals and performed higher catalytic work. Recently, an industrially significant different type's fungus was used to synthesize uniform-sized Au NPs, easier to handle than other types bacteria and yeast [16]. In addition, several forms of algae are currently being used for NPs synthesis due to their tremendous power of bioremediation of different toxic metals. Further, they are used very decently to fabricate various metal and metal oxide NPs [17].

Apart from these, plant root, stem, leaf, latex and seed have extensively been used for NPs synthesis, and they act as strong reducing or stabilizing agents [18]. More recently, different waste materials have been used for NPs synthesis. Accordingly, this chapter is therefore primarily focused on the synthesis of biogenic metal or metalloid NPs. Additionally, this chapter will address different biosynthesis mechanisms along with different influencing factors.

CAPPING AGENTS AND THEIR DIFFERENT TYPES

The capping agents perform a very central and versatile role in the synthesis of NPs. NPs can be used as capping agents to deliver fruitful properties by monitoring their morphology and size and protecting the total surface, thereby prohibiting the total quality. Many surfactants are used for capping agents for gating, but the most difficult problem is removing the surfactants because their properties are not easy to degrade. Liu *et al.*, and Gittins *et al.*reported that in their work, commercial surfactants are more hazardous for the environment [19, 20]. So, eco-friendly capping agents are more needed for industrial and laboratory-level NPs synthesis.

Biomolecules

The formulation of homogenous NPs using biomolecules has recently obtained prime interest because of its non-toxic character and not involving harsh synthetic origin. Different amino acids play a reducing role as well as capping materials to synthesize the NPs due to their unique shape and size. For instance, twenty different amino acids accept L-histidine, which reduces tetraauric acid to Au nanoparticles, but the formation of NPs is dose-dependent; the higher dose produces the smaller nanoparticle [21].

MODES OF BIOGENIC NPs SYNTHESIS AND PURIFICATION

Synthesis of NPs using Bacteria

Bacteria have the capacity to produce inorganic intracellular and extracellular materials, which makes them prospective bio-fermenters. Generally, bacteria-mediated NPs synthesis takes place in two ways: intracellular and extracellular approaches. Extracellular method has an advantage over the intracellular technique, since the extracellular method is less time-consuming as it does not require a downstream process for nanoparticle synthesis [22]. For example, *Pseudomonas stutzeri* AG259, the first silver-resistant bacteria, was used for noble AuNPs synthesis. *Pseudomonas stutzeri* was cultured in high silver nitrate concentration to accumulate silver in bulk amounts with a nanoscale 200 nm diameter [23]. *Proteus mirabilis* PTCC 1710 is more efficient for NPs synthesis as it promotes NPs synthesis (extracellular or intracellular) during the incubation of bacteria broth (nutrient broth, Muller-Hinton broth) [24].

In addition, bacteria contain reductase enzymes that catalyze the reduction of metal ions into metal NPs. For example, *D. radiodurans*, which has great antioxidant activity and is highly resistant to radiation and oxidative stress, is used for AuNPs green synthesis from its ionic form [25]. This fabricated AuNPs were more stable and had stronger antimicrobial activity than *D. radiodurans*. Recently, Kunoh*et al.* produced Au NPs from *Leptothrix* bacteria by reducing gold salt (by guanine residues of RNA molecules and 2-deoxy guanosine) in an aqueous medium for more stable NPs [26]. Therefore, it was clear that the bacteria-based synthesis of NPs is very cost-effective and highly efficient for large-scale production [27]. However, there is still a drawback of bacteria-based NPs, such as slow synthesis rate, and limited available size and shapes compared to the conventional process, which necessitates alternative biogenic procedures.

Synthesis of NPs using Fungi

Fungi are considered good sources of secondary metabolites and have high binding ability; intercellular uptake and active biomolecules make them more efficient for the green synthesis of NPs (Table **1**). However, the NPs synthesis protocol is diverse. At the beginning of the 20th century, Verticillium fungus was used to produce AgNPs with a diameter of 25±12 nm mainly through secreted enzymes, which help reduce silver ions and induce metal NP production [28]. NPs were formed below the fungal cell in comparison with bacteria [29] with diverse morphologies ranging from triangular, and spherical to hexagonal. Briefly, NPs are formed on the mycelia surface instead of in solution. Rauwel *et al.*opined that electrostatic interaction between positively charged silver ions and enzyme's (present in the cell wall of mycelia) negative charged carboxylate groups facilitates the silver ions to get adsorbed on the fungal surface [30]. Finally, the fungal enzyme reduces silver ions to silver nuclei. Fungal synthesis of NPs possesses advantages over bacterial green synthesis due to simpler bottom-up processing and easy biomass handling. For example, *F. oxysporum* secretes polymers, proteins, and enzymes that voluntarily help in metal NPs production [22]. These constituents enhance the yield and stability of NPs. Ahmad *et al.* observed that the reductase enzyme in the cytosol of F. oxysporum helps to reduce silver ions into silver metal in the presence of NADH+ during the formation of AgNPs. Generally, phytochelatin group of these compounds has a high capability to reduce silver ions into silver metal [22]. Sanghi and Verma used a culture supernatant of *Coriolus versicolor* to synthesize silver NPs (Ag NPs) [31]. They also confirmed that the presence of aliphatic and aromatic amines and some proteins in fungal extract act as capping agents to stabilize the formed AgNPs by binding with protein through amide bond. In another study, Tan *et al.* demonstrated the proteins containing SH group from fungal extract involved in the capping and stabilization of AgNPs [32].

Table 1. Biological entities used in the synthesis of metal and metal oxide NPs with their size, shape and brief biocidal activity.

Biogenic Origin	NPs	Morphology	Biocidal Effects	Refs.
Bacteria				
Escherichia coli, Exiguobacterium aurantiacumm, and Brevundimonas diminuta	AgNP	5 to 50 nm	Showed great potential as antimicrobial agent against MRSA	[38]
Escherichia coli	CdS	2-5 nm/spherical	Used to synthesize green solar cells and effective against *E. coli* (BW25113)	[39]
Bacillus subtilis	TiO_2 NP	10-30 nm/spherical	Bioremediation without producing toxic chemicals in the environment	[40]
Proteus vulgaris ATCC-29905	IONP	19.23 nm ad 30.51 nm	Showed good activity against methicillin-resistant Staphylococcus aureus (MRSA)	[41]
Aeromonas hydrophila	ZnO NP	57-72 nm/spherical	Exhibits antimicrobial activity against both bacteria (*Pseudomonas aeruginosa*) and fungi (*Aspergillus flavus*)	[42]

Biogenic Origin	NPs	Morphology	Biocidal Effects	Refs.
Fungus				
Penicillium oxalicum.	AgNP	60 -80 nm/spherical	Effective against *Staphylococcus aureus, S. dysenteriae*, and *Salmonella typh*i b	[43]
Trichoderma hamatum SU136.	AuNP	5-30 nm/spherical	Showed antimicrobial activity against four pathogenic bacterial strains	[44]
Aspergillus terreus IF0	AuNP	10-19 nm	Shows bactericidal activity against *E. coli*	[45]
Fusarium oxysporum	AuNP	20-40 nm/spherical, triangle	Antibacterial activity against burns bacterial growth, *E. coli, S. aureus*	[46]
Volvariella volvacea	AgNP&AuNP	20-150 nm/spherical/hexagonal	Shows antimicrobial activity	[47]
Verticillium sp.	Ag NP	25±12 nm/spherical	Shows antimicrobial activity	[29]
Aspergillus flavus	TiO$_2$ NPs	62–74 nm/spherical, oval in shape	Shows antimicrobial activity against *E. coli*	[48]
Yeast				
MKY3	Ag NP	2-5 nm/hexagonal	Activity against *E.coli, S. aureus*	[36]
Schizosacchromyces pombe	Cd NP	1-2 nm/hexagonal	-	[49]
Saccharomyces cerevisiae	TiO$_2$ NPs	6.7 ± 2.2 nm/spherical	Significant antimicrobial effect on gram-positive bacteria	[50]

The fungal green synthesis of NPs is more significant than bacteria as the fugal cell secretes a much higher amount of protein, hence employed large-scale NPs production. AsNPs are produced extracellularly, therefore they are easily purified and directly used in different applications. Additionally, fungal mycelia mesh can tolerate extra flow pressure and other conditions in bioreactors than plant material or bacteria [22]. Apart from these, most of the fungi have a high tolerance towards metals as well as high wall-binding capability and intracellular metal uptake capacity. As a result, natural nanofactories shifted from bacteria to fungi. For instance, the white rot fungus, *Phaenerochaetechrysosporium*, is nonpathogenic and used for large-scale production of AgNPs [33]. Das *et al.* utilized mycelia of *R. oryzae* for gold (Au) NPs synthesis through *in situ* reduction of chloroauric acid (HAuCl$_4$) in an acidic medium (pH 3) [34]. Verticillium fungus is used as a good mediator for the synthesis of silver NPs. *A. flavus* NPs combined with antibiotics enhanced the antibacterial activity against the MDR bacteria [35].

Recently, fungi-assisted NPs were produced intracellularly on $AgNO_3$ exposure, in an acidic medium (pH 5.5–6) [29].

Synthesis of NPs using Yeast

Yeast was also used for the production of NPs, such as silver NPs. For example, yeast strains (MKY3) are silver tolerant, accordingly, they are used for extracellular synthesis of NPs (Table **1**) [36]. The gold and AgNPs were synthesized using yeast supernatant broth of *Saccharomyces cerevisiae* and *Candida guilliermondii* [37].

Synthesis of NPs using Plant Materials

Plant-assisted NP synthesis is more advantageous as they are easily accessible, ecofriendly, safe to handle and contain several metabolites and biochemicals. Almost all plant parts, namely root, stem, leaves, latex, flowers and seeds, are used for NP synthesis [51, 52]. These metabolites and biochemicals act as capping agents that stabilize and govern NPs morphology and reduce agents during biogenic NP synthesis. The leading biomolecules that are employed as bio-reducing and capping agents during NPs synthesis include polysaccharides, flavones, phenols, terpenoids, proteins, alkaloids, amino acids, enzymes and alcoholic compounds. In addition, chlorophyll pigments, methyl chavicol, quinol, eugenol, caffeine, linalool, ascorbic acid, theophylline and other vitamins are also used as reducing agents during NPs synthesis [53]. Only hydroxyl and carboxyl groups of phenolic compounds bind to the surface of metals [27]. Apart from this, plant-assisted NPs were more stable than microbes and fungus-assisted NPs [52]. Generally, plant-mediated NP synthesis is done through three techniques, namely intracellular, extracellular and phytochemical pathways [35]. Intracellular NP synthesis occurs inside the cells by the utilization of cellular enzymes. Post synthesis, the NPs are recovered by rupturing the cell wall of plant cells. The technique is cheaper and produces a higher yield due to the presence of metabolites and biochemicals. Extracellular technique is employed when plant extract is used as raw material for NP synthesis. On the other hand, phytochemical-assisted NPs synthesis is rarely used as it requires knowledge about particular phytochemicals for stabilized NPs synthesis [35].

The first approach toward plant-assisted NPs synthesis was made with Alfalfa sprouts [54]. Alfalfa root can also absorb silver and transfer it to shoots in the same oxidation state. This translocated silver in shoots is accumulated and finally formed NPs faster than bacteria and fungi. The production of NPs with plants gives a faster rate of synthesis than bacteria and fungi. In another study, Shankar *et al.* reported that Geranium leaf extracts take only 9 h to reach 90% completion of NPs synthesis [55]. These findings open a new era in the green synthesis of

NPs. Since then, several authors have demonstrated that plant-assisted NPs synthesis in metal salt solution takes only a few minutes at room temperature, but the rate depends on plant extracts (Table **2**). For instance, Ahmed *et al.* synthesized spherical-shaped Ag NPs from *A. indica* leaf extract and exhibited potent antimicrobial activity [56].

Table 2. Plants/plant part-mediated synthesis of metal and metal oxide NPs with their size, shape and brief biocidal activity.

Biogenic Origin	NPs	Morphology	Biocidal Effects	Refs.
Alfalfa	AgNP	2-20 nm	Significantly increases root & stem growth, antibacterial	[57]
Aloe vera	AuNP and AgNP	50-350 nm/spherical, triangular	Bactericidal effects	[58]
Pongamia pinnata	ZnO NP	-	Effective against *Staphylococcus aureus* and *Escherichia coli*	[59]
Camellia sinensis	ZnO NP	16 nm	Strong antimicrobial effects	[60]
Catharanthus roseus	TiO$_2$ NP	25-110 nm/irregular	Effective again *Hippoboscamaculata* (flies) and *Bovicolaovis* (lice)	[61]
Parthenium hysterophorus	ZnO NP	10 nm	Strong antimicrobial activity against both bacterial and fungal strains	[62]
Lysiloma acapulcensis	AgNP	5 nm	Significant antimicrobial effect against *C. albicans, E. coli, S. aureus* and *P. aeruginosa*	[63]
Trigonella foenum-graecum	TiO$_2$ NP	20-90 nm	Antimicrobial activity against *S. aureus, E. faecalis, K. pneumoniae, S. faecalis, P. aeruginosa, E. coli, P. vulgaris, B. subtilis, Y. enterocolitica*	[64]
Gloriosa superba	CuO NP	5-10 nm/spherical	Effective against *S. aureus* and *Klebsiellaaerogenes*	[65]
Gum karaya	CuO NP	Avg. 4.8 nm 16-	Antimicrobial activity against *E. coli*	[66]
Geranium leaves	AgNP	40 nm/quasilinear superstructure	Antimicrobial activity	[55]
Hibiscus rosasinensis	AgNP and AuNP	14 nm/spherical	Strong antimicrobial activity	[67]
Ipomoea aquatica	AgNP	Prism 100-400 nm/spherical, cubic	-	[68]
Jatropha curcas	AgNP	15-50 nm/spherical	Biocidal effects	[69]
Malva sylvestris	CuONP	14 nm/spherical	Effective against both Gram-positive and negative bacteria	[70]

(Table 2) cont.....

Biogenic Origin	NPs	Morphology	Biocidal Effects	Refs.
Phyllanthus amarus	CuONP	20 nm/spherical	Effective than Rifampicin against *B. subtilis*	[71]
Curcuma longa	Cu NP	5–20 nm	Effective against Gram-positive and Gram-negative bacteria	[72]
Zingiber officinale	Fe_3O_4 NP	-	Antibacterial activity against Gram-negative bacteria *E. coli*	[73]
Extracts of neem, onion and tomato	AgNP	3–28 nm	Improved antimicrobial activity against *S. aureus*	[74]

Synthesis of NPs using Algae

Metallic and Non-metallic NPs

More recently, different algae species have been used for the synthesis of metal NPs nanotechnology field by different scientists due to their eco-friendly nature. For example, Ag nanoparticles derived from microalgae have acquired widespread care due to their potential antibacterial activities. The use of algae for NPs synthesis is advantageous in two ways: 1) nucleation and crystal formation are swift due to the appearance of negative charge on the cell surface, 2) synthesis of NPs on a large scale at a very small cost [75].

NP activities are reported as size-dependent. Algae species give a fast, green, cheap route and play as effective 'nanoreserves' for different sustainable NPs [76]. Moreover, these may be freely prepared in laboratories. The algal species diversity has played their exploitation and is used for the synthesis of metallic NPs. They perform super antimicrobial and antioxidant properties for functionalization on their surface [77, 78]. As a result, algal-NPs activities are very significant for the fruitful use of NPs in drug delivery in respective cells. On the other hand, metal oxide NPs are equally important for the generation of biogenic NPs. Several synthetic procedures viz., solvothermal, hydrothermal, vapour-deposition, microwave, wet-chemical, spray pyrolysis and seed-mediated, have been used for the formation of metal oxide NPs with different morphology, shape and size [79, 80]. However, few research articles have been published on biosynthesis approaches with special emphasis on algae. Apart from this, toxicology issues connected to the NPs are also a major concern. Accordingly, *in-vitro* and *in-vivo* studies have been carried out to calculate the toxicology levels of NPs using different models.

Mechanism of Biosynthesis of NPs using Algae

Across the presence of sophisticated biochemical methods, infrastructure and instruments, it is very easy to recognize the interaction and role of only a particular biomolecule viz., proteins, polysaccharides, and enzymes exist in the particular organism with NPs as a result of understanding the mechanism. The very simple technique in different enzymes and different functional groups exists in the cell walls of algae as a binding agent with the precursors, causing deposition and reduction of different metal/metal oxide nanoparticles at ambient conditions [81, 82].

Synthesis of NPs using Waste Materials

Metallic and Non-metallic NPs

The natural environment acts as a 'treasure' of different waste materials, particularly food which is successfully exploited for biosynthesis in NPs. For example, food waste product contains a huge number of organic compounds viz., flavonoids, polyphenols, vitamins and carotenoids [83] which is used as templating agents. In this way, the food waste performs as a 'biofactory' in nanoscience and nanotechnology. Au nanoparticles have been practically synthesized by using the mango peel extraction process. The extraction process rate was observed to be faster as compared with the other plant extraction process. In another example, the wine production industry generated a huge amount of grape waste which is a primary source of a sufficient amount of organic compounds that are directed to metals minimization to NPs. Chicken or duck eggshell membrane is a prime natural source used as waste. Eggshell membrane has been exploited for the synthesis of fluorescent Au NPs through a single step in ambient conditions [84].

Likely, different non-metallic NPs are synthesized using agricultural and processed food waste. For instance, waste product of tea has been synthesizing for the fabrication of iron oxide (Fe_3O_4) NPs with a size range between 5 to 25 nm with cuboid/pyramid in shape and size. As a result, NPs play vital roles in removing arsenic particles from the water and are used for five types of adsorption cycles [85]. According to statistics, 100 million tons of bananas are consumed every year worldwide [86], and banana peel is dumped into garbage, which is used to synthesize Mn_3O_4 NPs, which possess high-capacitive properties [87].

Mechanisms of Waste Material Mediated Synthesis of NPs

Cell walls of waste food materials contain hemicelluloses, cellulose, lignins, pectins, proteins and finally, biodegradable polysaccharides. These act as templating agents for the synthesis of NPs and determining their morphology, shape and size. So, the biomolecules act as a decrease of metal salts into the metal or metal oxide NPs [88].

Advantage of Waste Material in the Synthesis of NPs

Eco-friendly synthesis has led to the adulteration of huge inorganic NPs, particularly metal NPs. Various biomolecules present in the different waste act as templating agents, leading to fast, and, finally, eco-friendly and cost-effective approaches. Agro and food waste are easily accessible and do not require rigorous technical methods. So, it can be rightly processed for the synthesis of NPs and proper management of waste. On the other hand, proper maintenance of algae cultures is time-consuming; as a result, the production of algae requires much time and cost. Apart from this, some of the algae produced toxicological responses during the synthesis of NPs. Thus, the prime challenges related to algae mediated NPs synthesis are [89]: synthesis method needs to be improved, need a proper explanation of the mechanism of NPs formation and shape and size control for monodispersity of NPs to need to be improved.

Purification of Biogenic NPs

Purification is very vital after NPs synthesis before employing in different application sectors. Centrifugation has been a very common technique to purify nanoparticles for a long back due to easy operation cost and less time consumption. In addition to this, repetitive washing and high-speed centrifugation are highly necessary to eliminate unreacted bioactive molecules [22]. However, centrifugation techniques have some limitations, such as it caused NPs agglomeration, NPs destabilization and changed the intrinsic property of NPs. Dialysis is another important purification method that uses a suitable cut-off membrane. Generally, small organic molecules are extracted through this technique, but organic molecules remain intact and conjugated with NPs inside the dialysis membrane. Apart from this, this method is time-consuming (> 24 h). In addition to this, bio-fabricated nanoparticles are not purified through this method as they are water insoluble and, in the case of magnetic nanoparticles like Fe_2O_3 and Fe_3O_4, need external magnetic force.

FACTORS INFLUENCING BIOGENIC NPs SYNTHESIS

pH

NPs synthesis is directly dependent on the medium/reaction pH value as it regulates the formation of nucleation centres, which is directly proportional to pH value leading to enhanced formation of metallic NPs. It also played a critical role in regulating the morphology and size of NPs. For example, Armendariz *et al.*synthesized Au NPs from *A. sativa* at different pH conditions [90]. They observed that at pH 2 very small number of NPs were formed, but NPs size was considerably larger (25-85 nm) and opined that AuNPs produce lower nucleation centres at lower pH. Contrarily, Armendariz *et al.*synthesized small-sized NPs formed at slightly higher pH (pH 3-4). Okitsu *et al.* recorded pH effects on the size and average aspect ratio of gold nanorods; they opined that size and aspect ratio decreases with increasing pH value [91]. In another study, Tan *et al.* recorded that Camptothecin loaded NPs (NIPAm poly-(N-isopropylacrylamide)/ chitosan NPs) showed more Camptothecin release to target when NIPAm and chitosan ratio is 4:1(w/w), maximum release rate at pH 6.8 while it decreased on increasing or decreasing pH at 37 ^{0}C [92].

Temperature

Temperature is another important factor that influences the size and shape of NPs as well as their rate of synthesis. Generally, the temperature is primarily responsible for different shapes (spherical, triangle, octahedral and rod-like), size and NP synthesis. Reaction rate and nucleation centres formation are directly proportional to temperature [22]. For example, Sneha *et al.* synthesize AuNPs from Piper betel leaf extract in a temperature-dependent manner [93]. They also observed triangular-shaped NPs at 20 ^{0}C, octahedral-shaped NPs with sizes 5 to 500 nm at temperatures 30 to 40 ^{0}C and consistent spherical-shaped NPs at temperatures (50-60 ^{0}C). Islam *et al.* standardized AuNPs synthesis protocol from silver-ammonia complex solution of poly (ethylene oxide)-poly(propylene oxide) at different temperatures [94]. They observed that at a lower temperature, the size and NPs distribution are controlled by polymer morphology, but at a higher temperature, it is regulated by polymeric chemical composition. In another study, Tan *et al.* recorded the temperature effect on encapsulation and formation of AuNPs using PNIPAm/PEI [92]. The findings (TEM analysis) depicted that optimum AuNPs encapsulation and formation of stable Au-PNIPAm/PEI composite occurred at 25-30 ^{0}C, but at lower temperatures (15 ^{0}C), encapsulation was very poor. Iravani and Zolfaghari synthesized AgNPs from *P. eldarica* bark extract at different temperatures (25, 50, 100 and 150 ^{0}C) [95]. They observed that NP size decreases with increasing temperature. Fleitas-Salazar *et al.* recorded the impact of oxidizing atmosphere and temperature on Ag NPs synthesis using biocompatible polymer PEG. The findings showed that ability of polyethylene glycol (PEG) to reduce silver salt was high at 100 ^{0}C while at 60 ^{0}C, reduction of Ag ions occurred through oxidation of hydroxyl groups [96].

Reaction Time

Like temperature and pH, reaction time is another regulating factor that influences NPs morphology. For instance, Flor *et al.* observed the effect of reaction time ZnO and Cerium-doped ZnO particle size [97]. They reported a linear increase in particle size with increasing reaction time, and Cerium doped particle size is than ZnO. Ahmad *et al.* in their study, reported that the rate of reduction for AgNPs generation (from oil palm extract, *E. guineensis*) increases with an increase in reaction time [98]. Simultaneously, the effects of reaction time on the particle size of cadmium selenide NPs were conducted by Rose *et al.* [99] and concluded that the particle size of cadmium selenide NPs decreases with an increase in reaction time. Recently, Karade*et al.* studied the effect of reaction time on magnetic NPs, synthesized from Ferric nitrate solution containing green tea extract [100]. They observed that reaction time influences the structural as well as magnetic properties of magnetic FeNPs. They also noticed that increase in reaction time also increases particle size as well as enhances the saturation magnetization of NPs.

MECHANISM OF ANTIMICROBIAL ACTIVITY OF BIOGENIC NPs

Biogenic nanoparticles possess various modes of action for their antimicrobial activity [101]. The precise pathways for nanoparticle-mediated bactericidal activity are still emerging. Generally, the composition, size, surface modification, intrinsic properties as well as the bacterial species themselves play a role in the process of nanoparticle-mediated antimicrobial activity. It was reported that various nanoparticles can act upon the bacterial membrane causing loss of cell permeability. Likewise, nanoparticles of smaller size and positive zeta potential have been reported to be better antimicrobial agents.

Silver has been famous for its antimicrobial properties since ancient times. Silver nanoparticles use multifaceted factors simultaneously to combat bacterial cells. Silver, either in metallic form or in ionic form, reveals broad-spectrum antibacterial activity [102 - 104]. The antimicrobial activity of silver nanoparticles is shape and size-dependent as evidenced by previous studies [105]. Oxidation of metal or metal oxide nanoparticles with the release of positive ions such as Ag^+, Cu^{2+}, and Zn^{2+} can adsorb on the bacterial cell membrane and lead to cell wall damage by the formation of the pit in bacteria [106]. It has been studied earlier that silver nanoparticles generate free radicles and reactive oxygen species (ROS), that are responsible for the oxidation of cellular DNA and proteins, and damage the bacterial cell membranes *via* lipid peroxidation. The silver nanoparticles interact with the phosphorus and sulfur molecule of the bacterial DNA, which block the DNA replication process leading the cell death [107]. It has also been proposed that silver nanoparticles release silver ions, can interact with the thiol groups of enzymes and inactivate them, leading to damage to cells [108].

The antibacterial activity of ZnO nanoparticles is based on photocatalysis. The direct contact of the ZnO nanoparticles with the cell wall of the microorganisms through electrostatic interactions disrupted the cell membrane and resulted in the leakage of cell contents. Zn^{2+} ions released from the ZnO nanoparticles have been reported to interact with the various respiratory enzymes and inhibit DNA replication and cell division [109, 110].

The antibacterial activity of copper oxide nanoparticles (CuONPs) is revealed by damaging the bacterial cell membrane, inactivation of enzymes by interacting with them, exchange of essential ions and leakage of cytoplasmic content, and generation of free radicals (H_2O_2) [111, 112]. TiO_2 inactivates microorganisms *via* photocatalytic reactions. Superoxide anions, hydroxyl radical, and hydrogen peroxide are produced on the surface of irradiated TiO_2, which inactivate a wide spectrum of microbes by oxidizing the polyunsaturated phospholipid components of the cell membrane [113, 114]. An overview of various antibacterial modes of action biogenic nanoparticles is indicated as a cartoon in Fig. (**1**).

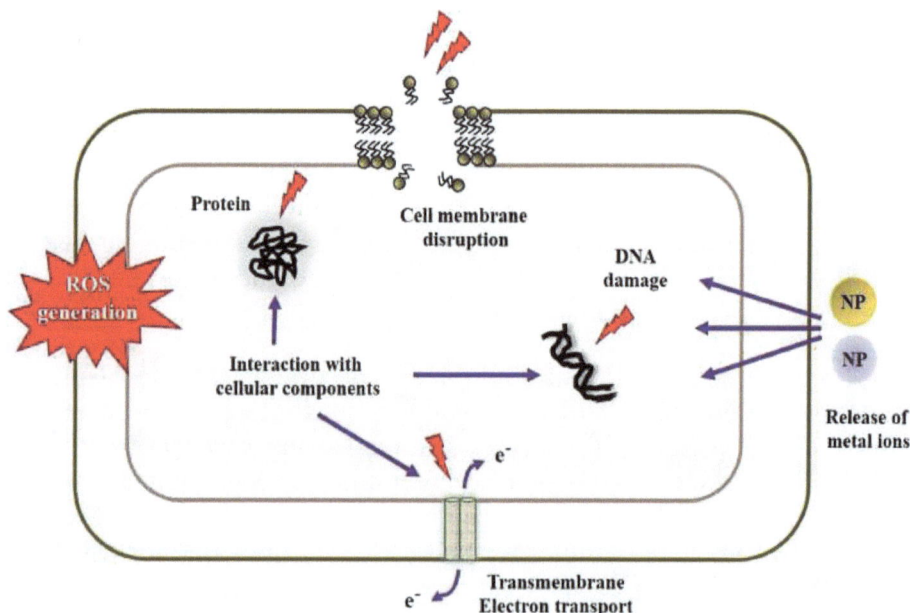

Fig. (1). The various antibacterial modes of action biogenic nanoparticles.

BIOGENIC NPs APPLICATIONS OR ACTIVITIES

Biogenic nanoparticles now became central focus in different fields for antimicrobial applications because of biocompatibility and long-term stability. Oxidative stress, cell wall rupture, metal ion release, and non-oxidative stress are the prime mechanisms behind the antimicrobial effect of these nanoparticles. A number of examples exist currently about the application of biogenic NPs against several pathogenic microorganisms. Herein, we will discuss the role of a few biogenic NPs against pathogenic microorganisms.

Antimicrobial Activity of Gold Nanoparticles (AuNPs)

Gold NPs are well known as the most studied biogenic NPs for their antimicrobial activity and biocompatibility. It needs capping agents for effective antimicrobial properties. Mainly collagen, chitosan or gelatin and even drugs are used as capping agents. Gu *et al.* observed that the antimicrobial effects of vancomycin enhanced 50 folds when it was conjugated with AuNPs against vancomycin-resistant *Enterococci* (VRE) and *E. coli*, mainly due to photothermal and photodynamic activation [115]. Generally, AuNPs are very biocompatible with microbial cells with no bacteriostatic or bactericidal activity. However, AuNPs integrated antibiotics showed a strong bactericidal effect against drug-resistant bacteria. For example, ampicillin-bound AuNPs damaged ampicillin-resistant bacteria, including *P. aeruginosa*, MRSA, *Enterobacteraerogenes*, and *E. coli* K-12 sub-strain DH5-alpha by different mechanisms such as inhibition of ATP synthase, changes in high surface/volume ratio etc. Li *et al.* observed the efficacy of AuNPs (2 nm size) having cationic surface properties against Gram-negative bacteria like *P. aeruginosa*, and *E. cloacae* complex, Gram-positive bacteria *S. aureus* and methicillin-resistant *S. aureus* [116]. Size and shape-dependent efficacy studies of the antibacterial and anti-cancerous activity of AuNPs have also been reported by several authors [117 - 119]. They also opined that smaller AuNPs showed better bactericidal effects than larger AuNPs.

Antimicrobial Activity of Silver Nanoparticles (AgNPs)

Silver is another widely used antimicrobial agent, particularly in paints, pharmaceuticals, ointments, fabrics, food, and packaging industries. But recently, it has been used as an antibacterial agent for wound dressing. Like AuNPs, antibacterial activity of AgNPs is size- and shape-dependent, mainly through disrupted energy metabolism, cell membrane damage, oxidative stress formation, and transcription inhibition [22]. In addition to this, Pal *et al.* observed that small-sized AgNPs led to enhanced activity due to their large surface area with higher antimicrobial activity by triangular-shaped AgNP than spherical or rod-shaped ones [105]. It was also observed that Gram-negative bacteria showed higher sensitivity than Gram-positive bacteria. This is mainly due to Gram-negative bacteria capsulated with lipopolysaccharides possessing negative charge, while Gram-positive bacteria are cross-linked by thick peptidoglycans and linear polysaccharides, providing rigidity to the cell [22]. AgNPs enhanced the antimicrobial activity of antibiotics such as amoxicillin, penicillin G, clindamycin, vancomycin, and especially erythromycin against *S. aureus* and *E. coli* [104]. Furthermore, the antibacterial activity of AgNPs against multidrug-resistant bacteria, namely *Streptococcus pyogenes, P. aeruginosa, S. aureus, E. coli, Klebsiellapneumoniae, Enterococcus* species and *Salmonella* species, have been reported by several authors [120, 121]. Additionally, silver carbene complex encapsulated by nanoparticles showed.

Strong antimicrobial activity against multidrug-resistant bacteria, including *A. baumannii* (MRAB), MRSA, *Burkholderiacepacia, P. aeruginosa*, and *K. pneumoniae* [122]. Kim *et al.* in their study reported that antimicrobial activity of AgNPs on *S. aureus* and *E. coli* are independent of temperature and pH [123].

Antimicrobial Activity of Iron NPs (Fe NPs)

Biogenic FeNPs have strong antimicrobial activity. For example, FeNPs extracted from biogenic sources (*T. procumbens, G. jasminoides, Skimmialaureola* and *L. inermis*) showed bactericidal potency against *S. aureus, P. mirabilis, S. enterica, E. coli, R. solanacearum* and *P. aeruginosa* [124, 125]. Khalil *et al.* observed that FeNPs synthesized from *Sageretiathea* showed inhibitory effects against the growth of *S. epidermidis, P. aeruginosa, E. coli, B. subtilis*, and *K. pneumonia* [126]. Shape-dependent antimicrobial activity of FeNPs synthesized from *E. crassipes* leaf extract against *S. aureus* and *P. fluorescens* has been reported by Jagathesan and Rajiv [127]. Recently, Hassan *et al.* synthesized magnetic FeNPs from *C. viminalis* and proved that biogenic NPs highly effective as antimicrobial agents in some Gram-negative and Gram-positive bacteria like *S. aureus, S. typhi, S. enterica, S. dysenteriae* and *K. pneumonia* [128].

Antimicrobial Activity of Zinc Oxide NPs (ZnONPs)

Zinc oxide NPs are used widely due to their unique antifungal and antibacterial properties. It possesses unique photo-oxidizing and photocatalytic capacity, enabling various applications, such as drug delivery and anticancer therapy. It is also effective against microorganisms and nanometers to micrometer ranges. It is equally effective for both Gram-negative and Gram-positive bacteria [22]. Jones *et al.* observed that ZnONPs possess higher bactericidal effects on *S. aureus* than MgO, Al_2O_3, TiO_2, CuO and CeO_2 NPs [109]. This is mainly because ZnO can absorb UV light heavily, which helps them hold higher conductivity and finally enhances interaction time between ZnONPs and bacteria. Talebian *et al.* reported shape-dependent bacterial activity: enhanced bacterial activity by flower-shaped NPs against *S. aureus* and *E. coli* than rod-shaped ones [129]. Sirelkhatim *et al.* demonstrated oxygen annealing of ZnO enhanced the higher production of ROS and enhanced antibacterial activity [130]. Lingaraju *et al.* recorded size and concentration-dependent antibacterial activity of ZnONPs; the larger the surface area (small size) and higher the concentration, the more the bactericidal effect [131]. They also opined that ZnONPs is also used as a substitute as they are equally effective against Gram-negative (*K. aerogenes, P. aeruginosa, E. coli,* etc.) and Gram-positive bacteria (*S. aureus*).

Antimicrobial Activity of Copper NPs (CuONPs)

Cupric oxide (CuO) nanoparticles have gained lots of importance in recent years not only because of their antimicrobial activity but also because of their effective utilization in the pharma industry, cosmetics, farming, etc [132]. The antibacterial mechanism of copper and CuONPs are believed to be due to electrostatic attraction between Cu^{+2} ion and a plasma membrane that helps to damage the membrane and finally kills the cell. CuONPs have strong antimicrobial activity against both Gram-positive and Gram-negative bacteria. For example, CuNPs synthesized from *Moringaoleifera* (drumstick) fruit pulp aqueous extracts showed strong antibacterial activity against *Pseudomonas putida, Bacillus subtilis, Staphylococcus aureus, Escherichia coli,* and *Klebsiella pneumonia* [133]. They are also used in wound healing (bactericidal plasters and bandages) because of their strong bactericidal effect and illegible sensibility of human tissues towards copper compounds.

Antimicrobial Activity of Titanium Dioxide NPs (TiO₂NPs)

TiO₂NPs have gained much attraction because of their exclusivity and improved properties compared to bulk counterparts. Apart from its use in the microbiological field, TiO₂NPs have been approved by Food and Drug Administration (FDA) for use in human food, drugs, cosmetics, and food contact materials. Likely, TiO₂NPs also exhibit antimicrobial activity by multiple mechanisms. Antibacterial nature of TiO₂NPs has been extensively demonstrated in *E. coli, S. aureus, P. aeruginosa,* and *E. faecium* [134, 135]. Combined interaction study between TiO₂NPs and various antibiotics showed maximum inhibition activity against penicillin and amikacin followed by ampicillin, oxacillin, amoxicillin, cephalexin, vancomycin, streptomycin, erythromycin, clindamycin and tetracycline [136]. Santoshkumar *et al.* reported that TiO₂NPs biosynthesized from *Psidium guajava* significantly prevented the growth of *E.*

coli and *S. aureus* [137]. Likely, *T. foenumgraecum* leaf-derived TiO_2NPs shower strong antimicrobial activity against *P. vulgaris, Y. enterocolitica, P. aeruginosa, E. faecalis, S. aureus, S. faecalis, E. coli, B. subtilis*, and *C. albicans*.

FACTORS AFFECTING BIOGENIC NPs APPLICATIONS

Nanoparticle Shape

Shape plays a crucial role in microbial activity, particularly the intensity of biocidal activity. Several studies demonstrated that different shapes (elongated rod, spherical, and truncated triangular) have different biocidal activity [27]. For example, media containing different shapes of NPs have different *E. coli* colony-forming unit [138]. Generally, NPs shape influencing the facets content: spherical NPs contain 100 facets, rod-shaped NPs contain 111 facets on the side surface and 100 on the end, and truncated triangular NPs contain 111 facets. Among different facets, facets 111 contain high atom density that favors higher antibacterial reactivity [27].

Chemical Composition

The chemical composition of NPs is the basis that regulates their variations in activities. Generally, NPs (TiO_2, ZnO_2 and SiO_2) themselves produce ROS against bacteria like *Bacillus subtilis* and *E. coli* [27, 139, 140]. The findings showed that *B. subtilis* growth was 90% inhibited by 10 ppm ZnONPs, while 90% *B. subtilis* growth inhibition occurred at 1000 and 2000 ppm TiO_2NPs and SiO_2NPs, respectively, which indicated that biocidal activity of these compounds is in ascending order from SiO_2 to TiO_2 to ZnO. On the other hand, Huang *et al.* reported that *E. coli* inhibition effect by NPs was partial at 10 and 500 ppm ZnONPs [140]. These findings indicated that bactericidal activity does not affect by light or dark; this inhibition is mainly due to involved mechanisms rather than ROS production.

Size and Concentration

The NPs size and concentration also played an influencing role in microbial activity. For example, Kim *et al.* in their study demonstrated that smallest-sized spherical AgNPs were more efficient to kill and destroy bacteria as compared to larger spherical AgNPs [141]. They also opined that this is basically due to the high surface-to-volume ratio. Likely, Liu *et al.* recorded that smaller NPs are more efficient in killing bacteria than large NPs as small-sized NPs released more silver cations [19].

Photo Activation

Photo activation is another parameter that regulates antimicrobial activity. For example, Liu *et al.* reported that the activity of TiO_2NPs against *E. coli* significantly increases after UV treatment compared with TiO_2NPs [19]. When TiO_2 nanoparticles were modified with *Garcinia zeylanica* aqueous extract and exposed to sunshine, they showed increased antibacterial activity against Methicillin-resistant *Staphylococcus aureus* [142]. In another study, Lipovsky*et al.* observed that photoactivation with the blue light of ZnONPs exhibited higher microbial activity by enhancing ROS production [143].

Target Microorganisms

It has been demonstrated that NPs showed greater biocidal activity against Gram-negative rod-shaped bacteria than Gram-positive cocci. For example, Qais*et al.* observed significantly higher antimicrobial activity of AgNPs against *E. coli* (MIC 3.3-3.6) than *S. aureus* (MIC more than 33 nM) [144]. They also demonstrated that this difference is mainly due to difference in cell wall organization, as well as cell wall composition of these bacteria. On the other hand, Huang *et al.* recorded higher ZnO NPs activity against *S. aureus* than *E. coli* and *P. aeruginosa* [140].

Isolation and Purification of NPs

Isolation and purification of NPs also play a vital role in microbial activity. Generally, after centrifugation at 10000 rpm for 10 min, metal NPs are settled at the bottom [27]. Depending on the preparation and centrifugation method, various impurities are found in NPs pellets. For example, simple filtration only removes polymer aggregates, and other impurities require other techniques. Methods like gel filtration, dialysis and ultracentrifugation are not capable of removing high molecular weight molecules. Accordingly, they showed varied antimicrobial activities [27].

FUTURE PROSPECTIVES

Metal NPs produced from biogenic origin support researchers to design eco-friendly nanomaterials and enhance the efficiency of the health and safety index of NPs. Biogenic NPs have wide applications in various sectors, including antimicrobial applications, particularly in biomedical industries. However, there are restrictions and certain holes in the effective production of these NPs, which should be figured out by the scientific community. One of the major limitations is the underlying mechanisms of the biogenic NPs mode of action. In addition, the synthesis and stabilization of biogenic NPs is very challenging regarding the biomineralization of metal NP ions. Particularly in terms of NP distribution, bioavailability and biocompatibility, it is significant to understand how active groups of biogenic materials are involved in NPs synthesis with greater efficacy. Further, in-depth biocompatibility and bioavailability evaluation of biogenic nanomaterials are still in the infancy stage. Therefore, a surfeit of biogenic NPs synthesis needs in-depth evaluation and considerable research efforts with regard to biocompatibility and bioavailability to meet the necessity for nanoproducts in several arenas. Toxicity evaluation of NPs is another important aspect for widespread application in the medical sector as the toxicity of biogenic NPs is not evaluated due to bio-origin. Therefore, ample investigation is very much needed between biogenic and chemically-synthesized NPs for successful future application of biogenic NPs. Finally, the true challenge for commercializing biogenic NPs is to find the right balance between production costs, scalability, and applicability.

CONCLUSION

Biogenic NPs synthesis has recently emerged as a noble research topic as an alternative to chemical-based NPs synthesis for future sustainability. Biogenic NPs synthesis routes are cleaner, environmentally and economically *via*ble, and more scalable than other routes like chemical and physical. Overall, this review paper summarized recent advances in biogenic NPs synthesis with mechanistic antimicrobial effects, particularly multi-drug resistant microorganisms. Although we have delineated the underlying mechanism of biogenic NPs, the resistance mechanism of microorganisms to these NPs is warranted and demands adequate attention in future studies. This review did not discuss biocompatibility, their distribution and bioavailability and toxic fate in vivo, which need utmost focus, and further studies for economical production of biogenic NPs are very much needed for wide applicability relevant to either antimicrobial era or other fields. Finally, the review outcomes will bring new inspiration to researchers and the scientific community for further research on this emerging topic.

CONSENT FOR PUBLICATION

Not applicable.

CONFLICT OF INTEREST

The authors declare no conflict of interest, financial or otherwise.

ACKNOWLEDGEMENTS

The authors MDA and SS would like to thank the HOD (Department of Biotechnology), Registrar, and Vice-Chancellor of the University of North Bengal, West Bengal, India, for their moral support. PS is thankful to Principal, Sukanta Mahavidyalaya, Dhupguri, West Bengal, India, for his support and encouragement. The author, SD would like to thank HOD (Department of Environmental Science), Vice-Chancellor of the University of Burdwan, West Bengal, India, for their moral support.

REFERENCES

[1]   Antonyraj CA, Jeong J, Kim B, *et al.* Selective oxidation of HMF to DFF using Ru/γ-alumina catalyst in moderate boiling solvents toward industrial production. J Ind Eng Chem 2013; 19(3): 1056-9.
[http://dx.doi.org/10.1016/j.jiec.2012.12.002]

[2]   Staniland SS. Magnetosomes: bacterial biosynthesis of magnetic nanoparticles and potential biomedical applications 2007.

[3]   Fu H, Yang X, Jiang X, Yu A. Bimetallic Ag-Au nanowires: synthesis, growth mechanism, and catalytic properties. Langmuir 2013; 29(23): 7134-42.
[http://dx.doi.org/10.1021/la400753q] [PMID: 23679079]

[4]   Roy N, Gaur A, Jain A, Bhattacharya S, Rani V. Green synthesis of silver nanoparticles: An approach to overcome toxicity. Environ Toxicol Pharmacol 2013; 36(3): 807-12.
[http://dx.doi.org/10.1016/j.etap.2013.07.005] [PMID: 23958974]

[5]   Xia Y, Yang H, Campbell CT. Nanoparticles for Catalysis. Acc Chem Res 2013; 46(8): 1671-2.
[http://dx.doi.org/10.1021/ar400148q] [PMID: 23957601]

[6]   Luo X, Morrin A, Killard AJ, Smyth MR. Application of Nanoparticles in Electrochemical Sensors and Biosensors. Electroanalysis 2006; 18(4): 319-26.
[http://dx.doi.org/10.1002/elan.200503415]

[7]   Shivaji S, Madhu S, Singh S. Extracellular synthesis of antibacterial silver nanoparticles using psychrophilic bacteria. Process Biochem 2011; 46(9): 1800-7.
[http://dx.doi.org/10.1016/j.procbio.2011.06.008]

[8]   Chan YS, Mat Don M. Biosynthesis and structural characterization of Ag nanoparticles from white rot fungi. Mater Sci Eng C 2013; 33(1): 282-8.
[http://dx.doi.org/10.1016/j.msec.2012.08.041] [PMID: 25428073]

[9]   Akhtar MS, Panwar J, Yun YS. Biogenic synthesis of metallic nanoparticles by plant extracts. ACS Sustain Chem& Eng 2013; 1(6): 591-602.
[http://dx.doi.org/10.1021/sc300118u]

[10]   Kanchi S, Kumar G, Lo AY, *et al.* Exploitation of de-oiled jatropha waste for gold nanoparticles synthesis: A green approach. Arab J Chem 2018; 11(2): 247-55.
[http://dx.doi.org/10.1016/j.arabjc.2014.08.006]

[11]   Hakim LF, Portman JL, Casper MD, Weimer AW. Aggregation behavior of nanoparticles in fluidized beds. Powder Technol 2005; 160(3): 149-60.
[http://dx.doi.org/10.1016/j.powtec.2005.08.019]

[12]   Tripp SL, Pusztay SV, Ribbe AE, Wei A. Self-assembly of cobalt nanoparticle rings. J Am Chem Soc 2002; 124(27): 7914-5.
[http://dx.doi.org/10.1021/ja0263285] [PMID: 12095331]

[13]   Huang J, Lin L, Sun D, Chen H, Yang D, Li Q. Bio-inspired synthesis of metal nanomaterials and

applications. Chem Soc Rev 2015; 44(17): 6330-74.
[http://dx.doi.org/10.1039/C5CS00133A] [PMID: 26083903]

[14]   Schröfel A, Kratošová G, Šafařík I, Šafaříková M, Raška I, Shor LM. Applications of biosynthesized metallic nanoparticles – A review. Acta Biomater 2014; 10(10): 4023-42.
[http://dx.doi.org/10.1016/j.actbio.2014.05.022] [PMID: 24925045]

[15]   Riddin T, Gericke M, Whiteley CG. Biological synthesis of platinum nanoparticles: Effect of initial metal concentration. Enzyme Microb Technol 2010; 46(6): 501-5.
[http://dx.doi.org/10.1016/j.enzmictec.2010.02.006] [PMID: 25919626]

[16]   Mishra A, Tripathy SK, Yun SI. Fungus mediated synthesis of gold nanoparticles and their conjugation with genomic DNA isolated from Escherichia coli and Staphylococcus aureus. Process Biochem 2012; 47(5): 701-11.
[http://dx.doi.org/10.1016/j.procbio.2012.01.017]

[17]   Patel V, Berthold D, Puranik P, Gantar M. Screening of cyanobacteria and microalgae for their ability to synthesize silver nanoparticles with antibacterial activity. Biotechnol Rep (Amst) 2015; 5: 112-9.
[http://dx.doi.org/10.1016/j.btre.2014.12.001] [PMID: 28626689]

[18]   Vanlalveni C, Lallianrawna S, Biswas A, Selvaraj M, Changmai B, Rokhum SL. Green synthesis of silver nanoparticles using plant extracts and their antimicrobial activities: A review of recent literature. RSC Advances 2021; 11(5): 2804-37.
[http://dx.doi.org/10.1039/D0RA09941D] [PMID: 35424248]

[19]   Liu FK, Ko FH, Huang PW, Wu CH, Chu TC. Studying the size/shape separation and optical properties of silver nanoparticles by capillary electrophoresis. J Chromatogr A 2005; 1062(1): 139-45.
[http://dx.doi.org/10.1016/j.chroma.2004.11.010] [PMID: 15679152]

[20]   Gittins DI, Bethell D, Schiffrin DJ, Nichols RJ. A nanometre-scale electronic switch consisting of a metal cluster and redox-addressable groups. Nature 2000; 408(6808): 67-9.
[http://dx.doi.org/10.1038/35040518] [PMID: 11081506]

[21]   Maruyama T, Fujimoto Y, Maekawa T. Synthesis of gold nanoparticles using various amino acids. J Colloid Interface Sci 2015; 447: 254-7.
[http://dx.doi.org/10.1016/j.jcis.2014.12.046] [PMID: 25591824]

[22]   Singh A, Gautam PK, Verma A, *et al.* Green synthesis of metallic nanoparticles as effective alternatives to treat antibiotics resistant bacterial infections: A review. Biotechnol Rep (Amst) 2020; 25: e00427.
[http://dx.doi.org/10.1016/j.btre.2020.e00427] [PMID: 32055457]

[23]   Al-Harbi MS, El-Deeb BA, Mostafa N, Amer SAM. Extracellular biosynthesis of AgNPs by the bacterium Proteus mirabilis and its toxic effect on some aspects of animal physiology. Adv Nanopart 2014; 3(3): 83-91.
[http://dx.doi.org/10.4236/anp.2014.33012]

[24]   Samadi N, Golkaran D, Eslamifar A, Jamalifar H, Fazeli MR, Mohseni FA. Intra/extracellular biosynthesis of silver nanoparticles by an autochthonous strain of Proteus mirabilis isolated from photographic waste. J Biomed Nanotechnol 2009; 5(3): 247-53.
[http://dx.doi.org/10.1166/jbn.2009.1029] [PMID: 20055006]

[25]   Li J, Li Q, Ma X, *et al.* Biosynthesis of gold nanoparticles by the extreme bacterium *Deinococcus radiodurans* and an evaluation of their antibacterial properties. Int J Nanomedicine 2016; 11: 5931-44.
[http://dx.doi.org/10.2147/IJN.S119618] [PMID: 27877039]

[26]   Kunoh T, Takeda M, Matsumoto S, *et al.* Green synthesis of gold nanoparticles coupled with nucleic acid oxidation. ACS Sustain Chem& Eng 2018; 6(1): 364-73.
[http://dx.doi.org/10.1021/acssuschemeng.7b02610]

[27]   Qidwai A, Pandey A, Kumar R, Shukla SK, Dikshit A. Advances in biogenic nanoparticles and the mechanisms of antimicrobial effects. Indian J Pharm Sci 2018; 80(4): 592-603.

[http://dx.doi.org/10.4172/pharmaceutical-sciences.1000398]

[28]   Khan SA, Ahmad A. Fungus mediated synthesis of biomedically important cerium oxide nanoparticles. Mater Res Bull 2013; 48(10): 4134-8.
[http://dx.doi.org/10.1016/j.materresbull.2013.06.038]

[29]   Mukherjee P, Ahmad A, Mandal D, *et al.* Fungus-mediated synthesis of silver nanoparticles and their immobilization in the mycelia matrix: A novel biological approach to nanoparticle synthesis. Nano Lett 2001; 1(10): 515-9. [a].
[http://dx.doi.org/10.1021/nl0155274]

[30]   Rauwel P, Küünal S, Ferdov S, Rauwel E. A review on the green synthesis of silver nanoparticles and their morphologies studied *via* TEM. Adv Mater Sci Eng 2015; 2015: 1-9.
[http://dx.doi.org/10.1155/2015/682749]

[31]   Sanghi R, Verma P. Biomimetic synthesis and characterisation of protein capped silver nanoparticles. Bioresour Technol 2009; 100(1): 501-4.
[http://dx.doi.org/10.1016/j.biortech.2008.05.048] [PMID: 18625550]

[32]   Tan Y, Wang Y, Jiang L, Zhu D. Thiosalicylic acid-functionalized silver nanoparticles synthesized in one-phase system. J Colloid Interface Sci 2002; 249(2): 336-45.
[http://dx.doi.org/10.1006/jcis.2001.8166] [PMID: 16290606]

[33]   Narayanan KB, Sakthivel N. Biological synthesis of metal nanoparticles by microbes. Adv Colloid Interface Sci 2010; 156(1-2): 1-13.
[http://dx.doi.org/10.1016/j.cis.2010.02.001] [PMID: 20181326]

[34]   Das SK, Das AR, Guha AK. Gold nanoparticles: microbial synthesis and application in water hygiene management. Langmuir 2009; 25(14): 8192-9.
[http://dx.doi.org/10.1021/la900585p] [PMID: 19425601]

[35]   Mohammadinejad R, Shavandi A, Raie DS, *et al.* Plant molecular farming: production of metallic nanoparticles and therapeutic proteins using green factories. Green Chem 2019; 21(8): 1845-65.
[http://dx.doi.org/10.1039/C9GC00335E]

[36]   Apte M, Sambre D, Gaikawad S, *et al.* Psychrotrophic yeast Yarrowia lipolytica NCYC 789 mediates the synthesis of antimicrobial silver nanoparticles *via* cell-associated melanin. AMB Express 2013; 3(1): 32.
[http://dx.doi.org/10.1186/2191-0855-3-32] [PMID: 23758863]

[37]   Kowshik M, Ashtaputre S, Kharrazi S, *et al.* Extracellular synthesis of silver nanoparticles by a silver-tolerant yeast strain MKY3. Nanotechnology 2003; 14(1): 95-100.
[http://dx.doi.org/10.1088/0957-4484/14/1/321]

[38]   Saeed S, Iqbal A, Ashraf MA. Bacterial-mediated synthesis of silver nanoparticles and their significant effect against pathogens. Environ Sci Pollut Res Int 2020; 27(30): 37347-56.
[http://dx.doi.org/10.1007/s11356-020-07610-0] [PMID: 32130634]

[39]   Mirzajani F, Ghassempour A, Aliahmadi A, Esmaeili MA. Antibacterial effect of silver nanoparticles on Staphylococcus aureus. Res Microbiol 2011; 162(5): 542-9.
[http://dx.doi.org/10.1016/j.resmic.2011.04.009] [PMID: 21530652]

[40]   Holt KB, Bard AJ. Interaction of silver(I) ions with the respiratory chain of Escherichia coli: An electrochemical and scanning electrochemical microscopy study of the antimicrobial mechanism of micromolar Ag+. Biochemistry 2005; 44(39): 13214-23.
[http://dx.doi.org/10.1021/bi0508542] [PMID: 16185089]

[41]   Majeed S, Danish M, Ibrahim MN, *et al.* Bacteria Mediated Synthesis of Iron Oxide Nanoparticles and Their Antibacterial, Antioxidant, Cytocompatibility Properties. J Cluster Sci 2020; 1-2.

[42]   Chaloupka K, Malam Y, Seifalian AM. Nanosilver as a new generation of nanoproduct in biomedical applications. Trends Biotechnol 2010; 28(11): 580-8.
[http://dx.doi.org/10.1016/j.tibtech.2010.07.006] [PMID: 20724010]

[43] Feroze N, Arshad B, Younas M, Afridi MI, Saqib S, Ayaz A. Fungal mediated synthesis of silver nanoparticles and evaluation of antibacterial activity. Microsc Res Tech 2020; 83(1): 72-80.
[http://dx.doi.org/10.1002/jemt.23390] [PMID: 31617656]

[44] Abdel-Kareem MM, Zohri AA. Extracellular mycosynthesis of gold nanoparticles using *Trichoderma hamatum* : optimization, characterization and antimicrobial activity. Lett Appl Microbiol 2018; 67(5): 465-75.
[http://dx.doi.org/10.1111/lam.13055] [PMID: 30028030]

[45] Priyadarshini E, Pradhan N, Sukla LB, Panda PK. Controlled synthesis of gold nanoparticles using Aspergillus terreus IF0 and its antibacterial potential against Gram negative pathogenic bacteria. J Nanotechnol 2014; 2014.

[46] Gudikandula K, Charya Maringanti S. Synthesis of silver nanoparticles by chemical and biological methods and their antimicrobial properties. J Exp Nanosci 2016; 11(9): 714-21.
[http://dx.doi.org/10.1080/17458080.2016.1139196]

[47] Jung WK, Koo HC, Kim KW, Shin S, Kim SH, Park YH. Antibacterial activity and mechanism of action of the silver ion in Staphylococcus aureus and Escherichia coli. Appl Environ Microbiol 2008; 74(7): 2171-8.
[http://dx.doi.org/10.1128/AEM.02001-07] [PMID: 18245232]

[48] Rajakumar G, Rahuman AA, Roopan SM, *et al.* Fungus-mediated biosynthesis and characterization of TiO$_2$ nanoparticles and their activity against pathogenic bacteria. Spectrochim Acta A Mol Biomol Spectrosc 2012; 91: 23-9.
[http://dx.doi.org/10.1016/j.saa.2012.01.011] [PMID: 22349888]

[49] Dos Santos CA, Seckler MM, Ingle AP, *et al.* Silver nanoparticles: therapeutical uses, toxicity, and safety issues. J Pharm Sci 2014; 103(7): 1931-44.
[http://dx.doi.org/10.1002/jps.24001] [PMID: 24824033]

[50] Peiris MMK, Gunasekara TDCP, Jayaweera PM, Fernando SSN. TiO$_2$ Nanoparticles from Baker□□s Yeast: A Potent Antimicrobial. J Microbiol Biotechnol 2018; 28(10): 1664-70.
[http://dx.doi.org/10.4014/jmb.1807.07005] [PMID: 30178650]

[51] Md Ishak N A I, Kamarudin SK, Timmiati SN. Green synthesis of metal and metal oxide nanoparticles *via* plant extracts: An overview. Mater Res Express 2019; 6(11): 112004.
[http://dx.doi.org/10.1088/2053-1591/ab4458]

[52] Singh P, Kim YJ, Zhang D, Yang DC. Biological synthesis of nanoparticles from plants and microorganisms. Trends Biotechnol 2016; 34(7): 588-99.
[http://dx.doi.org/10.1016/j.tibtech.2016.02.006] [PMID: 26944794]

[53] Sathishkumar M, Sneha K, Won SW, Cho CW, Kim S, Yun YS. Cinnamon zeylanicum bark extract and powder mediated green synthesis of nano-crystalline silver particles and its bactericidal activity. Colloids Surf B Biointerfaces 2009; 73(2): 332-8.
[http://dx.doi.org/10.1016/j.colsurfb.2009.06.005] [PMID: 19576733]

[54] Gardea-Torresdey JL, Gomez E, Peralta-Videa JR, Parsons JG, Troiani H, Jose-Yacaman M. Alfalfa sprouts: A natural source for the synthesis of silver nanoparticles. Langmuir 2003; 19(4): 1357-61.
[http://dx.doi.org/10.1021/la020835i]

[55] Shankar SS, Ahmad A, Pasricha R, Sastry M. Bioreduction of chloroaurate ions by geranium leaves and its endophytic fungus yields gold nanoparticles of different shapes. J Mater Chem 2003; 13(7): 1822-6.
[http://dx.doi.org/10.1039/b303808b]

[56] Ahmed S. Saifullah, Ahmad M, Swami BL, Ikram S. Green synthesis of silver nanoparticles using Azadirachtaindica aqueous leaf extract. Journal of radiation research and applied sciences. 2016; 9(1): 1-7.

[57] Órdenes-Aenishanslins NA, Saona LA, Durán-Toro VM, Monrás JP, Bravo DM, Pérez-Donoso JM.

Use of titanium dioxide nanoparticles biosynthesized by Bacillus mycoides in quantum dot sensitized solar cells. Microb Cell Fact 2014; 13(1): 90.
[http://dx.doi.org/10.1186/s12934-014-0090-7] [PMID: 25027643]

[58]   Rai M, Deshmukh SD, Ingle AP, Gupta IR, Galdiero M, Galdiero S. Metal nanoparticles: The protective nanoshield against virus infection. Crit Rev Microbiol 2014; 1-11.
[PMID: 24754250]

[59]   Sundrarajan M, Ambika S, Bharathi K. Plant-extract mediated synthesis of ZnO nanoparticles using Pongamia pinnata and their activity against pathogenic bacteria. Adv Powder Technol 2015; 26(5): 1294-9.
[http://dx.doi.org/10.1016/j.apt.2015.07.001]

[60]   Senthilkumar SR, Sivakumar T. Green tea (Camellia sinensis) mediated synthesis of zinc oxide (ZnO) nanoparticles and studies on their antimicrobial activities. Int J Pharm Pharm Sci 2014; 6(6): 461-5.

[61]   Velayutham K, Rahuman AA, Rajakumar G, *et al.* Evaluation of Catharanthus roseus leaf extract-mediated biosynthesis of titanium dioxide nanoparticles against Hippobosca maculata and Bovicola ovis. Parasitol Res 2012; 111(6): 2329-37.
[http://dx.doi.org/10.1007/s00436-011-2676-x] [PMID: 21987105]

[62]   Umavathi S, Mahboob S, Govindarajan M, *et al.* Green synthesis of ZnO nanoparticles for antimicrobial and vegetative growth applications: A novel approach for advancing efficient high quality health care to human wellbeing. Saudi J Biol Sci 2021; 28(3): 1808-15.
[http://dx.doi.org/10.1016/j.sjbs.2020.12.025] [PMID: 33732066]

[63]   Garibo D, Borbón-Nuñez HA, de León JND, *et al.* Green synthesis of silver nanoparticles using Lysiloma acapulcensis exhibit high-antimicrobial activity. Sci Rep 2020; 10(1): 12805.
[http://dx.doi.org/10.1038/s41598-020-69606-7] [PMID: 31913322]

[64]   Subhapriya S, Gomathipriya P. Green synthesis of titanium dioxide (TiO_2) nanoparticles by Trigonella foenum-graecum extract and its antimicrobial properties. Microb Pathog 2018; 116: 215-20.
[http://dx.doi.org/10.1016/j.micpath.2018.01.027] [PMID: 29366863]

[65]   Naika HR, Lingaraju K, Manjunath K, *et al.* Green synthesis of CuO nanoparticles using *Gloriosa superba* L. extract and their antibacterial activity. J Taibah Univ Sci 2015; 9(1): 7-12.
[http://dx.doi.org/10.1016/j.jtusci.2014.04.006]

[66]   Thekkae Padil VV, Černík M. Green synthesis of copper oxide nanoparticles using gum karaya as a biotemplate and their antibacterial application. Int J Nanomedicine 2013; 8: 889-98.
[PMID: 23467397]

[67]   Tyagi S, Kumar A, Tyagi PK. Comparative analysis of metal nanoparticles synthesized from Hibiscus rosasinesis and their antibacterial activity estimation against nine pathogenic bacteria. Asian J Pharm Clin Res 2017; 10(5): 323-9.
[http://dx.doi.org/10.22159/ajpcr.2017.v10i5.17458]

[68]   Khan MR, Hoque SM, Hossain KFB, Siddique MAB, Uddin MK, Rahman MM. Green synthesis of silver nanoparticles using *Ipomoea aquatica* leaf extract and its cytotoxicity and antibacterial activity assay. Green Chem Lett Rev 2020; 13(4): 303-15.
[http://dx.doi.org/10.1080/17518253.2020.1839573]

[69]   Chauhan N, Tyagi AK, Kumar P, Malik A. Antibacterial potential of Jatropha curcas synthesized silver nanoparticles against food borne pathogens. Front Microbiol 2016; 7: 1748.
[http://dx.doi.org/10.3389/fmicb.2016.01748] [PMID: 27877160]

[70]   Awwad AM, Albiss BA, Salem NM. Antibacterial activity of synthesized copper oxide nanoparticles using Malvasylvestris leaf extract. SMU Med J 2015; 2(1): 91-101.

[71]   Acharyulu NP, Dubey RS, Swaminadham V, Kollu P, Kalyani RL, Pammi SV. Green synthesis of CuO nanoparticles using Phyllanthusamarus leaf extract and their antibacterial activity against multidrug resistance bacteria. Int J Eng Res Technol (Ahmedabad) 2014; 3(4)

[72] Jayarambabu N, Akshaykranth A, Venkatappa Rao T, Venkateswara Rao K, Rakesh Kumar R. Green synthesis of Cu nanoparticles using Curcuma longa extract and their application in antimicrobial activity. Mater Lett 2020; 259: 126813.
[http://dx.doi.org/10.1016/j.matlet.2019.126813]

[73] Kirdat PN, Dandge PB, Hagwane RM, Nikam AS, Mahadik SP, Jirange ST. Synthesis and characterization of ginger (Z. officinale) extract mediated iron oxide nanoparticles and its antibacterial activity. Mater Today Proc 2021; 43: 2826-31.
[http://dx.doi.org/10.1016/j.matpr.2020.11.422]

[74] Chand K, Abro MI, Aftab U, *et al.* Green synthesis characterization and antimicrobial activity against *Staphylococcus aureus* of silver nanoparticles using extracts of neem, onion and tomato. RSC Advances 2019; 9(30): 17002-15.
[http://dx.doi.org/10.1039/C9RA01407A] [PMID: 35519862]

[75] Chugh D, Viswamalya VS, Das B. Green synthesis of silver nanoparticles with algae and the importance of capping agents in the process. J Genet Eng Biotechnol 2021; 19(1): 126.
[http://dx.doi.org/10.1186/s43141-021-00228-w] [PMID: 34427807]

[76] Baker S, Harini BP, Rakshith D, Satish S. Marine microbes: Invisible nanofactories. J Pharm Res 2013; 6(3): 383-8.
[http://dx.doi.org/10.1016/j.jopr.2013.03.001]

[77] Mikami Y, Dhakshinamoorthy A, Alvaro M, García H. Catalytic activity of unsupported gold nanoparticles. Catal Sci Technol 2013; 3(1): 58-69.
[http://dx.doi.org/10.1039/C2CY20068F]

[78] Zhao Y, Tian Y, Cui Y, Liu W, Ma W, Jiang X. Small molecule-capped gold nanoparticles as potent antibacterial agents that target Gram-negative bacteria. J Am Chem Soc 2010; 132(35): 12349-56.
[http://dx.doi.org/10.1021/ja1028843] [PMID: 20707350]

[79] Hayashi H, Hakuta Y. Hydrothermal synthesis of metal oxide nanoparticles in supercritical water. Materials (Basel) 2010; 3(7): 3794-817.
[http://dx.doi.org/10.3390/ma3073794] [PMID: 28883312]

[80] Leonelli C, Lojkowski W. Main development directions in the application of microwave irradiation to the synthesis of nanopowders. Chem Today 2007; 25: 34-8.

[81] Gade AK, Bonde P, Ingle AP, Marcato PD, Durán N, Rai MK. Exploitation of Aspergillus niger for synthesis of silver nanoparticles. J Biobased Mater Bioenergy 2008; 2(3): 243-7.
[http://dx.doi.org/10.1166/jbmb.2008.401]

[82] Crookes-Goodson WJ, Slocik JM, Naik RR. Bio-directed synthesis and assembly of nanomaterials. Chem Soc Rev 2008; 37(11): 2403-12.
[http://dx.doi.org/10.1039/b702825n] [PMID: 18949113]

[83] Kim H, Kim H, Mosaddik A, Gyawali R, Ahn KS, Cho SK. Induction of apoptosis by ethanolic extract of mango peel and comparative analysis of the chemical constitutes of mango peel and flesh. Food Chem 2012; 133(2): 416-22.
[http://dx.doi.org/10.1016/j.foodchem.2012.01.053] [PMID: 25683414]

[84] Devi PS, Banerjee S, Chowdhury SR, Kumar GS. Eggshell membrane: A natural biotemplate to synthesize fluorescent gold nanoparticles. RSC Advances 2012; 2(30): 11578-85.
[http://dx.doi.org/10.1039/c2ra21053c]

[85] Lunge S, Singh S, Sinha A. Magnetic iron oxide (Fe3O4) nanoparticles from tea waste for arsenic removal. J Magn Magn Mater 2014; 356: 21-31.
[http://dx.doi.org/10.1016/j.jmmm.2013.12.008]

[86] Venkateswarlu S, Rao YS, Balaji T, Prathima B, Jyothi NVV. Biogenic synthesis of Fe3O4 magnetic nanoparticles using plantain peel extract. Mater Lett 2013; 100: 241-4.
[http://dx.doi.org/10.1016/j.matlet.2013.03.018]

[87]   Yan D, Zhang H, Chen L, *et al.* Supercapacitive properties of Mn3O4 nanoparticles bio-synthesized from banana peel extract. RSC Advances 2014; 4(45): 23649-52.
[http://dx.doi.org/10.1039/c4ra02603a]

[88]   Heim KE, Tagliaferro AR, Bobilya DJ. Flavonoid antioxidants: chemistry, metabolism and structure-activity relationships. J Nutr Biochem 2002; 13(10): 572-84.
[http://dx.doi.org/10.1016/S0955-2863(02)00208-5] [PMID: 12550068]

[89]   Seabra A, Durán N. Nanotoxicology of metal oxide nanoparticles. Metals (Basel) 2015; 5(2): 934-75.
[http://dx.doi.org/10.3390/met5020934]

[90]   Armendariz V, Herrera I, peralta-videa JR, *et al.* Size controlled gold nanoparticle formation by Avena sativa biomass: use of plants in nanobiotechnology. J Nanopart Res 2004; 6(4): 377-82.
[http://dx.doi.org/10.1007/s11051-004-0741-4]

[91]   Okitsu K, Sharyo K, Nishimura R. One-pot synthesis of gold nanorods by ultrasonic irradiation: the effect of pH on the shape of the gold nanorods and nanoparticles. Langmuir 2009; 25(14): 7786-90.
[http://dx.doi.org/10.1021/la9017739] [PMID: 19545140]

[92]   Tan NPB, Lee CH, Li P. Influence of temperature on the formation and encapsulation of gold nanoparticles using a temperature-sensitive template. Data Brief 2015; 5: 434-8.
[http://dx.doi.org/10.1016/j.dib.2015.09.035] [PMID: 26594653]

[93]   Sneha K, Sathishkumar M, Kim S, Yun YS. Counter ions and temperature incorporated tailoring of biogenic gold nanoparticles. Process Biochem 2010; 45(9): 1450-8.
[http://dx.doi.org/10.1016/j.procbio.2010.05.019]

[94]   Islam AKMM, Mukherjee M. Effect of temperature in synthesis of silver nanoparticles in triblock copolymer micellar solution. J Exp Nanosci 2011; 6(6): 596-611.
[http://dx.doi.org/10.1080/17458080.2010.506518]

[95]   Iravani S, Zolfaghari B. Green synthesis of silver nanoparticles using Pinuseldarica bark extract 2013.

[96]   Fleitas-Salazar N, Silva-Campa E, Pedroso-Santana S, Tanori J, Pedroza-Montero MR, Riera R. Effect of temperature on the synthesis of silver nanoparticles with polyethylene glycol: new insights into the reduction mechanism. J Nanopart Res 2017; 19(3): 113.
[http://dx.doi.org/10.1007/s11051-017-3780-3]

[97]   Flor J, de Lima SM, Davolos MR. Effect of reaction time on the particle size of ZnO and ZnO: Ce obtained by a sol–gel method. Surface and Colloid Science. Berlin, Heidelberg: Springer 2004; pp. 239-43.

[98]   Ahmad T, Irfan M, Bustam MA, Bhattacharjee S. Effect of reaction time on green synthesis of gold nanoparticles by using aqueous extract of Elaiseguineensis (oil palm leaves). Procedia Eng 2016; 148: 467-72.
[http://dx.doi.org/10.1016/j.proeng.2016.06.465]

[99]   Rose IC, Sathish R, Rajendran AJ, Sagayaraj P. Effect of reaction time on the synthesis of cadmium selenide nanoparticles and the efficiency of solar cell. J Mater Environ Sci 2016; 7: 1589-96.

[100]   Karade VC, Dongale TD, Sahoo SC, *et al.* Effect of reaction time on structural and magnetic properties of green-synthesized magnetic nanoparticles. J Phys Chem Solids 2018; 120: 161-6.
[http://dx.doi.org/10.1016/j.jpcs.2018.04.040]

[101]   Nisar P, Ali N, Rahman L, Ali M, Shinwari ZK. Antimicrobial activities of biologically synthesized metal nanoparticles: An insight into the mechanism of action. J Biol Inorg Chem 2019; 24(7): 929-41.
[http://dx.doi.org/10.1007/s00775-019-01717-7] [PMID: 31515623]

[102]   Yamanaka M, Hara K, Kudo J. Bactericidal actions of a silver ion solution on Escherichia coli, studied by energy-filtering transmission electron microscopy and proteomic analysis. Appl Environ Microbiol 2005; 71(11): 7589-93.
[http://dx.doi.org/10.1128/AEM.71.11.7589-7593.2005] [PMID: 16269810]

[103] Lara HH, Ayala-Núñez NV, Ixtepan Turrent LC, Rodríguez Padilla C. Bactericidal effect of silver nanoparticles against multidrug-resistant bacteria. World J Microbiol Biotechnol 2010; 26(4): 615-21.
[http://dx.doi.org/10.1007/s11274-009-0211-3]

[104] Shahverdi AR, Fakhimi A, Shahverdi HR, Minaian S. Synthesis and effect of silver nanoparticles on the antibacterial activity of different antibiotics against Staphylococcus aureus and Escherichia coli. Nanomedicine 2007; 3(2): 168-71.
[http://dx.doi.org/10.1016/j.nano.2007.02.001] [PMID: 17468052]

[105] Pal S, Tak YK, Song JM. Does the antibacterial activity of silver nanoparticles depend on the shape of the nanoparticle? A study of the Gram-negative bacterium Escherichia coli. Appl Environ Microbiol 2007; 73(6): 1712-20.
[http://dx.doi.org/10.1128/AEM.02218-06] [PMID: 17261510]

[106] Slavin YN, Asnis J, Häfeli UO, Bach H. Metal nanoparticles: understanding the mechanisms behind antibacterial activity. J Nanobiotechnology 2017; 15(1): 65.
[http://dx.doi.org/10.1186/s12951-017-0308-z] [PMID: 28974225]

[107] Liu JL, Luo Z, Bashir S. A progressive approach on inactivation of bacteria using silver–titania nanoparticles. Biomater Sci 2013; 1(2): 194-201.
[http://dx.doi.org/10.1039/C2BM00010E] [PMID: 32481799]

[108] Dakal TC, Kumar A, Majumdar RS, Yadav V. Mechanistic basis of antimicrobial actions of silver nanoparticles. Front Microbiol 2016; 7: 1831.
[http://dx.doi.org/10.3389/fmicb.2016.01831] [PMID: 27899918]

[109] Jones N, Ray B, Ranjit KT, Manna AC. Antibacterial activity of ZnO nanoparticle suspensions on a broad spectrum of microorganisms. FEMS Microbiol Lett 2008; 279(1): 71-6.
[http://dx.doi.org/10.1111/j.1574-6968.2007.01012.x] [PMID: 18081843]

[110] Li M, Zhu L, Lin D. Toxicity of ZnO nanoparticles to Escherichia coli: mechanism and the influence of medium components. Environ Sci Technol 2011; 45(5): 1977-83.
[http://dx.doi.org/10.1021/es102624t] [PMID: 21280647]

[111] Rastogi L, Arunachalam J. Synthesis and characterization of bovine serum albumin–copper nanocomposites for antibacterial applications. Colloids Surf B Biointerfaces 2013; 108: 134-41.
[http://dx.doi.org/10.1016/j.colsurfb.2013.02.031] [PMID: 23531744]

[112] Zhu L, Elguindi J, Rensing C, Ravishankar S. Antimicrobial activity of different copper alloy surfaces against copper resistant and sensitive Salmonella enterica. Food Microbiol 2012; 30(1): 303-10.
[http://dx.doi.org/10.1016/j.fm.2011.12.001] [PMID: 22265316]

[113] Armelao L, Barreca D, Bottaro G, et al. Photocatalytic and antibacterial activity of TiO $_2$ and Au/TiO $_2$ nanosystems. Nanotechnology 2007; 18(37): 375709.
[http://dx.doi.org/10.1088/0957-4484/18/37/375709]

[114] Chauhan I, Mohanty P. In situ decoration of TiO2 nanoparticles on the surface of cellulose fibers and study of their photocatalytic and antibacterial activities. Cellulose 2015; 22(1): 507-19.
[http://dx.doi.org/10.1007/s10570-014-0480-3]

[115] Gu H, Ho PL, Tong E, Wang L, Xu B. Presenting vancomycin on nanoparticles to enhance antimicrobial activities. Nano Lett 2003; 3(9): 1261-3.
[http://dx.doi.org/10.1021/nl034396z]

[116] Li X, Robinson SM, Gupta A, et al. Functional gold nanoparticles as potent antimicrobial agents against multi-drug-resistant bacteria. ACS Nano 2014; 8(10): 10682-6.
[http://dx.doi.org/10.1021/nn5042625] [PMID: 25232643]

[117] Yang N, WeiHong L, Hao L. Biosynthesis of Au nanoparticles using agricultural waste mango peel extract and its in vitro cytotoxic effect on two normal cells. Mater Lett 2014; 134: 67-70.
[http://dx.doi.org/10.1016/j.matlet.2014.07.025]

[118]   Abdel-Raouf N, Al-Enazi NM, Ibraheem IBM. Green biosynthesis of gold nanoparticles using Galaxaura elongata and characterization of their antibacterial activity. Arab J Chem 2017; 10: S3029-39.
[http://dx.doi.org/10.1016/j.arabjc.2013.11.044]

[119]   Rajan A, Vilas V, Philip D. Studies on catalytic, antioxidant, antibacterial and anticancer activities of biogenic gold nanoparticles. J Mol Liq 2015; 212: 331-9.
[http://dx.doi.org/10.1016/j.molliq.2015.09.013]

[120]   Gopinath PM, Narchonai G, Dhanasekaran D, Ranjani A, Thajuddin N. Mycosynthesis, characterization and antibacterial properties of AgNPs against multidrug resistant (MDR) bacterial pathogens of female infertility cases. Asian Journal of Pharmaceutical Sciences 2015; 10(2): 138-45.
[http://dx.doi.org/10.1016/j.ajps.2014.08.007]

[121]   Jinu U, Jayalakshmi N. SujimaAnbu, A.; Mahendran, D.; Sahi, S.; Venkatachalam, P. Biofabrication of Cubic Phase Silver Nanoparticles Loaded with Phytochemicals from Solanumnigrum Leaf Extracts for Potential Antibacterial, Antibiofilm and Antioxidant Activities Against MDR Human Pathogens. J Cluster Sci 2017; 28: 489-505.
[http://dx.doi.org/10.1007/s10876-016-1125-5]

[122]   Leid JG, Ditto AJ, Knapp A, *et al. In vitro* antimicrobial studies of silver carbene complexes: Activity of free and nanoparticle carbene formulations against clinical isolates of pathogenic bacteria. J Antimicrob Chemother 2012; 67(1): 138-48.
[http://dx.doi.org/10.1093/jac/dkr408] [PMID: 21972270]

[123]   Kim SH, Lee HS, Ryu DS, Choi SJ, Lee DS. Antibacterial activity of silver-nanoparticles against Staphylococcus aureus and Escherichia coli. Microbiology and Biotechnology Letters 2011; 39(1): 77-85.

[124]   Naseem T, Farrukh MA. Antibacterial activity of green synthesis of iron nanoparticles using Lawsoniainermis and Gardenia jasminoides leaves extract. J Chem 2015; 2015.

[125]   Alam T, Khan RAA, Ali A, Sher H, Ullah Z, Ali M. Biogenic synthesis of iron oxide nanoparticles *via* Skimmia laureola and their antibacterial efficacy against bacterial wilt pathogen Ralstonia solanacearum. Mater Sci Eng C 2019; 98: 101-8.
[http://dx.doi.org/10.1016/j.msec.2018.12.117] [PMID: 30812984]

[126]   Khalil AT, Ovais M, Ullah I, Ali M, Shinwari ZK, Maaza M. Biosynthesis of iron oxide (Fe$_2$O$_3$) nanoparticles *via* aqueous extracts of *Sageretia thea* (Osbeck.) and their pharmacognostic properties. Green Chem Lett Rev 2017; 10(4): 186-201.
[http://dx.doi.org/10.1080/17518253.2017.1339831]

[127]   Jagathesan G, Rajiv P. Biosynthesis and characterization of iron oxide nanoparticles using Eichhornia crassipes leaf extract and assessing their antibacterial activity. Biocatal Agric Biotechnol 2018; 13: 90-4.
[http://dx.doi.org/10.1016/j.bcab.2017.11.014]

[128]   Hassan D, Khalil AT, Saleem J, *et al.* Biosynthesis of pure hematite phase magnetic iron oxide nanoparticles using floral extracts of Callistemon viminalis (bottlebrush): their physical properties and novel biological applications. Artificial Cells, Nanomedicine, and Biotechnology 2018; 46(sup1): 693-707.

[129]   Talebian N, Amininezhad SM, Doudi M. Controllable synthesis of ZnO nanoparticles and their morphology-dependent antibacterial and optical properties. J Photochem Photobiol B 2013; 120: 66-73.
[http://dx.doi.org/10.1016/j.jphotobiol.2013.01.004] [PMID: 23428888]

[130]   Sirelkhatim A, Mahmud S, Seeni A, *et al.* Review on zinc oxide nanoparticles: Antibacterial activity and toxicity mechanism. Nano-Micro Lett 2015; 7(3): 219-42.
[http://dx.doi.org/10.1007/s40820-015-0040-x] [PMID: 30464967]

[131] Lingaraju K, Raja Naika H, Manjunath K, *et al*. Biogenic synthesis of zinc oxide nanoparticles using Ruta graveolens (L.) and their antibacterial and antioxidant activities. Appl Nanosci 2016; 6(5): 703-10.
[http://dx.doi.org/10.1007/s13204-015-0487-6]

[132] El-Batal AI, El-Sayyad GS, El-Ghamery A, Gobara M. Response Surface Methodology Optimization of Melanin Production by Streptomyces cyaneus and Synthesis of Copper Oxide Nanoparticles Using Gamma Radiation. J Cluster Sci 2017; 28(3): 1083-112.
[http://dx.doi.org/10.1007/s10876-016-1101-0]

[133] DeAlba-Montero I, Guajardo-Pacheco J, Morales-Sánchez E, *et al*. Antimicrobial Properties of Copper Nanoparticles and Amino Acid Chelated Copper Nanoparticles Produced by Using a Soya Extract. Bioinorg Chem Appl 2017; 2017: 1-6.
[http://dx.doi.org/10.1155/2017/1064918] [PMID: 28286459]

[134] Foster HA, Ditta IB, Varghese S, Steele A. Photocatalytic disinfection using titanium dioxide: spectrum and mechanism of antimicrobial activity. Appl Microbiol Biotechnol 2011; 90(6): 1847-68.
[http://dx.doi.org/10.1007/s00253-011-3213-7] [PMID: 21523480]

[135] Li Y, Zhang W, Niu J, Chen Y. Mechanism of photogenerated reactive oxygen species and correlation with the antibacterial properties of engineered metal-oxide nanoparticles. ACS Nano 2012; 6(6): 5164-73.
[http://dx.doi.org/10.1021/nn300934k] [PMID: 22587225]

[136] Roy AS, Parveen A, Koppalkar AR, Prasad MA. Effect of nano-titanium dioxide with different antibiotics against methicillin-resistant Staphylococcus aureus. J Biomater Nanobiotechnol 2010; 1: 37.
[http://dx.doi.org/10.4236/jbnb.2010.11005]

[137] Santhoshkumar T, Rahuman AA, Jayaseelan C, *et al*. Green synthesis of titanium dioxide nanoparticles using Psidium guajava extract and its antibacterial and antioxidant properties. Asian Pac J Trop Med 2014; 7(12): 968-76.
[http://dx.doi.org/10.1016/S1995-7645(14)60171-1] [PMID: 25479626]

[138] Hsu S, Liu , Dai S, Fu . Antibacterial properties of silver nanoparticles in three different sizes and their nanocomposites with a new waterborne polyurethane. Int J Nanomedicine 2010; 5: 1017-28.
[http://dx.doi.org/10.2147/IJN.S14572] [PMID: 21187943]

[139] Adams LK, Lyon DY, Alvarez PJJ. Comparative eco-toxicity of nanoscale TiO2, SiO2, and ZnO water suspensions. Water Res 2006; 40(19): 3527-32.
[http://dx.doi.org/10.1016/j.watres.2006.08.004] [PMID: 17011015]

[140] Huang Z, Zheng X, Yan D, *et al*. Toxicological effect of ZnO nanoparticles based on bacteria. Langmuir 2008; 24(8): 4140-4.
[http://dx.doi.org/10.1021/la7035949] [PMID: 18341364]

[141] Kim JS, Kuk E, Yu KN, *et al*. Antimicrobial effects of silver nanoparticles. Nanomedicine 2007; 3(1): 95-101.
[http://dx.doi.org/10.1016/j.nano.2006.12.001] [PMID: 17379174]

[142] Senarathna ULNH, Fernando SSN, Gunasekara TDCP, *et al*. Enhanced antibacterial activity of TiO$_2$ nanoparticle surface modified with *Garcinia zeylanica* extract. Chem Cent J 2017; 11(1): 7.
[http://dx.doi.org/10.1186/s13065-017-0236-x] [PMID: 28123449]

[143] Lipovsky A, Gedanken A, Nitzan Y, Lubart R. Enhanced inactivation of bacteria by metal-oxide nanoparticles combined with visible light irradiation. Lasers Surg Med 2011; 43(3): 236-40.
[http://dx.doi.org/10.1002/lsm.21033] [PMID: 21412807]

[144] Qais FA, Shafiq A, Khan HM, *et al*. Antibacterial effect of silver nanoparticles synthesized using Murrayakoenigii (L.) against multidrug-resistant pathogens. Bioinorganic chemistry and applications. 2019 Oct 2019.

CHAPTER 8

2D Molybdenum Disulfide (MoS₂) Nanosheets: An Emerging Antibacterial Agent

Praveen Kumar[1] and **Amit Jaiswal**[1,*]

[1] School of Basic Sciences, Indian Institute of Technology Mandi, Kamand, Mandi-175005, Himachal Pradesh, India

Abstract: The development of resistance against antibiotics in microorganisms has led to the search for alternatives that can effectively kill microbes and will have a lesser probability of the generation of resistance. In this regard, nanomaterials have emerged as protagonists demonstrating efficient antibacterial activities against drug-resistant strains. Amongst nanomaterials, 2D nanosheets have attracted attention as an antibacterial agent due to their sheet-like features, having sharp edges and corners which can pierce through bacterial membranes, subsequently leading to membrane damage. The present chapter discusses the antibacterial potential of one such 2D material, transition metal dichalcogenides, specifically MoS$_2$ nanosheets and their composites. A brief discussion about the synthesis of MoS$_2$ nanosheets is presented, and a detailed overview of its application as an antibacterial agent is illustrated. The mechanism of action of antibacterial activity of 2D MoS$_2$ nanosheets is discussed, which shows that these nanosheets can cause bacterial cell death through membrane damage and depolarization, metabolic inactivation and generation of reactive oxygen species (ROS). Further, the photothermal property and the intrinsic peroxidase-like activity in certain conditions can also show antibacterial activity, which is summarized in the chapter along with the biocompatibility evaluation.

Keywords: 2D Nanomaterials, Antibacterial, MoS$_2$, photothermal, Nanomaterials, Nanosheets.

INTRODUCTION

The development of widespread antibiotic-resistant bacteria has proved to be a global health burden for antibiotics currently available in the market [1]. The IDSA has classified some of the bacterial strains as the most dangerous pathogens owing to their fast growth of antibiotic resistance [2, 3] ESKAPE [4] pathogens, which include *Enterococcus faecium, Staphylococcus aureus, Klebsiella pneumoniae, Acinetobacter baumannii, Pseudomonas aeruginosa,* and

---

* **Corresponding author Amit Jaiswal:** School of Basic Sciences, Indian Institute of Technology Mandi, Kamand, Mandi-175005, Himachal Pradesh, India; E-mail:j.amit@iitmandi.ac.in

Tilak Saha, Manab Deb Adhikari and Bipransh Kumar Tiwary (Eds.)

Enterobacter species and comprises both Gram-negative and Gram- positive pathogens. Gram-positive bacteria can successfully escape the bactericidal activity of maximum traditional antibiotics due to the presence of an extra lipopolysaccharide layer which acts as an additional barrier for antibiotic interactions or modification of the target site. Further, the presence of powerful efflux pumps in Gram-negative bacteria that easily pump out drugs from the cytoplasm (Fig. **1**) puts an additional burden on antibiotic therapy [5, 6]. Since ESKAPE pathogens may cause life-threatening infections, therefore,to overcome the problem of antibiotic resistance, it is important to create an alternative therapeutic approach. In recent years, small-molecule-based antibiotics, such as daptomycin, fidaxomicin, and retapamulin, were confirmed to have resistant bacterial species [7]. As a result, researchers are looking for novel antibacterial agents that can destroy pathogenic strains selectively and effectively while causing no negative side effects to the host [8].

Several nanomaterials have emerged as potential antibacterial agent that has shown strong antibacterial action towards ESKAPE pathogens without developing resistance andare now considered a potential therapeutic option for treating drug-resistant infections. Silver nanoparticles are by far the most investigated and used antibacterial agents in clothing and food packaging, water disinfection, wound healing, and antibacterial coatings [9 - 15]. The toxicity of silver nanoparticles to human or hosts have raised concern about their transition to antibacterial therapy [16]. Further, for photo-thermal inactivation of bacteria, the conventional gold nanoparticles with near-infrared responsive photo-thermal activity have been investigated. Besides metal nanoparticles, the class of 2-D (two-dimensional) nanomaterials have gained a lot of recognition as possible antibacterial agents in the last decade. Amongst 2D materials, graphene has been widely investigated for antibacterial applications [17]. Recently, 2D material, transition metal dichalcogenides (TMDs), has arisen as a potential candidate for antibacterial agents. TMDs are X-M-X sandwich layered materials in which chalcogen atoms are linked to metals through covalent bonds and the layers are attached by weak van der Waals forces [18]. Molybdenum disulfide (MoS_2) is an S-Mo-S sandwich layered material belonging to the TMDs family that has drawn increasing interest as a 2D layered nanomaterial due to its intriguing properties, such as excellent mechanical properties arising from the in-plane stiffness of monolayer [19, 20], having suitable band gap for light harvesting thereby exhibiting strong photo-catalytic and photo-thermal properties [21 - 24], and potential antibacterial activity against bacterial pathogens. The main mechanism behind the MoS_2 killing or inhibiting microorganisms is oxidative stress and inducing contact-mediated membrane damage. It is important to use an antimicrobial agent that possesses biocompatibility along with the efficient antimicrobial activity. In this regard,

MoS$_2$ has been reported to show low cytotoxicity and genotoxicity [25 - 27], making it a suitable antibacterial nanomaterial.

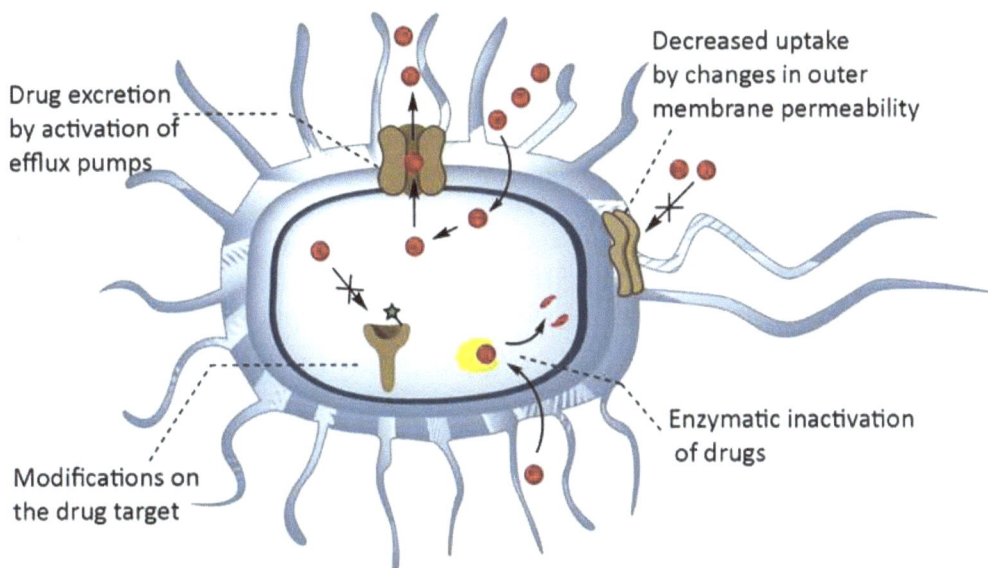

Fig. (1). Schematic illustration of the bacterial mechanisms of antibiotic resistance With permission, this image has been reproduced [28]. Copyright 2017, Elsevier.

Methods of Preparation of MoS$_2$ Nanomaterials

MoS$_2$ nanomaterial exists in various forms categorized as 0D, 1D, and 2D, and for the synthesis of these nanocomposites, a range of approaches have been developed. Bottom-up and Top-down are two widely used strategies for nanomaterial synthesis. The top-down method involves the breakdown of bulk material into its compositional components. This process involves the exfoliation of MoS$_2$ multi-layered bulk into mono or few layers by applying physical/mechanical force or intercalation and exfoliation by solvents and chemicals. Solvent-assisted exfoliation, mechanical exfoliation, thermal decomposition, electrochemical exfoliation and chemical-assisted exfoliation are a few examples of top-down approaches which are presently used for the synthesis of MoS$_2$ nanomaterials [29, 30]. Physical processes such as mechanical exfoliation and laser thinning, as shown in Figs. (**2a** & **2b**), generally use micromechanical force to synthesize MoS$_2$ monolayer from bulk [29]. The mechanism of electrochemical exfoliation of MoS$_2$ is represented in Fig. (**2c**). The first step in this process is the oxidation of water. The resulting formation of ·OH and ·O radicals around the bulk MoS$_2$ layer then lead to the expansion of the interlayers resulting in the exfoliation of MoS$_2$ nanosheets [31]. The bottom-up approach, on the other hand, includes assembling compositional components to form a dynamic alternative. Chemical vapor deposition (CVD), hydro thermal/solvothermal reaction, thermal anisotropy (sputtering), and physical vapor deposition are the processes used in this method, where synthesis of nanosheets directly takes place from precursors [32 - 35]. All the above methods have their own benefits and drawbacks, and the decision of preparation is primarily based on the choice of application for which it is sought for. For biomedical applications, the synthesis method, which is safe, non-toxic, and easy to manufacture, is generally selected. There have been major developments in synthesis techniques that use a range of, processes, such as chemical-assisted exfoliation and intercalation, liquid, or solvent-assisted exfoliation and hydrothermal/ solvothermal preparation methods [36], which provides better biocompatibility and sizes of MoS_2 nanostructures for use in biomedical applications [37 - 40].

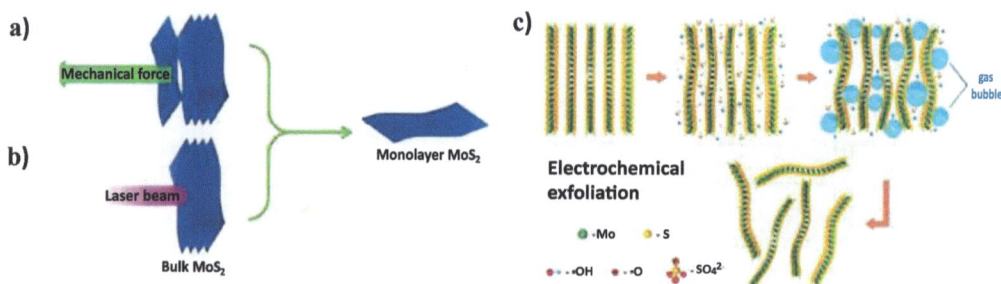

Fig. (2). Schematic of the synthesis procedure of MoS_2nanosheets a) Mechanical force-based exfoliation and b) laser beam thinning. With permission, this image has been reproduced [29]. Copyright 2015, Royal Society of Chemistry. c) Electrochemical exfoliation of MoS_2nanosheets [31]. With permission, this image has been reproduced copyright 2014, American Chemical Society.

MoS_2-Based Materials and their Antibacterial Activity

MoS_2 nanosheets possess sharp edges and defects, which leads to excellent antimicrobial activity that is similar or better to that of commonly used nano-agents like metallic nanoparticles [41]. Yang *et al.* reported that MoS_2 nanosheets have better antibacterial activity than bulk MoS_2 due to the sharp edges of MoS_2 nanosheets and act as nanoknives or nanoscissors. These sharp edges and corners damage the bacterial membrane by puncturing the lipid bilayers, extracting phospholipids, and rupturing the cells [42, 43]. Zang *et al.* found that MoS_2 nanosheet has antibacterial effects, which are attributed to the generation of superoxide anion, which causes the formation of reactive oxygen species (ROS) generation, such as $\cdot OH$, H_2O_2, $O_2^{\cdot-}$ and $^1O_2$, leading to both membrane damage and oxidation tension [44]. In another study, Liu *et al.* coated the few-layered vertically MoS$_2$on copper film, which resulted in a six-fold increase in sterilization efficiency in the presence of visible light [45].

Bacterial membrane and exfoliated MoS$_2$ nanosheets have surface negative charge [46], because of which they repel each other [27]. To improve the interaction between bacterial membrane and the MoS$_2$ nanosheet thus needs surface functionalization. Functionalization of MoS$_2$ nanosheet increases the antibacterial activity against bacterial species [26, 27, 47]. Several reports demonstrated that the functionalization of MoS$_2$ nanosheet with antimicrobial peptides [48], antibiotics [45], or hydrophobic ligands [27, 49] increases the water disinfection and wound healing properties. Coupled action of peroxidase-like activity and near-infrared stimuli responsiveness of MoS$_2$ nanosheets was used for the development of stimulus-responsive photo-catalytic and photo-thermal disinfection systems [44, 50 - 52]. The functionalization with thiolated ligands of different hydrophobicity and charge showed excellent antibacterial action towards ESKAPE pathogens [27, 49]. The thiolated ligand functionalized MoS$_2$ nanosheets inhibit the beta-lactamase enzyme of bacterial cells and terminate the chance of development of resistance towards modified MoS$_2$ [53]. ce-MoS$_2$nanosheets (Chemically exfoliated) modified by chitosan and loaded with antibiotics have been shown to be very effective in inhibiting the growth of *S. aureus* and its biofilm formation [48]. The study stated that there was no contribution of chitosan towards antibacterial action. The antibacterial activity was achieved due to the trapping of bacteria by the MoS$_2$ nanosheets, and then antibiotics being released into the trapped pathogen's environment. Feng *et al.* investigated the photo-thermal and photo-dynamic behavior of chitosan functionalized MoS$_2$ nanosheets coated upon titanium surface against both gram-positive and gram-negative bacterial cells using visible and near-infrared radiation as shown in Fig. (**3i**). The functionalized surface exhibited excellent antibacterial effectiveness of 99.84% and 99.65% against *E. coli* and *S. aureus* in the presence of 660 nm visible light for 10 sec. Malondialdehyde (MDA) content of both Gram-positive and negative of different samples was shown in Fig. (**3ii**) .The higher concentration of MDA content found in the dual lights group CS@MoS$_2$-Ti + 660 nm VL and CS@MoS$_2$-Ti + indicate that the bacterial cell membrane lipids are significantly oxidized under these two models [51]. Similarly, Yin *et al.* functionalized MoS$_2$ with peroxidase and checked its photothermal activity for wound healing (Fig. **3iii**), where they investigated the healing ability of functionalized MoS$_2$ in wound-infected bacterial skin of mice. MoS$_2$ + H$_2$O$_2$+ NIR was able to decrease the CFU count as compared to other samples [50]. Analogously, the hybrid MoS$_2$ nanomaterial with graphene oxide, Ti$_3$C$_2$MXene, and other nanomaterials has been explored to create increasingly powerful antimicrobial nanocomposite [54]. The better antimicrobial effect of these hybrid MoS$_2$ nanocomposites than that of the single-component MoS$_2$ nanosheets may be attributed to an improvement of the separation of photogenerated electrons and holes [55]. Awasthi *et al.* reported the synthesis of MoS_2/ZnO nanocomposites coated with flower-like ZnO on the MoS_2nanosheets surface. They found that MoS_2/ZnO nanocomposites show higher antibacterial activity in comparison with pristine MoS_2 and ZnO [56]. In Fig. (**3iv**), MoS_2@polydopamine-Ag nanosheets (MPA NSs) are able to increase the antibacterial activity towards biofilms and wound infections. By applying NIR laser, the temperature of modified MoS_2 rapidly increases to 50 degrees celsius within 2 min. The wound area of mice treated with MPA NSs under NIR laser promotes faster wound healing as compared to control. Other MoS_2-based hybrid nanocomposites have also exhibited similar behavior [57, 58]. Using a bath sonication process, Meng *et al.* found that a MoS_2 QDs-interspersed Bi_2WO_6 heterostructure exhibits synergetic effects on photocatalytic disinfection [59]. The photothermal property of the MoS_2 nanosheets was also exploited for the development of reusable antibacterial fabric, which showed the excellent antibacterial property. The fabricated fabric was also demonstrated to be used as an additional layer in face mask for better protection against microbial agents [60]. The antibacterial activity of MoS_2 nanosheets are summarized in Table **1**.

Table 1. Antibacterial activity of MoS_2-based nanomaterials.

Strains of microorganism	MoS_2/MoS_2 based nanomaterials	Methods used for synthesis	Concentration of MoS_2-based materials	Incubation duration in hrs	Observation methods	Inhibition in %	References
E. coli	MoS_2	Chemical exfoliation	80μg/ml	2	Cell *viability* assay	91.8	[43]
S aureus and *P aeruginosa*	MoS_2	Chemical exfoliation	-	-	Growth kinetics	-	[27]
E. coli	MoS_2	-	1mg/ml	12	Growth kinetics	99	[62]
E. coli	MoS_2	Chemical exfoliation	100 ppm	3	Plate count	92	[63]
S aureus and *E. coli*	Chitosan -MoS_2	Ultrasonication	60μg/ml	6	Growth kinetics	99 & 98	[26]
S aureus and *E. coli*	Chitosan -MoS_2	Ultrasonication	10μg/ml	5	Plate count	100 & 98.1	[64]
S aureus and *E. coli*	Chitosan -MoS_2	Hydrothermal	100μg/ml	24	Inhibition zone	-	[65]
E. coli and B. subtilis	PEG- MoS_2	Hydrothermal	1 mg/ml	12	Plate count	97 & 100	[52]
S aureus and *E. coli*	Chitosan magnetic MoS_2	Hydrothermal	0.2 mg/ml	1.5 min	Growth kinetics	100	[48]
S aureus and *E. coli*	PDDA-Ag^+-Cys-MoS_2	Solvo-thermal	30μg/ml	30 min		100	[66]
S aureus	MoS_2-PDA-Ag	Microwave irradiation	20 mg/ml	6	Plate count	86.31	[67]
S aureus and *E. coli*	Chitosan -MoS_2-Ti	Electrophoretic deposition	-	10 min	Plate count	99.8 & 99.6	[53]

(Table 1) cont.....

Strains of microorganism	MoS₂/MoS₂ based nanomaterials	Methods used for synthesis	Concentration of MoS₂-based materials	Incubation duration in hrs	Observation methods	Inhibition in %	References
S aureus, E. faecalis and *E. coli*	MoS₂□BNN6	Grinding	250μg/ml	10 min	Plate count	97.2	[47]
S aureus and *E. coli*	MoS₂	Chemical exfoliation	6mg/ml	24	Plate count	99.9	[68]
S aureus and *E. coli*	MoS₂, MoS₂/rGO, and MoS₂/MXene	Chemical exfoliation	100μg/ml	3	Flow cytometry	-	[54]
S aureus	MoS₂@PDA-Ag	Microwave irradiation	125μg/ml	5 min	Plate count	99	[63]
S aureus and *E. coli*	MoS2-ZnO-rGO	Hydrothermal	100μg/ml	12	Inhibition zone	100	[69]
S aureus, E. coli and *K pneumoniae*	PEG-MoS₂-AMP	Solvothermal	-	24	Growth kinetics	100	[49]
E. coli	GO-MoS₂	Hummer's method and Chemical exfoliation	-	6	Plate count	96.4	[70]
S aureus	MoS₂	Lithium-intercalation	-	16	Growth kinetics	-	[55]
S aureus and *E. coli*	MoS₂/polydopamine - arginine-glycine-aspartic acid	Hydrothermal	-	10 min	Plate count	94.6	[71]
S aureus and *E. coli*	AgBr@MoS₂-Ti	Hydrothermal	-	20 min	Plate count	99.6 & 99.5	[72]
S aureus	MoS₂ @PDA-PEG/Ig	Lithium intercalation	160μg/ml	6	Growth kinetics	99.99	[73]
S aureus and *E. coli*	Chitosan /Ag/MoS₂-Ti	Hydrothermal	-	20 min	Plate count	98.6 & 99.7	[74]
S aureus	Bi-doped MoS₂	Chemical exfoliation	1.0 mg/50 μl/ml	12	Inhibition zone	-	[75]
S aureus	Ag–Cu/MoS₂	Hydrothermal	1.0 mg/50 μl/ml	12	Inhibition zone	-	[76]
S aureus and *E. coli*	Zr-doped MoS₂	Chemical exfoliation	1.0 mg/50 μl/ml	12	Inhibition zone	-	[77]
S aureus and E. coli	rGO–MoS₂–Ag	Hydrothermal	100, 200μg/ml	6	Plate count	100	[78]
E. coli and B. subtilis	MoS₂□Lys	Ultrasonication	400μg/ml	12	Plate count	93	[79]
S aureus and *E. coli*	Polyethylenimine modified MoS2	Ultrasonication	80μg/ml	5	Plate count	100	[80]
E. coli	Fe3O4@MoS₂-Ag	Hydrothermal	100μg/ml	1	Plate count	100	[81]
S. aureus and *E. coli*	β-cyclodextrin (β-CD) functionalized MoS2 (β-CD/MoS2) hybrid	Hydrothermal	100μg/ml	12	Inhibition zone	-	[82]
E. coli and B. subtilis	Van-MoS₂–Au	Solvothermal	100 ppm	20 min	Growth kinetics	-	[83]
S aureus and *E. coli*	PEG-MoS₂/rGO-streptomycin sulfate	Hydrothermal	150μg/ml	24	Growth kinetics	-	[84]
S aureus and *Salmonella*	Antibiotic-loaded - chitosan -MoS₂	Ionic liquid-assisted grinding	80 μg/ml	18	Inhibition zone	-	[45]

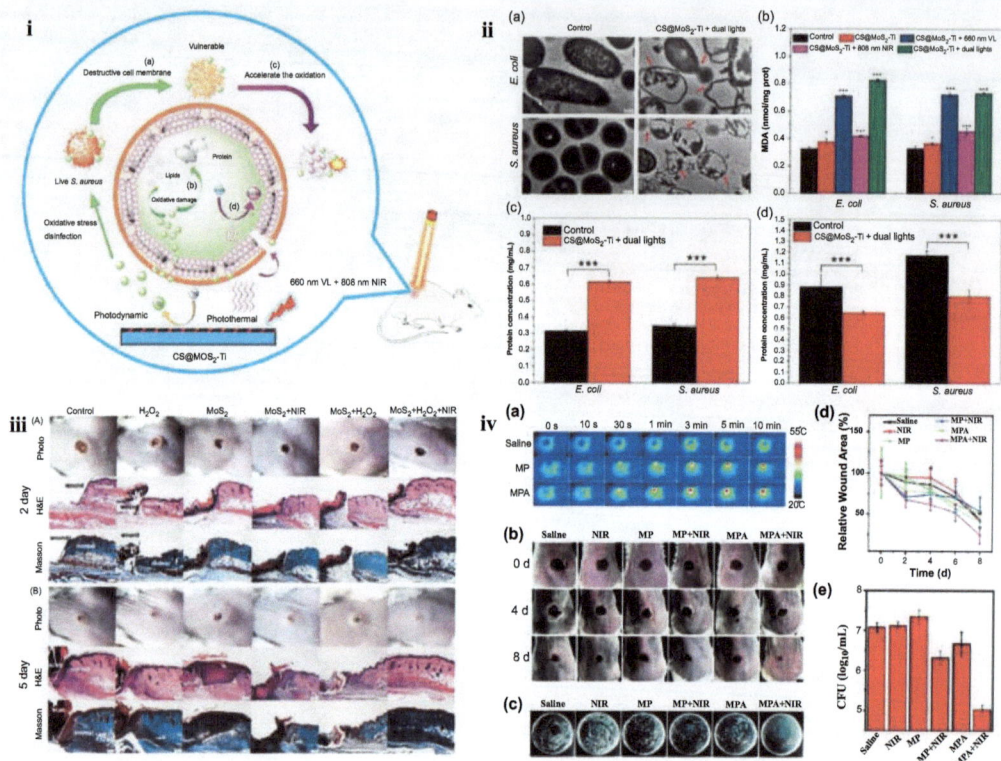

Fig. (3). i) Disinfection schematic *in vivo*, disruptive behavior on the bacterial cell membrane, and inactivating bacteria through photodynamic and photothermal actions of CS@MoS$_2$ hybrid coating on Ti implant under the dual lights (660 nm VL + 808 nm NIR). ii) a) a) Bacterial section images observed by TEM (scale bar: 200 nm). b) The MDA content of E. coli and S. aureus of samples. Protein concentration of E. coli and S. aureus on pure Ti without light (control) and CS@MoS$_2$ + dual lights for 10 min c) before, and d) after cell crushing processing. With permission, this image has been reproduced [51]. Copyright 2018, Wiley online library. iii) Photographs of Ampr E. coli infected wound treated with PBS (control), H$_2$O$_2$, MoS$_2$, MoS$_2$+NIR,MoS$_2$+H$_2$O$_2$, and MoS$_2$+H$_2$O$_2$+NIR at (A) the second day and (B) the fifth day and their corresponding histologic analyses. (Three mice in each group). With permission, this image has been reproduced [50]. Copyright 2016, American Chemical Society. iv) a) Infrared thermal images of S. aureus infected wounds in mice treated with saline (50 µL), MP NSs (MoS$_2$: 200 µg mL^{-1}, 50 µL), and MPA NSs (MoS$_2$: 200 µg mL^{-1} and Ag: 55 µg mL^{-1}, 50 µL) under NIR laser irradiation (0.5 W cm^{-2}) at different times. (b) The photographs of infected wounds in mice at different times after various treatments: saline, NIR, MP NSs, MPA NSs, MP NSs + NIR, and MPA NSs + NIR. (c) The photographs of bacterial colonies formed on LB agar plates from the infected wounds after different treatments. (d) The change of the relative wound area of the mice shown in (b). (e) The numbers of *via*ble S. aureus in infected wounds after different treatments determined by the plate counting method. With permission, this image has been reproduced [61]. Copyright 2018, Royal Society of Chemistry.

Mechanism of Antibacterial Action of MoS$_2$ Nanosheet

The mode of action for the antibacterial activity of MoS$_2$ nanosheets was reported to be a multistep process (Fig. **4**). The first step involves direct physical interaction between nanosheets surface and bacterial membranes. This is the most crucial step, which depends upon two factors (i) size and (ii) orientation of nanosheets. When the size of MoS_2 nanosheets is in the nano range, they can easily penetrate the plasma membrane of the bacterial cells, and this process is termed as edge mediated physical interaction. This involves direct puncture or piercing of bacterial membrane by sharp edges of nanosheets leading to pore formation in the membrane. As the size of nanosheets increases from nano to micro range, sheets start adhering to the bacterial membrane without direct penetration; this mode of interaction is known as surface area mediated physical interaction. In this interaction, nanosheets wrap around the entire bacterial cells and trap them in an isolated environment. The orientation of MoS_2 nanosheets is another factor that involves direct physical interaction. It was found that the vertical alignment of the nanosheets provides a surface with a high density of sharp edges, which interact with the bacterial cell through edge-mediated physical interaction (Fig. **4ii**) [54]. Attachment or adsorption of nanosheets on the membrane surface and its subsequent penetration results in creating membrane stress on the bacterial cells (Fig. **4iv**). The key events involved in membrane stress are the interaction of nanosheets with phospholipid headgroups of the bacterial membrane creating dents or troughs on the bacterial membrane surface. This membrane perturbation leads to the generation of patches of upturned phospholipid on the surface marking the initiation of membrane disruption. The strong interaction between nanosheets and exposed hydrophobic lipid tails results in the destructive extraction of phospholipid molecules from the membrane onto the surface of nanosheets(Fig. **4i**) [62]. The event of phospholipid extraction disturbs the membrane integrity to a large extent causing rapid depolarization of the membrane. The membrane losses its permeability barrier and become more permeable and vulnerable to further damage (Fig. **4ivb**). In the final event of membrane stress, the physical disruption of lipid bilayers takes place in which it becomes easier for the embedded nanosheets to penetrate further into the membrane, ultimately causing the membrane to rupture and disintegrate. Consequently, the membrane directed antibacterial action interrupted the respiratory chain of bacterial cells. Inhibition of respiratory dehydrogenases of the bacterial cells leads to metabolic inactivation by separating respiration from oxidative phosphorylation. Additionally, nanosheets have been found to induce both ROS-dependent and ROS-independent oxidative stress in bacterial cells. This acts as another prominent mode of action in the antibacterial activity of MoS_2 nanosheets. The ROS-generating ability of MoS_2 nanosheets is attributed to the discontinuous crystal planes and defects present on the surface of the nanosheets [85, 86]. Along with exposed edges, these surface defects act as the reactive sites due to the presence of active surface electrons [85], which react with molecular oxygen to form free radicals such as superoxide anions ($O_2\cdot^-$) and eventually generate further ROS *via* disproportionation which subsequently overcomes the bacterial cell antioxidant defense mechanism and leads to oxidative damage of cells [62]. Therefore, inhibition of bacterial respiratory metabolic action mediated by membrane disruption in a combination of oxidative damage was mutually involved in bactericidal action against both Gram-negative and Gram-positive bacteria (Fig. **4iii & iv**) [26].

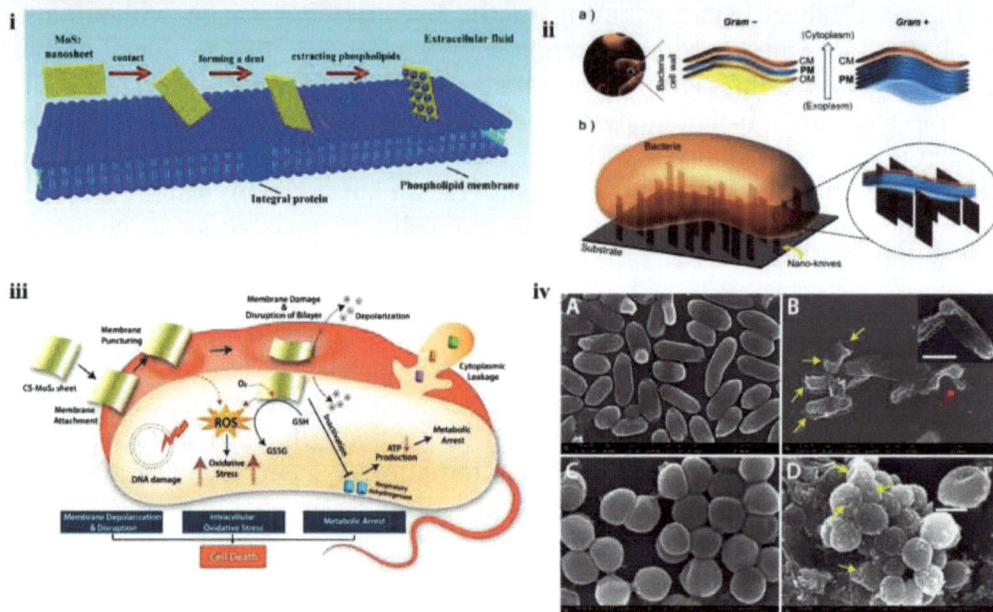

Fig. (4). i) Scheme of the molecular antibacterial mechanism of the MoS_2 nanosheet interacting with the bacterial cell membrane. These yellow blocks represent the three steps during the interaction process. With permission, this image has been reproduced [42]. Copyright 2014, Royal Society of Chemistry. ii). a) General membrane ultrastructure of Gram-negative and Gram-positive bacterial species. Gram-negative bacteria contain a pair of lipoprotein membranes separated by a rigid plasma membrane. Gram-positive bacteria have a significantly thicker plasma membrane and a single lipoprotein membrane. (b) Schematic representation of direct physical interaction of the bacterial surface with sharp edges of vertically aligned nanosheets onto a substrate. With permission, this image has been reproduced [54]. Copyright 2018, American Chemical Society. iii) Schematics represent the proposed mechanism of antibacterial action of MoS_2 Nanosheets. iv) SEM analysis of bacterial membrane damage. Representative SEM images of (A) untreated and (B) CS-MoS_2 treated *E. coli*. Inset shows the extent of membrane damage. Representative SEM images of (C) untreated and (D) CS-MoS_2 treated *S. aureus*. With permission, this image has been reproduced [26]. American Chemical Society.

Cellular Toxicity of the MoS_2- Nanosheets

In order to explore the antibacterial activity of MoS_2-based nanomaterial, it is important to determine its safety and biocompatibility. In this regard, the cytotoxic effects of the MoS_2 nanosheet were examined on mammalian cells by Fan *et al.* [65]. They demonstrated that exposing NIH 3T3 fibroblast cells to 50 and 100 ppm concentrations of MoS_2 nanosheets which exhibited the highest antimicrobial activity, did not show any significant reduction in cell *via*bility (Fig. **5a**) [65].Using an XTT-based cell *via*bility assay, the *via*bility of MCF-7 breast cancer cells and HEK-293A standard kidney cells treated with increasing concentration of CS-MoS$_2$ nanosheets (25-200 g/mL) was evaluated, which showed no significant decrease in the cell *via*bility in the tested concentration range as shown in Fig. (**5b** and **c**) [26]. The cytotoxicity experiment of MoS$_2$-Lysozyme was carried out on human umbilical vein endothelial cells (HUVECs) incubated with various concentrations of MoS$_2$-Lys nanosheets for 24 h. Results showed 100% cell *via*bility below 100 µg/mL [79]. These results clearly indicated that the MoS$_2$ nanosheets were not toxic to mammalian cells and thus can suitably be utilized as an antibacterial agent.

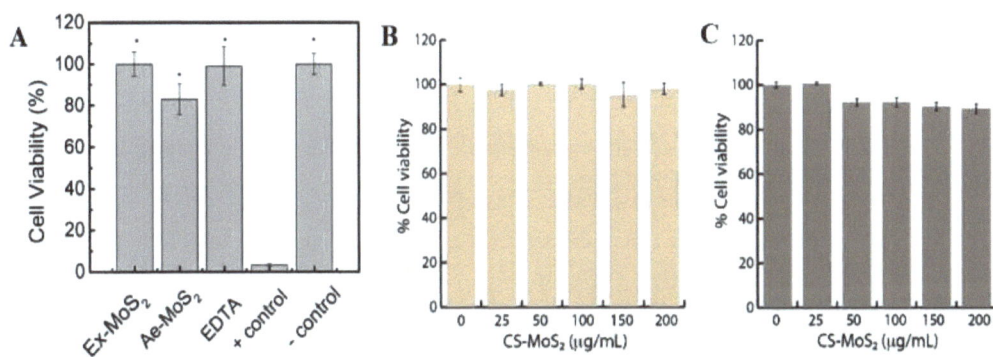

Fig. (5). A) Percent cytotoxicity of NIH 3T3 fibroblast cells after 3 h of exposure to 100 ppm Ex-MoS2, 50 ppm Ae-MoS$_2$, and 40 ppm of EDTA [65]. With permission, this image has been reproduced copyright 2015, Royal Society of Chemistry. Percentage cell *via*bility of **(B)**MCF-7 and **(C)** HEK-293A cells treated with CS-MoS$_2$nanosheets (0–200 µg/mL) for 24 h. With permission, this image has been reproduced [26]. Copyright 2018, American Chemical Society.

CONCLUSION

In this chapter, we provided an overview of the properties of MoS$_2$ nanosheets and their nanocomposites as an alternative to antibiotics for antibacterial action. We have presented the recent development on the antibacterial aspect of MoS$_2$ and how this material has been utilized in the development of a next-generation antibacterial agent for fighting against drug-resistant bacteria. In addition to this, we provide a detailed description of the mechanism involved in the antibacterial action of MoS$_2$.MoS$_2$ nanosheets, due to the presence of sharp edges and corners, are able to damage the bacterial cell through the involvement of membrane depolarization, membrane damage, metabolic inactivation, and oxidative stress generation. The biocompatibility of the MoS$_2$ nanosheets is also discussed briefly to demonstrate that the MoS$_2$ nanosheets did not show any significant toxicity at the concentration range showing efficient antibacterial activity. We also believe that as the membrane damage is the major cause for the antibacterial action of MoS_2, the possibility of generation of resistance towards these nanosheets will be negligible.

CONSENT FOR PUBLICATION

Not applicable.

CONFLICT OF INTEREST

The authors declare no conflict of interest, financial or otherwise.

ACKNOWLEDGEMENT

Declared none.

REFERENCES

[1]   Levy SB, Marshall B. Antibacterial resistance worldwide: causes, challenges and responses. Nat Med 2004; 10(S12) (Suppl.): S122-9.
[http://dx.doi.org/10.1038/nm1145] [PMID: 15577930]

[2]   Ventola CL. The antibiotic resistance crisis: part 1: causes and threats. P&T 2015; 40(4): 277-83.
[PMID: 25859123]

[3]   McKenna M. Antibiotic resistance: The last resort. Nature 2013; 499(7459): 394-6.
[http://dx.doi.org/10.1038/499394a] [PMID: 23887414]

[4]   Zhou Z, Li B, Liu X, *et al.* Recent Progress in Photocatalytic Antibacterial. ACS Appl Bio Mater 2021; 4(5): 3909-36.
[http://dx.doi.org/10.1021/acsabm.0c01335] [PMID: 35006815]

[5]   Nikaido H. Multidrug efflux pumps of gram-negative bacteria. J Bacteriol 1996; 178(20): 5853-9.
[http://dx.doi.org/10.1128/jb.178.20.5853-5859.1996] [PMID: 8830678]

[6]   Walsh CT, Wencewicz TA. Prospects for new antibiotics: A molecule-centered perspective. J Antibiot (Tokyo) 2014; 67(1): 7-22.
[http://dx.doi.org/10.1038/ja.2013.49] [PMID: 23756684]

[7]   Remanan S, Padmavathy N, Rabiya R, *et al.* Converting Polymer Trash into Treasure: An Approach to Prepare MoS_2 Nanosheets Decorated PVDF Sponge for Oil/Water Separation and Antibacterial Applications. Ind Eng Chem Res 2020; 59(45): 20141-54.
[http://dx.doi.org/10.1021/acs.iecr.0c03069]

[8]   Brooks BD, Brooks AE. Therapeutic strategies to combat antibiotic resistance. Adv Drug Deliv Rev 2014; 78: 14-27.
[http://dx.doi.org/10.1016/j.addr.2014.10.027] [PMID: 25450262]

[9]   Echegoyen Y, Nerín C, Toxicology C. Nanoparticle release from nano-silver antimicrobial food containers. Food Chem Toxicol 2013; 62: 16-22.
[http://dx.doi.org/10.1016/j.fct.2013.08.014] [PMID: 23954768]

[10]   von Goetz N, Fabricius L, Glaus R, Weitbrecht V, Günther D, Hungerbühler K. Migration of silver from commercial plastic food containers and implications for consumer exposure assessment. Food Addit Contam Part A Chem Anal Control Expo Risk Assess 2013; 30(3): 612-20.
[http://dx.doi.org/10.1080/19440049.2012.762693] [PMID: 23406534]

[11] Huang Y, Chen S, Bing X, Gao C, Wang T, Yuan B. Nanosilver migrated into food☐simulating solutions from commercially available food fresh containers. Packag Technol Sci 2011; 24(5): 291-7. [http://dx.doi.org/10.1002/pts.938]

[12] Zhang F, Wu X, Chen Y, Lin H. Application of silver nanoparticles to cotton fabric as an antibacterial textile finish. Fibers Polym 2009; 10(4): 496-501. [http://dx.doi.org/10.1007/s12221-009-0496-8]

[13] Tian X, Jiang X, Welch C, *et al.* Bactericidal effects of silver nanoparticles on lactobacilli and the underlying mechanism. ACS Appl Mater Interfaces 2018; 10(10): 8443-50. [http://dx.doi.org/10.1021/acsami.7b17274] [PMID: 29481051]

[14] Dankovich TA, Gray DG. Bactericidal paper impregnated with silver nanoparticles for point-of-use water treatment. Environ Sci Technol 2011; 45(5): 1992-8. [http://dx.doi.org/10.1021/es103302t] [PMID: 21314116]

[15] Kumar SSD, Houreld NN, Kroukamp EM, Abrahamse H. Cellular imaging and bactericidal mechanism of green-synthesized silver nanoparticles against human pathogenic bacteria. J Photochem Photobiol B 2018; 178: 259-69. [http://dx.doi.org/10.1016/j.jphotobiol.2017.11.001] [PMID: 29172133]

[16] Stensberg MC, Wei Q, McLamore ES, Porterfield DM, Wei A, Sepúlveda MS. Toxicological studies on silver nanoparticles: challenges and opportunities in assessment, monitoring and imaging. Nanomedicine (Lond) 2011; 6(5): 879-98. [http://dx.doi.org/10.2217/nnm.11.78] [PMID: 21793678]

[17] Szunerits S, Boukherroub R. Antibacterial activity of graphene-based materials. J Mater Chem B Mater Biol Med 2016; 4(43): 6892-912. [http://dx.doi.org/10.1039/C6TB01647B] [PMID: 32263558]

[18] Li X, Zhu H. Two-dimensional MoS2: Properties, preparation, and applications. Journal of Materiomics 2015; 1(1): 33-44. [http://dx.doi.org/10.1016/j.jmat.2015.03.003]

[19] Castellanos-Gomez A, Poot M, Steele GA, van der Zant HSJ, Agraït N, Rubio-Bollinger G. Elastic properties of freely suspended MoS2 nanosheets. Adv Mater 2012; 24(6): 772-5. [http://dx.doi.org/10.1002/adma.201103965] [PMID: 22231284]

[20] Bertolazzi S, Brivio J, Kis A. Stretching and breaking of ultrathin MoS2. ACS Nano 2011; 5(12): 9703-9. [http://dx.doi.org/10.1021/nn203879f] [PMID: 22087740]

[21] Li Z, Meng X, Zhang Z. Recent development on MoS2-based photocatalysis: A review. J Photochem Photobiol Photochem Rev 2018; 35: 39-55. [http://dx.doi.org/10.1016/j.jphotochemrev.2017.12.002]

[22] Nan F, Li P, Li J, Cai T, Ju S, Fang L. Fang LJTJoPCC. Experimental and theoretical evidence of enhanced visible light photoelectrochemical and photocatalytic properties in MoS2/TiO2 nanohole arrays. J Phys Chem C 2018; 122(27): 15055-62. [http://dx.doi.org/10.1021/acs.jpcc.8b01574]

[23] Abareshi A, Pirlar MA, Houshiar MJMRE. Photothermal property in MoS2 nanoflakes: theoretical and experimental comparison. Mater Res Express 2019; 6(10): 105050.

[24] Zhang Y, Jia G, Wang P, *et al.* Size effect on near infrared photothermal conversion properties of liquid-exfoliated MoS_2 and MoSe 2. Superlattices Microstruct 2017; 105: 22-7. [http://dx.doi.org/10.1016/j.spmi.2016.11.058]

[25] Appel JH, Li DO, Podlevsky JD, *et al.* Low cytotoxicity and genotoxicity of two-dimensional MoS2 and WS2. ACS Biomater Sci Eng 2016; 2(3): 361-7. [http://dx.doi.org/10.1021/acsbiomaterials.5b00467] [PMID: 33429540]

[26]　Roy S, Mondal A, Yadav V, *et al.* Mechanistic insight into the antibacterial activity of chitosan exfoliated MoS2 nanosheets: membrane damage, metabolic inactivation, and oxidative stress. ACS Appl Bio Mater 2019; 2(7): 2738-55.
[http://dx.doi.org/10.1021/acsabm.9b00124] [PMID: 35030809]

[27]　Pandit S, Karunakaran S, Boda SK, Basu B, De M. De MJAam, interfaces. High antibacterial activity of functionalized chemically exfoliated MoS2. ACS Appl Mater Interfaces 2016; 8(46): 31567-73.
[http://dx.doi.org/10.1021/acsami.6b10916] [PMID: 27933975]

[28]　González-Bello C. Antibiotic adjuvants – A strategy to unlock bacterial resistance to antibiotics. Bioorg Med Chem Lett 2017; 27(18): 4221-8.
[http://dx.doi.org/10.1016/j.bmcl.2017.08.027] [PMID: 28827113]

[29]　Zhang W, Zhang P, Su Z, Wei G. Synthesis and sensor applications of MoS $_2$ -based nanocomposites. Nanoscale 2015; 7(44): 18364-78.
[http://dx.doi.org/10.1039/C5NR06121K] [PMID: 26503462]

[30]　Yao Y, Lin Z, Li Z, Song X, Moon KS, Wong C. Large-scale production of two-dimensional nanosheets. J Mater Chem 2012; 22(27): 13494-9.
[http://dx.doi.org/10.1039/c2jm30587a]

[31]　Liu N, Kim P, Kim JH, Ye JH, Kim S, Lee CJ. Large-area atomically thin MoS2 nanosheets prepared using electrochemical exfoliation. ACS Nano 2014; 8(7): 6902-10.
[http://dx.doi.org/10.1021/nn5016242] [PMID: 24937086]

[32]　Li BL, Chen LX, Zou HL, Lei JL, Luo HQ, Li NB. Electrochemically induced Fenton reaction of few-layer MoS $_2$ nanosheets: preparation of luminescent quantum dots *via* a transition of nanoporous morphology. Nanoscale 2014; 6(16): 9831-8.
[http://dx.doi.org/10.1039/C4NR02592J] [PMID: 25027566]

[33]　Muratore C, Hu JJ, Wang B, *et al.* Continuous ultra-thin MoS $_2$ films grown by low-temperature physical vapor deposition. Appl Phys Lett 2014; 104(26): 261604.
[http://dx.doi.org/10.1063/1.4885391]

[34]　Nath M, Govindaraj A, Rao CNR. Simple synthesis of MoS2 and WS2 nanotubes. Adv Mater 2001; 13(4): 283-6.
[http://dx.doi.org/10.1002/1521-4095(200102)13:4<283::AID-ADMA283>3.0.CO;2-H]

[35]　Wu S, Huang C, Aivazian G, Ross JS, Cobden DH, Xu X. Vapor-solid growth of high optical quality MoS $_2$ monolayers with near-unity valley polarization. ACS Nano 2013; 7(3): 2768-72.
[http://dx.doi.org/10.1021/nn4002038] [PMID: 23427810]

[36]　Yadav V, Roy S, Singh P, Khan Z, Jaiswal A II. MoS2□based nanomaterials for therapeutic, bioimaging, and biosensing applications. Small 2019; 15(1): 1803706.
[http://dx.doi.org/10.1002/smll.201803706] [PMID: 30565842]

[37]　Lee YH, Zhang XQ, Zhang W, *et al.* Synthesis of large-area MoS2 atomic layers with chemical vapor deposition. Adv Mater 2012; 24(17): 2320-5.
[http://dx.doi.org/10.1002/adma.201104798] [PMID: 22467187]

[38]　Ye L, Xu H, Zhang D, Chen S. Synthesis of bilayer MoS2 nanosheets by a facile hydrothermal method and their methyl orange adsorption capacity. Mater Res Bull 2014; 55: 221-8.
[http://dx.doi.org/10.1016/j.materresbull.2014.04.025]

[39]　Yu Y, Li C, Liu Y, Su L, Zhang Y, Cao L. Controlled scalable synthesis of uniform, high-quality monolayer and few-layer MoS 2 films. Sci Rep 2013; 3(1): 1866.
[http://dx.doi.org/10.1038/srep01866]

[40]　Song I, Park C, Hong M, Baik J, Shin HJ, Choi HC. Patternable large-scale molybdenium disulfide atomic layers grown by gold-assisted chemical vapor deposition. Angew Chem Int Ed 2014; 53(5): 1266-9.
[http://dx.doi.org/10.1002/anie.201309474] [PMID: 24420501]

[41] Yang Q, Zhang L, Ben A, *et al.* Effects of dispersible MoS$_2$ nanosheets and Nano-silver coexistence on the metabolome of yeast. Chemosphere 2018; 198: 216-25.
[http://dx.doi.org/10.1016/j.chemosphere.2018.01.140] [PMID: 29421733]

[42] Wu R, Ou X, Tian R, *et al.* Membrane destruction and phospholipid extraction by using two-dimensional MoS$_2$ nanosheets. Nanoscale 2018; 10(43): 20162-70.
[http://dx.doi.org/10.1039/C8NR04207A] [PMID: 30259040]

[43] Yang X, Li J, Liang T, *et al.* Antibacterial activity of two-dimensional MoS$_2$ sheets. Nanoscale 2014; 6(17): 10126-33.
[http://dx.doi.org/10.1039/C4NR01965B] [PMID: 25042363]

[44] Zhang W, Shi S, Wang Y, *et al.* Versatile molybdenum disulfide based antibacterial composites for *in vitro* enhanced sterilization and *in vivo* focal infection therapy. Nanoscale 2016; 8(22): 11642-8.
[http://dx.doi.org/10.1039/C6NR01243D] [PMID: 27215899]

[45] Liu C, Kong D, Hsu PC, *et al.* Rapid water disinfection using vertically aligned MoS$_2$ nanofilms and visible light. Nat Nanotechnol 2016; 11(12): 1098-104.
[http://dx.doi.org/10.1038/nnano.2016.138] [PMID: 27525474]

[46] Gupta A, Arunachalam V, Vasudevan S. Vasudevan SJTjopcl. Water dispersible, positively and negatively charged MoS2 nanosheets: surface chemistry and the role of surfactant binding. J Phys Chem Lett 2015; 6(4): 739-44.
[http://dx.doi.org/10.1021/acs.jpclett.5b00158] [PMID: 26262496]

[47] Gao Q, Zhang X, Yin W, *et al.* Functionalized MoS2 nanovehicle with near□infrared laser□mediated nitric oxide release and photothermal activities for advanced bacteria□infected wound therapy. Small 2018; 14(45): 1802290.
[http://dx.doi.org/10.1002/smll.201802290] [PMID: 30307703]

[48] Begum S, Pramanik A, Gates K, Gao Y, Ray PC. Antimicrobial Peptide-Conjugated MoS$_2$-Based Nanoplatform for Multimodal Synergistic Inactivation of Superbugs. ACS Appl Bio Mater 2019; 2(2): 769-76.
[http://dx.doi.org/10.1021/acsabm.8b00632] [PMID: 35016281]

[49] Karunakaran S, Pandit S, Basu B, De M. De MJJotACS. Simultaneous exfoliation and functionalization of 2H-MoS2 by thiolated surfactants: Applications in enhanced antibacterial activity. J Am Chem Soc 2018; 140(39): 12634-44.
[http://dx.doi.org/10.1021/jacs.8b08994] [PMID: 30192533]

[50] Yin W, Yu J, Lv F, *et al.* Functionalized nano-MoS2 with peroxidase catalytic and near-infrared photothermal activities for safe and synergetic wound antibacterial applications. ACS Nano 2016; 10(12): 11000-11.
[http://dx.doi.org/10.1021/acsnano.6b05810] [PMID: 28024334]

[51] Feng Z, Liu X, Tan L, *et al.* Electrophoretic deposited stable chitosan@ MoS2 coating with rapid *in situ* bacteria□killing ability under dual□light irradiation. Small 2018; 14(21): 1704347.
[http://dx.doi.org/10.1002/smll.201704347] [PMID: 29682895]

[52] Begum S, Pramanik A, Davis D, *et al.* 2D and heterostructure nanomaterial based strategies for combating drug-resistant bacteria. ACS Omega 2020; 5(7): 3116-30.
[http://dx.doi.org/10.1021/acsomega.9b03919] [PMID: 32118128]

[53] Ali SR, Pandit S, De M. 2D-MoS2-based β-Lactamase inhibitor for combination therapy against drug-resistant bacteria. ACS Appl Bio Mater 2018; 1(4): 967-74.
[http://dx.doi.org/10.1021/acsabm.8b00105] [PMID: 34996138]

[54] Alimohammadi F, Sharifian Gh M, Attanayake NH, *et al.* Antimicrobial properties of 2D MnO2 and MoS2 nanomaterials vertically aligned on graphene materials and Ti3C2 MXene. Langmuir 2018; 34(24): 7192-200.
[http://dx.doi.org/10.1021/acs.langmuir.8b00262] [PMID: 29782792]

[55]   Xu Z, Lu J, Zheng X, *et al*. A critical review on the applications and potential risks of emerging MoS_2 nanomaterials. J Hazard Mater 2020; 399: 123057.
[http://dx.doi.org/10.1016/j.jhazmat.2020.123057] [PMID: 32521321]

[56]   Awasthi GP, Adhikari SP, Ko S, Kim HJ, Park CH, Kim CS. Facile synthesis of ZnO flowers modified graphene like MoS2 sheets for enhanced visible-light-driven photocatalytic activity and antibacterial properties. J Alloys Compd 2016; 682: 208-15.
[http://dx.doi.org/10.1016/j.jallcom.2016.04.267]

[57]   Tang K, Cao H, Liu X, Chu PK, Eds. Molybdenum disulfide nanosheets vertically coated on titanium for disinfection in the dark. 2019 Chinese Biomaterials Congress & International Symposium on Advanced Biomaterials.

[58]   Pal A, Jana TK, Roy T, *et al*. MoS2☐TiO2 nanocomposite with excellent adsorption performance and high antibacterial activity. ChemistrySelect 2018; 3(1): 81-90.
[http://dx.doi.org/10.1002/slct.201702618]

[59]   Meng X, Li Z, Zeng H, Chen J, Zhang Z. MoS 2 quantum dots-interspersed Bi 2 WO 6 heterostructures for visible light-induced detoxification and disinfection. Appl Catal B 2017; 210: 160-72.
[http://dx.doi.org/10.1016/j.apcatb.2017.02.083]

[60]   Kumar P, Roy S, Sarkar A, Jaiswal A. Reusable MoS 2 -Modified Antibacterial Fabrics with Photothermal Disinfection Properties for Repurposing of Personal Protective Masks. ACS Appl Mater Interfaces 2021; 13(11): 12912-27.
[http://dx.doi.org/10.1021/acsami.1c00083] [PMID: 33715350]

[61]   Yuwen L, Sun Y, Tan G, *et al*. MoS 2 @polydopamine-Ag nanosheets with enhanced antibacterial activity for effective treatment of *Staphylococcus aureus* biofilms and wound infection. Nanoscale 2018; 10(35): 16711-20.
[http://dx.doi.org/10.1039/C8NR04111C] [PMID: 30156245]

[62]   Wu N, Yu Y, Li T, *et al*. Investigating the influence of mos2 nanosheets on e. coli from metabolomics level. PLoS One 2016; 11(12): e0167245.
[http://dx.doi.org/10.1371/journal.pone.0167245] [PMID: 27907068]

[63]   Fan J, Li Y, Nguyen HN, Yao Y, Rodrigues DF. Toxicity of exfoliated-MoS 2 and annealed exfoliated-MoS 2 towards planktonic cells, biofilms, and mammalian cells in the presence of electron donor. Environ Sci Nano 2015; 2(4): 370-9.
[http://dx.doi.org/10.1039/C5EN00031A]

[64]   Cao W, Yue L, Wang Z. High antibacterial activity of chitosan – molybdenum disulfide nanocomposite. Carbohydr Polym 2019; 215: 226-34.
[http://dx.doi.org/10.1016/j.carbpol.2019.03.085] [PMID: 30981349]

[65]   Kasinathan K, Murugesan B, Pandian N, Mahalingam S, Selvaraj B, Marimuthu K. Synthesis of biogenic chitosan-functionalized 2D layered MoS_2 hybrid nanocomposite and its performance in pharmaceutical applications: *In-vitro* antibacterial and anticancer activity. Int J Biol Macromol 2020; 149: 1019-33.
[http://dx.doi.org/10.1016/j.ijbiomac.2020.02.003] [PMID: 32027897]

[66]   Cao F, Ju E, Zhang Y, *et al*. An efficient and benign antimicrobial depot based on silver-infused MoS2. ACS Nano 2017; 11(5): 4651-9.
[http://dx.doi.org/10.1021/acsnano.7b00343] [PMID: 28406604]

[67]   Zeng G, Huang L, Huang Q, *et al*. Rapid synthesis of MoS2-PDA-Ag nanocomposites as heterogeneous catalysts and antimicrobial agents *via* microwave irradiation. Appl Surf Sci 2018; 459: 588-95.
[http://dx.doi.org/10.1016/j.apsusc.2018.07.144]

[68]   Chen T, Zou H, Wu X, *et al*. Nanozymatic antioxidant system based on MoS2 nanosheets. ACS Appl

Mater Interfaces 2018; 10(15): 12453-62.
[http://dx.doi.org/10.1021/acsami.8b01245] [PMID: 29595050]

[69]   Priyadharsan A, Shanavas S, Vasanthakumar V, Balamuralikrishnan B, Anbarasan P. Synthesis and investigation on synergetic effect of rGO-ZnO decorated MoS2 microflowers with enhanced photocatalytic and antibacterial activity. Colloids Surf A Physicochem Eng Asp 2018; 559: 43-53.
[http://dx.doi.org/10.1016/j.colsurfa.2018.09.034]

[70]   Kim TI, Kwon B, Yoon J, *et al.* Antibacterial activities of graphene oxide–molybdenum disulfide nanocomposite films. ACS Appl Mater Interfaces 2017; 9(9): 7908-17.
[http://dx.doi.org/10.1021/acsami.6b12464] [PMID: 28198615]

[71]   Yuan Z, Tao B, He Y, *et al.* Biocompatible MoS_2/PDA-RGD coating on titanium implant with antibacterial property *via* intrinsic ROS-independent oxidative stress and NIR irradiation. Biomaterials 2019; 217: 119290.
[http://dx.doi.org/10.1016/j.biomaterials.2019.119290] [PMID: 31252244]

[72]   Zhu W, Liu X, Tan L, *et al.* AgBr nanoparticles *in situ* growth on 2D MoS2 nanosheets for rapid bacteria-killing and photodisinfection. ACS Appl Mater Interfaces 2019; 11(37): 34364-75.
[http://dx.doi.org/10.1021/acsami.9b12629] [PMID: 31442020]

[73]   Zhang Y, Xiu W, Gan S, *et al.* Antibody-functionalized mos2 nanosheets for targeted photothermal therapy of staphylococcus aureus focal infection. Front Bioeng Biotechnol 2019; 7: 218.
[http://dx.doi.org/10.3389/fbioe.2019.00218] [PMID: 31552242]

[74]   Zhu M, Liu X, Tan L, *et al.* Photo-responsive chitosan/Ag/MoS_2 for rapid bacteria-killing. J Hazard Mater 2020; 383: 121122.
[http://dx.doi.org/10.1016/j.jhazmat.2019.121122] [PMID: 31518801]

[75]   Qumar U, Ikram M, Imran M, *et al.* Synergistic effect of Bi-doped exfoliated MoS_2 nanosheets on their bactericidal and dye degradation potential. Dalton Trans 2020; 49(16): 5362-77.
[http://dx.doi.org/10.1039/D0DT00924E] [PMID: 32255457]

[76]   Ikram M, Abbasi S, Haider A, *et al.* Bimetallic Ag/Cu incorporated into chemically exfoliated MoS_2 nanosheets to enhance its antibacterial potential: *In silico* molecular docking studies. Nanotechnology 2020; 31(27): 275704.
[http://dx.doi.org/10.1088/1361-6528/ab8087] [PMID: 32182604]

[77]   Ikram M, Tabassum R, Qumar U, *et al.* Promising performance of chemically exfoliated Zr-doped MoS_2 nanosheets for catalytic and antibacterial applications. RSC Advances 2020; 10(35): 20559-71.
[http://dx.doi.org/10.1039/D0RA02458A] [PMID: 35517731]

[78]   Li J, Zheng J, Yu Y, Su Z, Zhang L, Chen X. Facile synthesis of rGO–MoS_2–Ag nanocomposites with long-term antimicrobial activities. Nanotechnology 2020; 31(12): 125101.
[http://dx.doi.org/10.1088/1361-6528/ab5ba7] [PMID: 31770730]

[79]   Ma D, Xie C, Wang T, *et al.* Liquid☐Phase Exfoliation and Functionalization of MoS_2 Nanosheets for Effective Antibacterial Application. ChemBioChem 2020; 21(16): 2373-80.
[http://dx.doi.org/10.1002/cbic.202000195] [PMID: 32227558]

[80]   Cao W, Yue L, Khan IM, Wang Z. Polyethylenimine modified MoS2 nanocomposite with high stability and enhanced photothermal antibacterial activity. J Photochem Photobiol Chem 2020; 401: 112762.
[http://dx.doi.org/10.1016/j.jphotochem.2020.112762]

[81]   Wei F, Cui X, Wang Z, Dong C, Li J, Han X. Recoverable peroxidase-like Fe_3O_4@MoS_2-Ag nanozyme with enhanced antibacterial ability. Chem Eng J 2021; 408: 127240.
[http://dx.doi.org/10.1016/j.cej.2020.127240] [PMID: 33052192]

[82]   Kasinathan K, Marimuthu K, Murugesan B, Pandiyan N, Pandi B, Mahalingam S, *et al.* Cyclodextrin functionalized multi-layered MoS2 nanosheets and its biocidal activity against pathogenic bacteria and MCF-7 breast cancer cells: Synthesis, characterization and *in-vitro* biomedical evaluation. J Mol Liq

2020; 114631.

[83]   Cai Y, Wang L, Hu H, Bing W, Tian L, Zhao J. A synergistic antibacterial platform: combining mechanical and photothermal effects based on Van-MoS $_2$ –Au nanocomposites. Nanotechnology 2021; 32(8): 085102.
[http://dx.doi.org/10.1088/1361-6528/abc98e] [PMID: 33176290]

[84]   Zhao X, Chen M, Wang H, *et al.* Synergistic antibacterial activity of streptomycin sulfate loaded PEG-MoS$_2$/rGO nanoflakes assisted with near-infrared. Mater Sci Eng C 2020; 116: 111221.
[http://dx.doi.org/10.1016/j.msec.2020.111221] [PMID: 32806251]

[85]   Muscuso L, Cravanzola S, Cesano F, Scarano D, Zecchina A. Zecchina AJTJoPCC. Optical, vibrational, and structural properties of MoS2 nanoparticles obtained by exfoliation and fragmentation *via* ultrasound cavitation in isopropyl alcohol. J Phys Chem C 2015; 119(7): 3791-801.
[http://dx.doi.org/10.1021/jp511973k]

[86]   Nel A, Xia T, Mädler L, Li N. Toxic potential of materials at the nanolevel. Science 2006; 311(5761): 622-7.
[http://dx.doi.org/10.1126/science.1114397] [PMID: 16456071]

CHAPTER 9

Metallic and Non-Metallic Quantum Dots as Potent Antibacterial Agents

Areeba Khayal¹, Kabirun Ahmed², Amaresh Kumar Sahoo³,* and **Md Palashuddin Sk¹,***

¹ *Department of Chemistry, Aligarh Muslim University, Uttar Pradesh, India*

² *Department of Chemical Sciences, Tezpur University, Assam, India*

³ *Department of Applied Sciences, Indian Institute of Information Technology, Allahabad, Uttar Pradesh 211012, India*

Abstract: The emergence of antibiotic-resistant bacteria poses a critical public health issue worldwide, which demands the development of novel therapeutic agents as *via*ble alternatives to antibiotics. The advent of nanoscience and technology offers the synthesis of several potential anti-microbial agents that are effective against both Gram-positive and Gram-negative bacterial strains. One such nanoscale material that fascinated researchers due to its unique optoelectronic properties is Quantum Dots (QDs). Moreover, these are found to be highly bactericidal, even against resistant bacterial infections. Thus, a significant number of researches have been going on globally to employ QDs as potent bactericidal agents alone or in combination with antibiotics. Studies demonstrated that intracellular uptakes of QDs elevate the level of reactive oxygen species (ROS) inside the cells, which turns-on cascades of intracellular events that cause damage to DNA and proteins. However, the inherent reactive nature of these metallic and semiconductor QDs raises huge concern for translational research as these are found to be cytotoxic and non-biocompatible. Moreover, the human body does not have a proper sequester mechanism to remove these metallic ions from the body, which limits its direct applications. Recent progress in this line of interest has focused on developing non-metallic quantum dots, such as carbon dots (CQDs) and Black Phosphorus quantum dots (BP QDs) which showed less toxicity and immunogenicity suitable for real-life applications. Therefore, in the present chapter, we are going to discuss the recent development of bactericidal QDs and various types of surface functionalization illustrated recently to increase biocompatibility.

---

* **Corresponding authors Md Palashuddin Sk:** Department of Chemistry, Aligarh Muslim University, Uttar Pradesh, India; E-mail: palashuddin.ch@amu.ac.in
Amaresh Kumar Sahoo: Indian Institute of Information Technology, Allahabad, India; E-mail: asahoo@iiita.ac.in

Tilak Saha, Manab Deb Adhikari and Bipransh Kumar Tiwary (Eds.)

Keywords: Antibacterial mode of action, Antibiotic resistant strains, Antibacterial agents, BP, CdTe QDs, CdSe QDs, Combination therapy, CQDs, *E. coli*, MDR, Metallic quantum dots, Non-metallic quantum dots, NPs, Nanocomposite, Photothermal, Photosensitized, Quantum dots, ROS, TiO_2 QDs, ZnO QDs.

INTRODUCTION

Antibiotic-resistant bacterial infections are found to be one of the most alarming threats to mankind as the conventional antibiotics found to be futile against it, which have been employed as useful gold standard medicines for more than fifty years [1, 2]. All parts of the world are now reported cases of resistant bacterial infections; however, the situation is very serious in third-world countries, including India, possibly due to a lack of awareness and poor healthcare systems. The incomplete dose of antibiotics and misuse of antibiotics in humans, as well as animals, lead to the occurrence of multidrug-resistant (MDR) and extensively drug-resistant (XDR) strains, which are not possible to treat by conventional therapeutics. For example, multi-drug-resistant *Mycobacterium tuberculosis* (MDR-TB), vancomycin-resistant *Enterococcus* (VRE) and methicillin-resistant *Staphylococcus aureus* (MRSA) are the most frequent among the reported drug-resistant cases. This causes higher medical costs for treatment, and longer hospital stays for patients of any age. Moreover, antibiotic resistance in bacterial infections also caused co-morbidity as well as increased mortality rate. This is a critical health crisis that has been going on globally because of an increasing number of cases of bacterial infections and the occurrence of drug-resistant strains. Hence, there is an urgent need to develop advanced therapeutic agents and/or strategies to address antibiotic-resistant bacterial infections.

The advent of nanotechnology offers several innovative solutions to address these contemporary issues. Nanoscale materials having the size of a few nanometers (1 nm = 10^{-9} m) provide the scope for easy penetration through the bacterial cell wall and leads to bacterial cell death [3 - 6]. In this regard, metallic nanoparticles (NPs) provide huge promises to kill different bacterial strains. Studies showed that size, shape, and surface chemistry play a vital role in bactericidal activity [5]. One of the most explored nano-scale materials is silver nanoparticles (Ag NPs) and their composites, which have been at the forefront owing to their highly bactericidal activity against a broad range of bacterial strains, more importantly, antibiotic-resistant bacterial strains [5]. Generally, Ag NPs damage bacterial cell walls after interactions with them; however, intracellular uptake leads to the elevation of reactive oxygen species (ROS) that damage various important biomacromolecules, including DNA and enzymes. Metallic NPs may also increase the concentration of intracellular metal ions, which directly affects several cellular functions that ultimately cause bacterial cell death. Apart from the Ag NPs, other metal NPs, such as copper, gold, and platinum, NPs also showed promising outcomes [5, 6]. Additionally, there are reports of the use of metal oxide NPs, such as copper oxide, zinc oxide, titanium oxide, and iron oxide NPs, which are found to be equally potent against bacterial infections [7, 8]. However, in most cases, higher doses of metal and metal oxide NPs are cytotoxic, which limits their direct use as an antibacterial agent for real-life applications. Thus, in the last few decades, researchers all over the world proposed the use of these metal and metal oxide NPs along with conventional antibiotics for combination therapy that offered reasonable success against several resistant bacterial strains due to synergistic bactericidal effects. Moreover, the use of other antibacterial agents such as amphiphilic molecules, cationic polymer (*e.g.,* chitosan), and supramolecules, in tandem with these metal and metal oxide NPs have been explored widely that reduces the effective concentration of the metal ions.

Another nanoscale material that has drawn significant consideration from the scientific community for the last few decades is quantum dots (QDs), a new form of fluorescent nanostructures. QDs are semiconductor nanocrystals (usually 2-10 nm in diameter), which possess size-dependent optical properties. They have been proven effective in several biophotonic and biomedical implementations [3, 4]. The dimension of this nanoscale material is that it shows a "quantum confinement" effect, which causes discretization of the energy levels responsible for size-tunable luminescence. QDs have specific physical and chemical characteristics compared to conventional materials, such as narrow emission bands, stability, and high quantum yield. Therefore, QDs are generally employed in various biomedical applications such as biosensors, cell labelling, drug delivery, photodynamic therapy and bioimaging [3, 4]. In addition, QDs are gaining significant attention due to their antibacterial activity, which suggests that QDs could be used as an antibacterial agent as an alternative to conventional antibiotics [4]. Different types of metallic and semiconductor quantum dots (ZnO, TiO_2, and CdTe) have been reported as potential bactericidal agents in the last decades. However, the cytotoxicity of metallic QDs has been proved in several *in vitro* researches that have restricted their use, particularly in medical applications. Presently, the researcher focuses on developing non-metallic QDs such as carbon dots (CQDs) and phosphorus QDs that have shown less toxicity suitable for biomedical applications. In this chapter, we have demonstrated the recent progress in the synthesis of various kinds of metallic and non-metallic QDs (Fig. **1**). Emphasis has been given to illustrate the recent outcomes and the scope of non-metallic QDs as potential bactericidal agents against both Gram-positive and Gram-negative bacterial strains.

ANTIBACTERIAL MODE OF ACTION OF QUANTUM DOTS

The development of a new class of antibacterial agents hugely relies on the basic understanding of its mode of action that offers the scope of monitoring its properties to get the best possible activity with minimal side effects on humans. Several studies have reported using QDs as bactericidal against Gram-positive and Gram-negative bacteria. However, the molecular mechanism of the bactericidal activity is not clear yet. It has been found that the size, charge, and surface functional groups play a significant role in its antibacterial action. Recent studies established that the smaller the size better the bactericidal activity, likewise, positively charged particles showed better activity as compared to their counterparts [4]. The bactericidal action of QDs, occurs primarily due to the following reason.

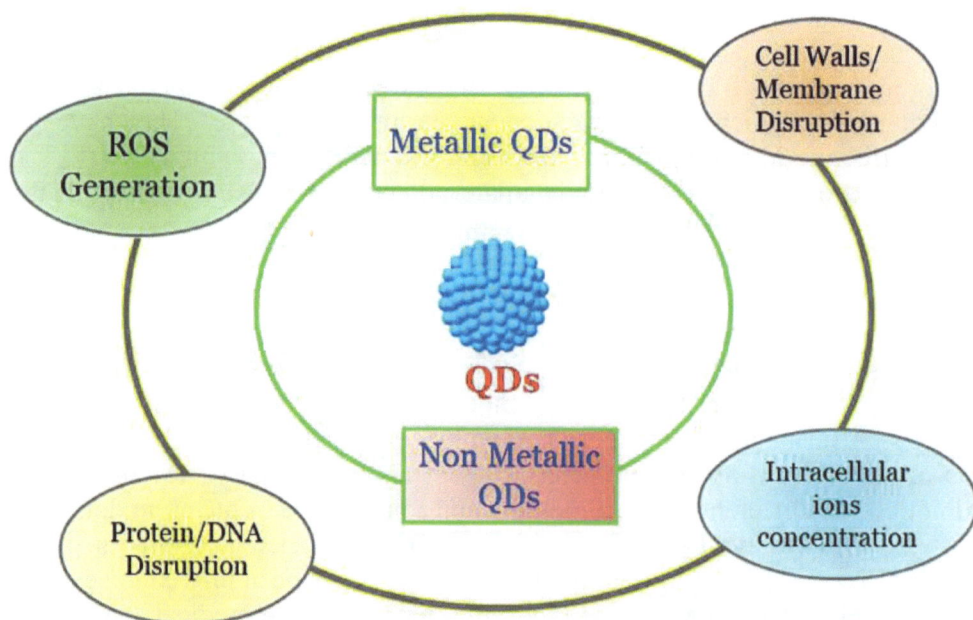

Fig. (1). Various antibacterial modes of actions of Metallic/Non-Metallic Quantum Dots.

Perforation of Bacterial Cell Walls and Cell Membranes

The size of the nanoscale materials is such that their passage through the bacterial cell barrier is not the same as small drug molecules. It offers a 'particle effect' while passing through the bacterial cell wall, which results in mechanical damage to the bacterial cell wall. It also causes osmotic imbalance and thus releases intracellular components, which leads to bacterial cell death [4 - 8]. Apart from this, QDs cause roughening and shrinking of the membrane of cells. Besides, the surface charge of QDs allows electrostatic interactions with a negatively charged bacterial cell wall, leading to a change in the cell wall integrity.

Generation of Reactive Oxygen Species (ROS)

The generation of ROS within the cells seems to be the most commonly applied mechanism of action for QDs. QDs constitute a significant number of free electrons and holes attributed to the increased transmission of electrons, which cause elevation of cellular ROS level. During the synthesis of the energy-carrying molecule adenosine triphosphate (ATP), a terminal electron acceptor of the electron transport chain (ETC), is oxygen, which is present at the bacterial cytoplasmic membrane. Generation of ATP involves the association of two electrons that combines with oxygen and H^+ ions and produces water molecule. However, exposure to QDs may cause single-electron oxidation of oxygen during this process that initiates a cascade of reactions, which generates highly reactive free radicals, such as peroxides (H_2O_2), superoxide ($O_2^{\cdot-}$) and hydroxyl radical ($HO^{\cdot}$). Cell respiration and replication get interrupted by the accumulation of ROS, which eventually results in cell death [4]. A high level of ROS is also responsible for lipid peroxidation and damage to vital macromolecules, including DNA and proteins.

Interactions with Biomacromolecules

Interactions of QDs with biomacromolecules, such as nucleic materials (*e.g.,* DNA, RNA) and/or proteins (*e.g.,* enzymes), may affect their functions that ultimately prevent cell proliferation. This also prevents DNA replication and transcription. Few studies demonstrated that QDs might also interact with the ribosome and influence protein synthesis.

Change in Intracellular Ions Concentration

An increase in the intracellular ions concentration causes issues in the function of several cellular pathways, including metabolic and cell signaling. The intercellular ion imbalance affects the intracellular charge and ion efflux. This incongruity in the ionic homeostatic also affects cellular activities, including changes in ROS level, gene expression and metabolism, which leads to cell death.

The recent studies provided evidence from various biochemical assays, genomics and proteomic analysis that the aforementioned intracellular and extracellular changes are the main reasons for the antibacterial action of QDs (Table **1**) [4]. Also, there were changes observed in bacterial cellular morphology and integrity due to the interaction of QDs, which was noted by advanced optical microscopy, and electron microscopy such as TEM & SEM. Notably, the potency of the antibacterial action varies depending on the characteristics of the ligand, size, shape, surface charge, and charge transfer effect in the case of functionalized QDs [4]. It is evident that QDs attack multiple pathways at a time, which makes it very potential and difficult for bacteria to develop resistance against it.

Table 1. Various types of Metallic and Nonmetallic QDs and their mode of bactericidal action.

Type of QDs	Name of Bacteria	Mode of Action
ZnO QDs	*Escherichia coli*	Photo excited ZnO QDs produce electron-hole pair, and the oxygen-trapped electron then causes the formation of excessive ROS [9].
PVP-capped ZnO QDs	*Listeria monocytogenes, Salmonella enteritis, Escherichia coli*	The QDs enter through the cell membrane and cause damage to cell organelles [10].
Ag/In/S QDs	*Candida albicans*	Inhibits bacteria generating ROS [11].
CdS/Ag$_2$S QDs	*Staphylococcus aureus, Escherichia coli, Pseudomonas aeruginosa*	The QDs bind to DNA infiltrating through the cell wall, which causes DNA condensation that destabilizes the structure of DNA [12].
CdSe QDs/TiO$_2$/nano graphene sheets	*Escherichia coli*	Oxidative stress is caused by the photo-induced π electrons, which are delocalized [13].
CdTe QDs	*Escherichia coli*	CdTe QDs induces membrane stress after interactions with cell. Also, the heavy metal ions which are released into the cells, inhibit the gene expression of superoxide dismutase (SOD) [14].
CdTe-Rocephin QD complex	*Escherichia coli*	Rochepins destroy the cell wall and produce membrane pits. CdTe QDs subsequently enter the cytoplasmic cell and then attach to the nucleic material to avoid anti-oxidase activation of the gene [15].
3-mercaptopropionic acid (MPA)-capped CdTe QDs	*Pseudomonas aeruginosa, Salmonella typhimurium, Acinetobacterbaumanni*	CdTe QDs bind to the phospholipid layer of bacteria; and Cd^{2+} inhibits respiration [16].
BP QDs	*Methicillin-resistant Staphylococcus aureus (MRSA)*	Photon-controlled antibacterial inactivation of drug-resistant bacteria [17].
CQDs-AMP (Ampicillin)	*Escherichia coli*	A moderate amount of reactive oxygen species were developed in combination with ampicillin in QDs under visible light illumination, which was rather efficient in inactivating *Escherichia coli* [18].

(Table 1) cont.....

Type of QDs	Name of Bacteria	Mode of Action
CQDs (Positive CQDs, Negative CQDs, Uncharged CQDs)	*Escherichia coli*	The CQDs induced the number of reactive oxygen species to increase, potentially contributing to the death of bacterial cells [19].
Nitrogen-doped CQDs	*Escherichia coli and Salmonella*	As a photothermal antibacterial chemotherapy of CQDs [20].

METALLIC QUANTUM DOTS

Due to their potent bactericidal activity, metallic QDs are regarded as a *viable* option to replace conventional organic antibacterial agents [21, 22]. Various studies demonstrate the efficiency of metallic QDs as a potential antibacterial agent [5 - 8]. Moreover, luminescent characteristics of QDs employed in the study of diverse microbial communities and bacteria identification. One of the major research fields in biological applications is single bacterial imaging by probe-conjugated QDs [23]. Studies showed that the release of metal ions from the QDs triggers bacteria toxicity [5 - 8]. The previous study has approved the development of free radicals to play a major role in antibacterial activity [24 - 26]. Antibacterial activity of different types of metallic quantum dots is discussed here below.

Antibacterial Activity of Cadmium Telluride Quantum Dots (CdTe QDs)

One of the most widely explored quantum dots is CdTe QDs. Several synthesis protocols have already been proposed by various research groups, which include ultrasonic-assisted extraction [27], aqueous synthesis [28], microwave [29] and green synthesis methods [30]. The synthesis method provided the scope of precise control of its diameter that precisely delivers the opportunity of monitoring its optical properties. Interestingly, CdTe QDs showed significant bactericidal activity against multidrug-resistant strains [31]. The bactericidal activity of CdTe QDs is assigned to the release of Cd^{2+} ions, which triggers bacterial toxicity by following several pathways. It also leads to the elevation of the ROS level, which is a consequence of the Cd^{2+} interactions with genomes [31]. Besides, various reports demonstrated that CdTe QDs affect gene expression. It downregulates the expression of superoxide dismutase (SOD) genes and endonuclease, essential for the cellular system's anti-oxidative capacity [31]. It is also found that CdTe QDs bind to the lipophilic tail of phospholipid bilayers depending on its hydrophobic characteristic, and destroy the outer peptidoglycan layer of the bacterial cell wall [31]. The research demonstrates that the procedure with CdTe QDs of smooth rod-shaped *B. subtilis* induced the bacterial cell wall to disintegrate within 4h of incubation, which was clear from either SEM or TEM images [31]. The dose-dependent inhibition of bacterial growth was reported after treatment with CdTe QDs. Interestingly, the Rosenzweig group reported that the poly(oxanorbornene) coated CdTe QDs exhibited better activity against *E. coli* as compared to free poly(oxanorbornene) or CdTe QDs [29].

In particular, the functionalized CdTe QDs delivered improved bactericidal activity contrasted with CdTe QDs alone. Potent bactericidal activity of the CdTe QD-Rocephin complex against *E. coli* bacteria has been shown. Results showed that the functional group of rocephin and bare CdTe QDs interacts with organic biomolecules, which enhances the antibacterial property compared to the QDs only [15]. The production of *E. coli* bacteria was also substantially impeded by rutin-conjugated TGA-CdTe QDs, where it was indicated that hydroxyl groups (-OH) of rutin were correlated with CdTe QDs. The QDs produce a huge amount of ROS inside the cell by facilitating the generation of hydroxyl radicals [32].

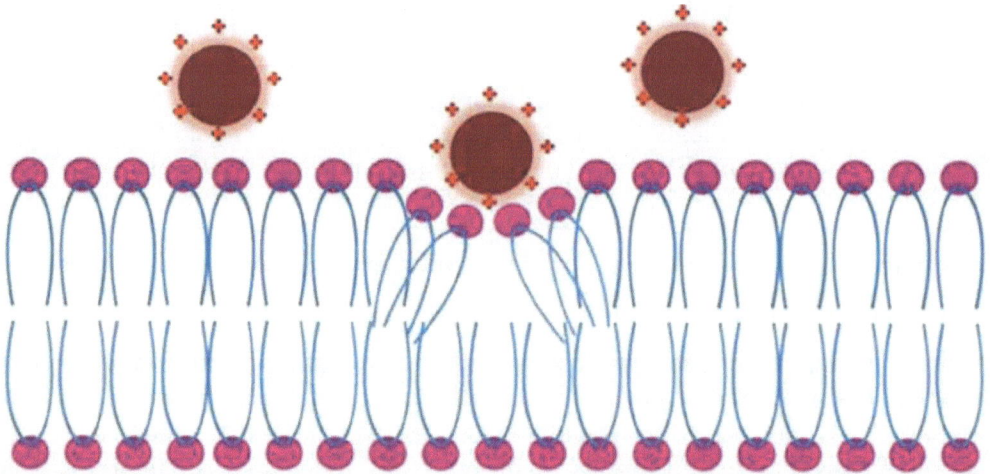

Membrane disruption mechanism of positive QDs.

Fig. (2). A pictorial overview of the membrane disruption due to the surface charge of QDs. Reprinted with permission from the reference [35].

A combination of biocompatible, cationic and antibacterial polymers like Poly-L-lysine (PLL) enhanced the bactericidal activity of CdTe QDs [33]. Results showed that the use of 8 - 10 bilayers of PLL on QDs significantly decreased the growth rate of *E. coli* bacteria. The inhibition mechanism was due to cell wall disruption depending on the electrostatic interactions of bilayers of PLL coated on CdTe QD [33]. Surface factionalized CdTe QDs with suitable molecules produce photo-generated electrons charge, which is very effective for photodynamic therapy (PDT). It would be mentioned here that low-power laser light with an appropriate wavelength can effectively eradicate bacterial cells while treated with photosensitizer drugs/agents such as QDs. This produces highly reactive singlet oxygen that could efficiently destroy bacterial cells. Thus, recent outcomes in this line of interest demonstrated that QDs-based PDT would be one of the promising ways of killing a wide range of multidrug-resistant bacterial populations [34]. Further, studies revealed that the surface charge of CdTe QDs could also impact the antibacterial activity (Fig. **2**) [35]. The membrane disruption of bacteria has occurred through the interaction between QDs and lipid molecules (hydrophobic head groups). In this regard, the surface charge of QDs plays a significant role.

It is worth mentioning that CdTe QDs are reported to be cytotoxic against human cells, which limits their *in-vivo* applications. However, the toxicity level varies depending on the concentration and surface functional groups of the CdTe QDs. Interestingly, reports demonstrated that functionalizing CdTe QDs with biomolecules reduces its toxicity level. Though CdTe QDs, while coated with PLL bioavailability, were seen on human cells, exhibiting no suspected cytotoxicity after 12 h of incubation on human cells [33].

Antibacterial Activity of Titanium Dioxide Quantum Dots (TiO$_2$ QDs)

Another type of QDs that showed huge promise in various biomedical applications is titanium dioxide (TiO$_2$) QDs. This is chemically inert, less toxic, economical and effective for photocatalytic activity. Globally, several research groups have been working on TiO$_2$ QDs, which could develop various wet chemical synthesis methods like solvothermal, hydrothermal, and sol-gel methods [36 - 40].

TiO$_2$ QDs exhibit high bactericidal activity. Hence, it is considered a potentially significant candidate for various therapeutic applications, including wound healing. Recent reports demonstrated that TiO$_2$ QDs have potent antibacterial activity against *E. coli, Bacillus megaterium, Enterococcus faecalis* and *Micrococcus luteus* [37 - 39]. Recently, anatase TiO$_2$ QDs were employed to inactivate *E. coli* under UV light irradiation [37]. Interestingly, it has been found that TiO$_2$ QDs (concentration 60 μg mL^{-1}) efficiently inhibited 91% of *E. coli* and showed better antibacterial properties as compared to the commercially available TiO$_2$ nanoparticles. Fakhri *et al.* also explored the anti-microbial activity of TiO$_2$ QDs against *E. faecalis*(Gram-positive) and *M. luteus*(Gram-negative) [38]. Their study revealed that bactericidal activity was due to the electrostatic interaction between TiO$_2$ QDs and bacterial cell walls. The minimum inhibitory concentration (MIC) was found to be 13.4mM and 1.2mM for *E. faecalis* and *M. luteus*, respectively. In addition to the above-mentioned study, the antibacterial activity of anatase TiO$_2$ nanocomposites was investigated by the Lin group also

[39]. They observed that TiO_2 nanocomposites can inhibit the growth of *E. coli* and *Bacillus megaterium* bacteria. However, a better bactericidal result was noted in the case of *E. coli* as compared to *Bacillus megaterium* bacteria. It is proposed that the small size, larger surface area, higher band gap, and active site of TiO_2 are the reason behind the higher bactericidal effect.

Another study was carried out using the composite photocatalystMn-$CdS/ZnCuInSe/CuInS_2/TiO_2$ for the inhibition of *E. coli* bacteria under visible light [41]. It was found that 96% of bacteria were eliminated within around 50 minutes in 50 mL of 10^5 colony developing systems (CFU/ mL) route [41]. Notably, Mn-CdS QDs, ZnCuInSe QDs, $CuInS_2$ QDs harvest visible light and TiO_2 nanowires played an important role in transferring photo-generated electrons.

Antibacterial Activity of Cadmium Selenide Quantum Dots (CdSe QDs)

The synthesis of cadmium selenide (CdSe) QDs was carried out by electrochemical deposition [42], solutions [43], sonochemical synthesis [44] and wet-chemical [45] methods. Holden group demonstrated that CdSe QD efficiently inhibited the *P. aeruginosa* PG201 bacteria [46]. It was proposed that Cd^{2+} ions reached into the cell cytoplasm in the impaired cell wall *via* pits and pores, which eventually promoted Cd^{2+} ions internalization and intracellular accumulation. The MIC dose of metal ions (Cd^{2+}) has raised the high level of bacterial cell toxicity. Detailed molecular mechanisms also demonstrated that cell toxicity restricted cell respiration and multiplication, including the proliferation of the bacteria. ROS is generated in the cellular environment due to the presence of oxygen molecules, which enhances oxidative stress leading to cell rupture, deformation, and cell death, as given in Fig. (**3**). Interestingly, MIC concentration was 5 µg / mL CdSe QDs [46]. Another study showed that combination therapy of CdSe QDs, TiO_2 and graphene nanosheet illustrated effective bactericidal potency against *E. coli* [47]. The bactericidal activity was increased manifold due to its photo-catalytic features produced by irradiation in visible light. Photo-radiation tended to be closely linked to the proliferation of *E. coli*, which greatly inhibited bacterial growth depending on the intensity and duration of exposure to light. The CdSe QDs used for photo-irradiation effectively eliminated approximately 90% of *E. coli*. Along with the rise of O_2^- and hydroxyl radicals, the light irradiation produced electrons- holes that increased the amount of ROS, leading to cell death [13].

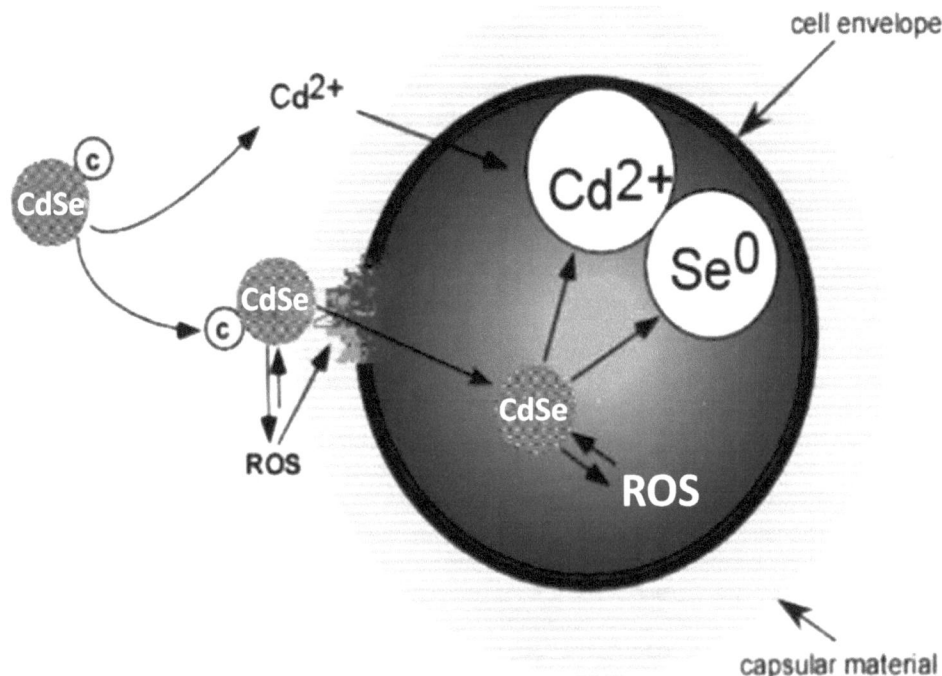

Fig. (3). Interaction between citrate-stabilized CdSe QD and planktonic *P. aeruginosa*. QDs dissolve extracellularly partially, but in the absence of cells, dissolution is faster. Extracellular ROS is produced by QDs that permeate membranes, facilitating cellular entry, where improved ROS output per QD is detected. Reprinted with permission from the reference [46].

Antibacterial Activity of Zinc Oxide Quantum Dots (ZnO QDs)

One of the most promising bactericidal QDs, which gained huge interest owing to its less toxicity and less immunogenic nature, is zinc oxide quantum dots (ZnO QDs). There are several gels and wound healing materials proposed based on the ZnO QDs. These are generally synthesized by Sol-gel [47], green synthesis [48], hydrolysis sol-gel hydrothermal [49], and wet chemical synthetic [9] methods. ZnO QDs can kill a wide range of bacterial strains while treating alone and/or with other bactericidal agents. It is worth mentioning that ZnO QDs produce electron-hole pairs that produce ROS species, including hydroxyl radical (OH·), superoxide anion (O^{2-}) and per-hydroxyl radical (HO^{2-}). Reports also suggested that ZnO QDs specifically damage the phospholipid layer, which causes the release of intracellular content. Furthermore, few studies indicate that the cell absorbs Zn^{2+} ions that damage DNA and proteins also [50].

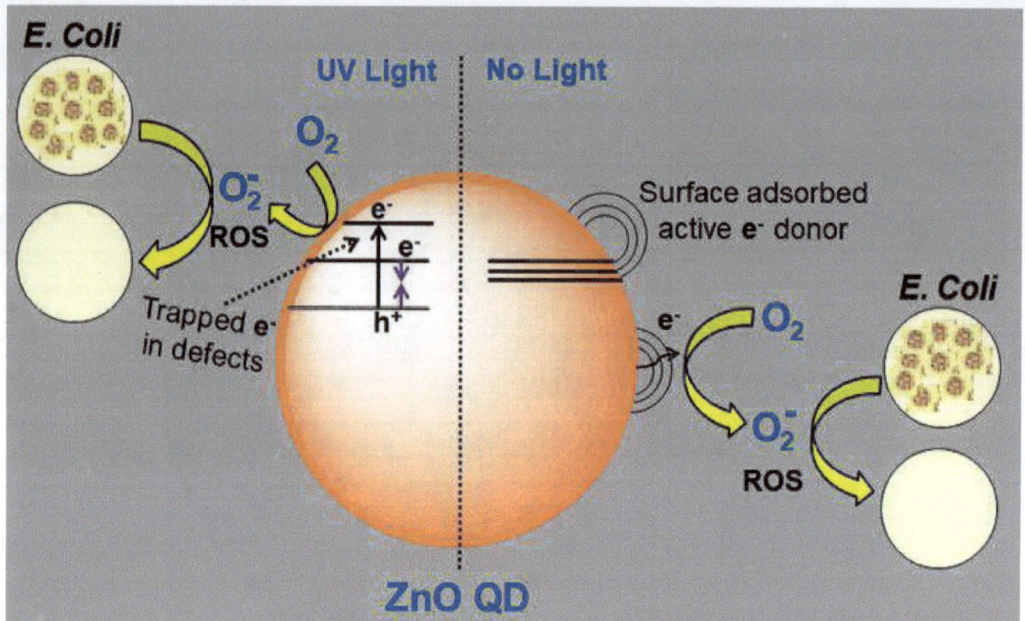

Fig. (4). Depiction of antibacterial activity of ZnO QDs under UV light and without UV light.

Interestingly, multidrug-resistant bacterial strains could be prevented by functionalized ZnO QDs (Fig. **4**) [9]. As in the antibacterial potency of ZnO QDs, limited size and surface enhancements, including suitable ligands, serve a critical role. The research was conducted on multi-drug resistant *E.coli* by treating ZnO QDs, including acetate (CH_3COO-) as well as nitrate (NO_3-) anion surface binding sites [9]. The MIC dosage of ZnO QDs (4-7 nm size) has been reported to yield 6 µL /mL in the appearance of nitrate. In the involvement of acetate ZnO QDs (3-5 nm size), the MIC exposure was lowered by 2.5 µL /mL [9]. With an estimated size of 10 nm, Gram-positive (*S. aureus*), as well as Gram-negative (E. coli) bacteria, have been administered with PEG-capped ZnO nanostructures that triggered increased antibacterial activities, which was approved to generate a significant amount of ROS [51]. Besides that, polyvinylpyrrolidone (PVP) was employed to generate ZnO QDs complex due to modulation of ZnO QDs. *Listeria monocytogenes, Salmonella enteritidis*, also *E. coli* O157:H7 have been used to evaluate the bactericidal capability of the complex [10]. In comparison to the uncapped form of ZnO QD, the procedure of the ZnO-PVP complex reduced bacterial cellular proliferation to 5.3 log for *L. monocytogenes* and 6.0 log for *E. coli* followed by 48 h of incubation.

NON-METALLIC QUANTUM DOTS

Antibacterial Activity of Black Phosphorus (BP)

Black phosphorus quantum dot (BP QD) is one of the promising biocompatible non-metallic photo thermal materials [52, 53]. One of the major advantages of BP QDs is non-cytotoxicity in human cells [52]. Thus, it could be a promising alternative option other than metallic QDs. Under NIR light irradiation, BP QDs exhibit high photothermal conversion efficiency of around 28.4% [52]. BP QDs are synthesized by the liquid exfoliation method using probe sonication or bath sonication. It has fairly photostability along with an average size of around 2.6 nm and a thickness of 1.5 nm [52]. Due to its superior physicochemical properties, BP has drawn significant attention in numerous biomedical applications. Fascinatingly, pathogenic bacterial infections that are difficult to diagnose and functionalized BP may be strategic nanostructures that showed great outcomes to kill these pathogenic strains selectively.

The BP QDs offer the possibility of spatiotemporal regulation of photothermal impact. Thus, it shows huge promise in photo-induced heat generation, which is useful in biomedical applications [52]. Consequently, by contemplating BP QDs number of researchers tried to integrate a new antibacterial system that can accomplish controlled pharmaceutical advent, and a productive complementary approach would contribute as a productive strategy to fight drug-resistant bacteria.

Proposed Mechanism of Antibacterial Action of Black Phosphorus Based Nanomaterials

Treatment of BP-based nanomaterial leads to the formation of singlet oxygen ($^1O_2$), peroxide (O_2^-), superoxide anion (O^{2-}), hydrogen peroxide (H_2O_2) and hydroxyl radicals (OH•), which ultimately contributed to the rise of ROS [54, 55]. Various study has already shown that the primary cause for bacterial cell destruction is ROS-induced oxidative stress [66, 74]. As side products of certain biological methods, ROS species are continually found to be very effective to influence various cellular functions. Application of BP based nanomaterial would result in an imbalance in the equilibrium of ROS production and its degradation by free radical scavengers like antioxidant molecules [54, 55].This interferes with various intracellular biomolecules like DNA, RNA, and proteins. The high amount of ROS also eliminates pathogenic bacteria by destroying the cell membrane *via* lipid peroxidation.

Photothermal Conversion of BP QDs for Improved Bactericidal Activity

Photothermal antibacterial activity is considered to be a safe and effective way to combat bacterial infections [17]. NIR light penetrates deeply without (or at minimum) damaging normal human cells. Thus, photothermal treatment is found to be one of the very impactful methods for clinical practices. It would be mentioned here that NIR light raises the heat that can be used to disrupt bacterial cells. Studies demonstrated that it would be one of the possible alternative strategies to replace the existing antibiotics [17].

A photon-responsive antibacterial system (regarded as BP QDs banco liposome) has been established by Zhang *et al*. The research team proposed a heat-resistant liposome as the transporter of BP QDs or vancomycin that displayed high drug delivery and efficient photothermal activity, which was found to be very lethal against bacterial infections [17]. Notably, the findings of subcutaneous abscess therapy showed that stimulus-responsive liposomes provide exemplary output in destroying *in vivo* drug-resistant bacteria. As displayed in Fig. (**5**), the thermal sensitive liposome is disrupted under NIR irradiation owing to the photodynamic impact of BP QDs, contributing to the release of embedded vancomycin antibiotics that inhibit bacteria. Simultaneously, an increase in local temperature triggered by BP QD led to an effective bacterial ablation against drug-resistant bacteria. During the diagnosis of skin abscess-related antibiotic-responsive bacteria, the photon-responsive antibacterial framework was also used. *In vivo* analysis showed that the skin abscesses induced by drug-resistant bacteria could be easily dealt with through the insertion of NIR- sensitive liposomes (Fig. **5**). A NIR-mediated photothermal antibacterial therapy may also be done to kill a wide range of Gram-positive and Gram-negative bacteria which are indeed paving several new pathways for fields of nanomedicine [17]. It would be mentioned here that the cytotoxicity of the BP QDs is less as compared to conventional QDs. However, toxicity hugely relies on the physiochemical properties of the BP QDS. For example, a new updated ultrasonic approach was mentioned for the preparation of BP QDs having an average size of 10 nm and height of 8.7 nm. These BP QDs showed *in vitro* cytotoxicity contrasted to graphene with graphene-oxide QDs while treated with HeLa, COS-7 including CHO-K1 cells [56].

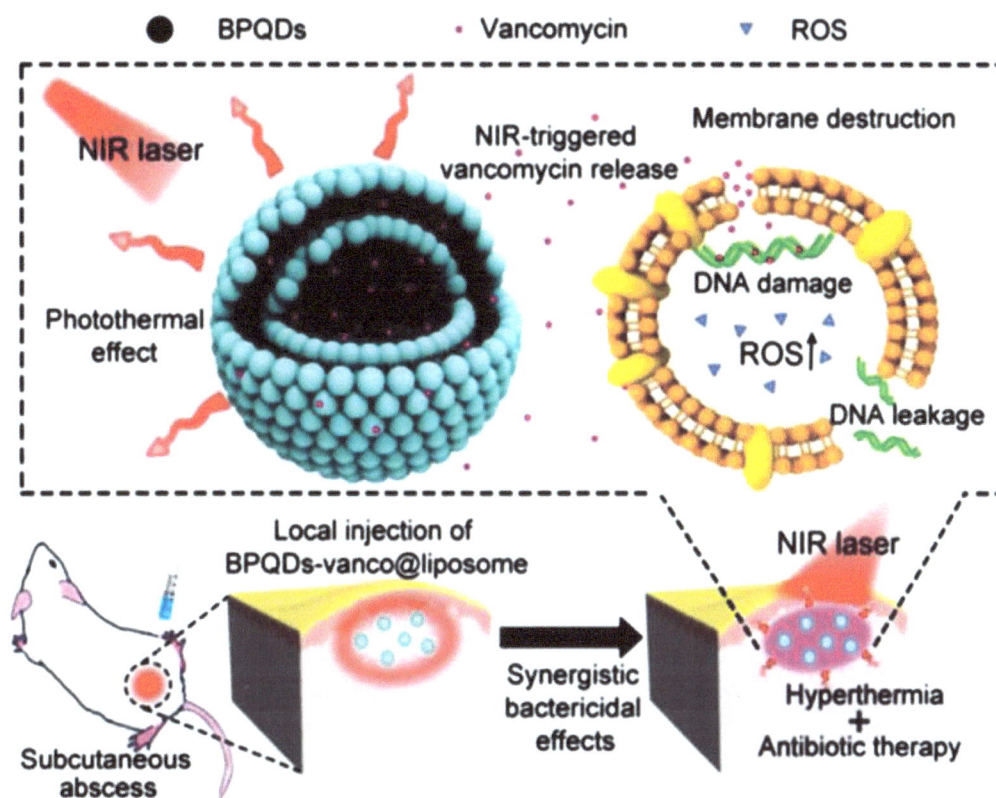

Fig. (5). Diagrammatic representation of the antibacterial framework driven by photons for the synergistic handling of subcutaneous abscess infected by bacteria. Reprinted with permission from the reference [17].

Antibacterial Activity of Carbon Quantum Dots (CQDs)

One of the most recent advent nanoscale materials is carbon quantum dots (CQDs) which fascinated researchers owing to their unique photo-physical properties [57 - 59]. These are found to be highly water soluble, biocompatible and bactericidal against both Gram-positive and Gram-negative bacteria. The bottom-up and top-down methods are utilized to synthesize CQDs. Pyrolysis, combustion, or hydrothermal methods which build up nanostructures from tiny organic molecular substrates are included in a bottom-up approach, however the top-down pathway is focused on reducing tiny particles into nanostructures of the desired size using physical, chemical or electrochemical approaches [58 - 60]. The basic content and the follow-up functionalization processes often lead to the key attributes within each kind of CQDs, in addition to the preference for a synthetic path. Therefore, synthesizing anti-microbial substances from CQDs is also feasible [60]. CQDs are stated to have antibacterial activity and low toxicity depending on their size and surface charge that could be conveniently controlled by surface modification (*e.g.,* carboxyl, hydroxyl, amino, epoxy, amides, *etc.*)

[62]. By facilitating CQDs-mediated bactericidal activity, membrane disruption and the development of reactive oxygen species (ROS) are significant mechanisms [62]. The bacterial membrane adsorption of CQDs involves alterations in the permeability of the membrane, resulting in leakage of the cytoplasm [61, 63]. The structure of DNA and proteins is affected by enhanced ROS levels and oxidative stress [64]. The antibacterial properties of carbon-based nanostructures were investigated, and further comprehensive studies were carried out to understand the interdependence of CQD surface chemistry and bactericidal efficiency.

Verma *et al.* employed microwave-mediated synthesis using biodegradable starch as a precursor of catalytic fluorescent CQDs [65]. This was used to perform combination therapy along with Ag NPs. Notably, in the photo-reduction of Ag^+ to silver nanoparticles (Ag NPs), CQDs demonstrated high catalytic activity. There was neither any requirement for excessive surface passivating agents to regulate the Ag NPs throughout the photo-reduction process. Furthermore, CQDs and Ag NPs combination treatment incorporating CQD-Ag NP nano-composite was conducted, which suggested enhanced antibacterial activity against recombinant *E. coli* bacteria resistant to antibiotics (Fig. **6**).

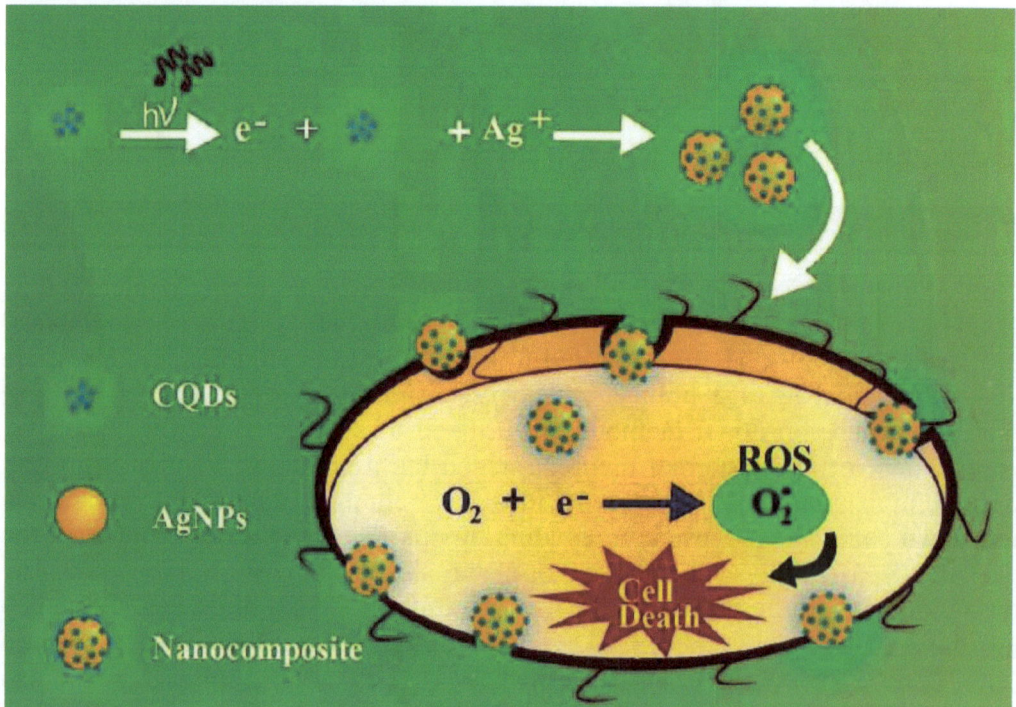

Fig. (6). Schematic representation of the synthesis of CQD-Ag nanocomposite and its antibacterial mechanism against recombinant *E. coli* bacteria resistant.

Since comparison includes its individual components, the composite increased the amount of ROS. The analysis of the flow cytometer showed that the combination technique induced blockage of the bacterial cell wall, which may have led to antibacterial action against both Gram-positive and Gram-negative bacteria. Owing to the ground state complexation, the existence of CQDs on the surface of the Ag NPs may have facilitated electrons transfer from Ag NPs to CQDs that improved the outcomes of ROS compared with just Ag NPs. The composite seemed capable of penetrating the cell wall of the bacteria and inducing elevated levels of growth of ROS. This could be potential in managing the problem of antibiotic-induced multiple drug resistance infections [65].

Antibacterial examination of four distinct strains of bacteria, namely *Escherichia coli, Staphylococcus aureus, Bacillus cereus,* and *Pseudomonas aeruginosa,* indicated that it increased antibacterial effect of thin films under blue light irradiation from hydrophobic CQDs [66]. Moreover, hydrophobic QDs display a non-cytotoxic impact on the line of mouse fibroblast cells. Such characteristics allow for the future use of thin films of hydrophobic CQDs as outstanding antibacterial coatings for various biomedical purposes [66]. Recently, photodynamic anti-microbial chemotherapy using CQDs has been reported by the Wu group [20]. They synthesized a series of water-soluble nitrogen-doped CQDs. These CQDs exhibit excellent photosensitized oxygen activation (Fig. 7).

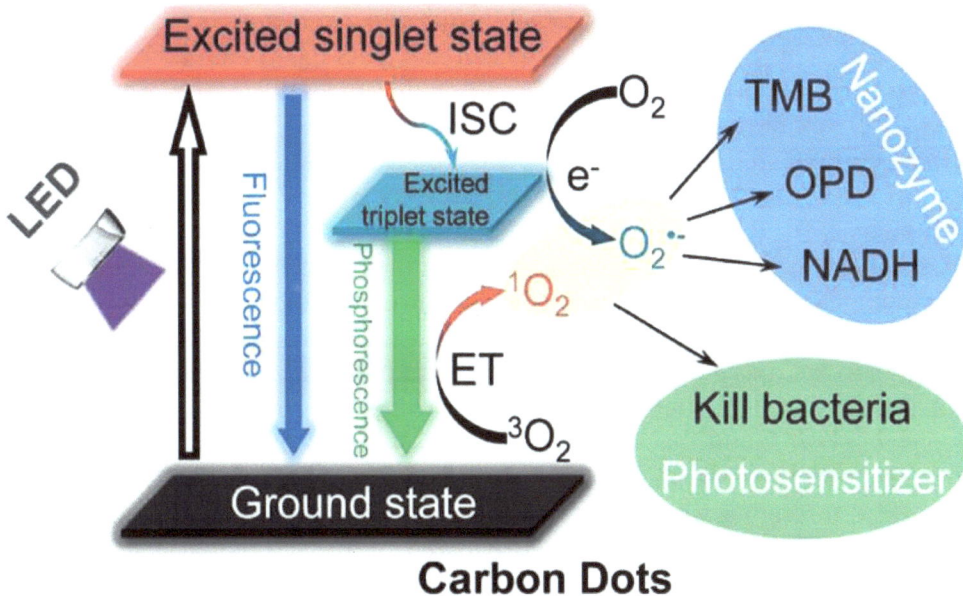

Fig. (7). Photosensitized Oxygen Activation with CQDs and photodynamic anti-microbial activity. Reprinted with permission from the reference [20].

Verma and coworkers used a microwave synthesis technique, fast and simple, to synthesize various charged CQDs to the similar initial material and the synthesized CQDs, which is very stable by employing diethylene glycol as a carbon source with numerous amines comprising surface passivating agents, the various surface charged CQDs (such as positive, negative and neutral) were produced [19]. The relation between *E. coli* bacteria and three different forms of CQDs was investigated, and reported that considerable antibacterial activities were shown by these CQDs. The bactericidal function of these CQDs, the recombinant *E. coli* has been used as a conceptual model. In contrast to negative charged CQDs, time-dependent bacterial growth and FACS studies showed that both uncharged CQDs and positive charged CQDs exhibit excellent bactericidal activity. The CQDs induced the amount of reactive oxygen species to increase, potentially contributing to the death of bacterial cells. The effective control of CQDs surface charge seemed to be the prominent trigger of the antibacterial activity that could be used for generating chemically functionalized CQDs for successful biomedical operations [19].

The antibiotic metronidazole was previously employed by Liu *et al.* to generate CQDs *via* a simple hydrothermal process [67]. Through this process, by photoluminescence, they synthesized a kind of CQDs that would detect required anaerobes and are potentially effective against such bacteria [68]. Thakur *et al.* used a microwave-assisted synthesis, the conjugated ciprofloxacin hydrochloride (an antibiotic) to the surface of CQDs produced from gum Arabic to generate Cipro CQDs (Cipro @ CQDs conjugate) [69]. Thereby it was found that CQDs can also function as a drug carriers. The results of their analysis indicated that Cipro CQDs exhibited increased antibacterial activities against a wide range of bacteria. Animal studies have also revealed that, like other metal and metal oxide nanostructures, the CQDs were easily removed from tissues and provided less cytotoxicity [69]. Recently, the antibacterial activity of photoactive hydrophobic CQDs/polyurethane nanocomposite was examined. For the development of these nanocomposites, the swell-encapsulation-shrink method was implemented. Hydrophobic CQDs/polyurethane nanocomposites have been regarded as a very efficient singlet oxygen generator when irradiated by low-power blue light. Antibacterial experimental research carried out on *Staphylococcus aureus* and *Escherichia coli* found that these nanocomposites had a 5-log bactericidal effect within 60 minutes of irradiation [70]. Small, uncharged CQDs had been synthesized through a simple thermal decomposition path, although amines were employed for surface modification. It showed high bactericidal activity against various bacterial species (*e.g.,Staphylococcus aureus, Escherichia coli,* and *Klebsiella pneumonia*). For *S. aureus*, MICs was found to be from 3.4 to 6.9 mg / mL [71].

FUTURE PROSPECTIVES

Metallic and non-metallic QDs are established as a novel class of nanoscale materials having superior optical and physicochemical properties. This is considered an effective alternative to replacing conventional organic dyes suitable for various biomedical applications. In addition, these nanoscale materials have size-dependent emission, high quantum, and photo-stability. Thus, for the last few decades, extensive research activities have been directed to explore several new aspects of QDs. It is found to be a very efficient candidate for the treatment of bacterial infections. Furthermore, to make it more suitable as an antibacterial agent, recent studies focused on finding out various methods of functionalization that make it very officious in biological fluid and enhance biocompatibility. One of the earliest commercialized nanotechnology-based products is QDs, which showed a great opportunity for tracking biomacromolecules and cell imaging. Despite its several remarkable features, metallic QDs are not found to be very useful for real-life applications as these are primarily made up of heavy metals. The major issue of metallic QDs is biocompatibility, which limits their direct use in *in-vivo* applications and translational research. Moreover, the release of heavy metal ions sometimes causes huge concern for users and the environment. Fortunately, the functionalization of QDs reduced the concurrent demerits of metallic QDs and increased the feasibility of biomedical applications. Also, the combination therapy of QDs with conventional antibacterial agents, including antibiotics, showed the great scope of using these as a novel class of agents for the treatment of drug-resistant bacterial strains. On the other hand, non-metallic QDs have shown huge prospects as they are biocompatible and biodegradable. This class of non-metallic QDs, including CQDs, is more suitable for real-life applications. Owing to its huge possibilities and prospects QDs research about the pharmacology and toxicity are not explored up to the mark to use these as novel therapeutic agents. A comprehensive understanding is very much essential to explore ADME properties (*i.e.,* absorption, distribution, metabolism, and excretion) of QDs, which is the basic pharmacological understanding important for real-life applications. Another very crucial aspect that needs to address properly is the blood-QDs interactions which are commonly coined as hemocompatibility. There are not enough studies carried out in this direction to get a detailed understanding of blood–QD interactions. Although there are few reports which demonstrate that QDs interacts with blood proteins, further research is important to get a proper idea about the immunogenicity of QDs, which is one of the vital parameters for *in-vivo* applications of QDs.

CONCLUSION

Metallic quantum dots (QDs) are one of the most explored new classes of an inorganic fluorophore, having unique optical and physicochemical properties. It has a size of a few nanometers, which determines the luminescent behavior of the QDs. Bright luminescence, high photostability and chemical inertness prompt us to use this as novel bioimaging and bio labeling agents. Apart from this, it showed strong bactericidal activity against both Gram-positive and Gram-negative depending on their size, surface charge, and surface ligands. The QDs are one of the most promising candidates for photodynamic therapy (PDT) as, in the presence of external light sources, it acts as a photosensitizer (PS). That is very officious for the killing of drug-resistant bacterial strains. However, metallic and semiconductor QDs generally lack the issues of biocompatibility and biodegradability, as enzymatic metabolism sometimes causes the leaching of metal ions, which are found to be very toxic for human health and the environment. This may also result in oxidative stress for mammalian cells, which raises huge concern for its direct use as an antibacterial agent. In this case, non-metallic QDs like CQDs may be an alternative choice as they showed strong bactericidal activity while administered alone or with other antibacterial agents, including conventional antibiotics. Moreover, the combination therapy of functionalized QDs is found to be most promising and holds huge prospects to use for the treatment of MDR and XDR bacterial strains. Thus, it is anticipated that in the near future, functionalized QDs would certainly enrich the repertoire of nanomedicines and might be a *via*ble choice for an alternative to antibiotics.

CONSENT FOR PUBLICATION

Not applicable.

CONFLICT OF INTEREST

The authors declare no conflict of interest, financial or otherwise.

ACKNOWLEDGEMENTS

AK and MPS acknowledge the Department of Chemistry (DRS-II (SAP), DST (FIST and PURSE) funded), AMU, Aligarh. AKS acknowledges the financial support from DBT, Govt. of India (SAN. No. BT/PR40544/COD/139/14/2020). KA acknowledges University GrantCommission, New Delhi, for providing her with a Moulana Azad National Fellowship. AK acknowledges Virtual Internship with Science Leaders (VISL) organized by the Young Academy of India.

REFERENCES

[1]   OnHealth. Bacterial infections 101: Types, symptoms, and treatments. 2020. Available from: https://www.onhealth.com/content/1/bacterial_infections (Accessed on: 18 September, 2020).

[2]   [Sugden R, Kelly R, Davies S. Combatting anti-microbial resistance globally. Nat Microbiol 2016; 1(10): 1-2.

[3]   Jahangir MA, Gilani SJ, Muheem A, *et al.* Quantum dots: next generation of smart nano-systems. Pharm Nanotechnol 2019; 7(3): 234-45.
[http://dx.doi.org/10.2174/2211738507666190429113906] [PMID: 31486752]

[4]   Rajendiran K, Zhao Z, Pei DS, Fu A. Antimicrobial activity and mechanism of functionalized quantum dots. Polymers (Basel) 2019; 11(10): 1670.
[http://dx.doi.org/10.3390/polym11101670] [PMID: 31614993]

[5]   Slavin YN, Asnis J, Häfeli UO, Bach H. Metal nanoparticles: understanding the mechanisms behind antibacterial activity. J Nanobiotechnology 2017; 15(1): 65.
[http://dx.doi.org/10.1186/s12951-017-0308-z] [PMID: 28974225]

[6]   Rice KM, Ginjupalli GK, Manne NDPK, Jones CB, Blough ER. A review of the antimicrobial potential of precious metal derived nanoparticle constructs. Nanotechnology 2019; 30(37): 372001.
[http://dx.doi.org/10.1088/1361-6528/ab0d38] [PMID: 30840941]

[7]   Kadiyala U, Kotov NA, VanEpps JS. Antibacterial metal oxide nanoparticles: challenges in interpreting the literature. Curr Pharm Des 2018; 24(8): 896-903.
[http://dx.doi.org/10.2174/1381612824666180219130659] [PMID: 29468956]

[8]   Raghunath A, Perumal E. Metal oxide nanoparticles as antimicrobial agents: A promise for the future. Int J Antimicrob Agents 2017; 49(2): 137-52.
[http://dx.doi.org/10.1016/j.ijantimicag.2016.11.011] [PMID: 28089172]

[9]   Joshi P, Chakraborti S, Chakrabarti P, *et al.* Role of surface adsorbed anionic species in antibacterial activity of ZnO quantum dots against Escherichia coli. J Nanosci Nanotechnol 2009; 9(11): 6427-33.
[http://dx.doi.org/10.1166/jnn.2009.1584] [PMID: 19908545]

[10]   Jin T, Sun D, Su JY, Zhang H, Sue HJ. Antimicrobial efficacy of zinc oxide quantum dots against Listeria monocytogenes, Salmonella Enteritidis, and Escherichia coli O157:H7. J Food Sci 2009; 74(1): M46-52.
[http://dx.doi.org/10.1111/j.1750-3841.2008.01013.x] [PMID: 19200107]

[11]   Mir IA, Radhakrishanan VS, Rawat K, Prasad T, Bohidar HB. Bandgap tunable AgInS based quantum dots for high contrast cell imaging with enhanced photodynamic and antifungal applications. Sci Rep 2018; 8(1): 9322.
[http://dx.doi.org/10.1038/s41598-018-27246-y] [PMID: 29921973]

[12]   Neelgund GM, Oki A, Luo Z. Antimicrobial activity of CdS and Ag2S quantum dots immobilized on poly(amidoamine) grafted carbon nanotubes. Colloids Surf B Biointerfaces 2012; 100: 215-21.
[http://dx.doi.org/10.1016/j.colsurfb.2012.05.012] [PMID: 22766300]

[13]   Ma X, Xiang Q, Liao Y, Wen T, Zhang H. Visible-light-driven CdSe quantum dots/graphene/TiO2 nanosheets composite with excellent photocatalytic activity for E. coli disinfection and organic pollutant degradation. Appl Surf Sci 2018; 457: 846-55.
[http://dx.doi.org/10.1016/j.apsusc.2018.07.003]

[14]   Lu Z, Li CM, Bao H, Qiao Y, Toh Y, Yang X. Mechanism of antimicrobial activity of CdTe quantum dots. Langmuir 2008; 24(10): 5445-52.
[http://dx.doi.org/10.1021/la704075r] [PMID: 18419147]

[15]   Luo Z, Wu Q, Zhang M, Li P, Ding Y. Cooperative antimicrobial activity of CdTe quantum dots with rocephin and fluorescence monitoring for Escherichia coli. J Colloid Interface Sci 2011; 362(1): 100-6.
[http://dx.doi.org/10.1016/j.jcis.2011.06.039] [PMID: 21757199]

[16]   Geraldo DA, Arancibia-Miranda N, Villagra NA, Mora GC, Arratia-Perez R. Synthesis of CdTe QDs/single-walled aluminosilicate nanotubes hybrid compound and their antimicrobial activity on bacteria. J Nanopart Res 2012; 14(12): 1286.
[http://dx.doi.org/10.1007/s11051-012-1286-6]

[17]   Zhang L, Wang Y, Wang J, *et al.* Photon-responsive antibacterial nanoplatform for synergistic photothermal-/pharmaco-therapy of skin infection. ACS Appl Mater Interfaces 2019; 11(1): 300-10.
[http://dx.doi.org/10.1021/acsami.8b18146] [PMID: 30520301]

[18]   Jijie R, Barras A, Bouckaert J, Dumitrascu N, Szunerits S, Boukherroub R. Enhanced antibacterial activity of carbon dots functionalized with ampicillin combined with visible light triggered photodynamic effects. Colloids Surf B Biointerfaces 2018; 170: 347-54.
[http://dx.doi.org/10.1016/j.colsurfb.2018.06.040] [PMID: 29940501]

[19]   Verma A, Arshad F, Ahmad K, *et al.* Role of surface charge in enhancing antibacterial activity of fluorescent carbon dots. Nanotechnology 2020; 31(9): 095101.
[http://dx.doi.org/10.1088/1361-6528/ab55b8] [PMID: 31703210]

[20]   Zhang J, Lu X, Tang D, *et al.* Phosphorescent carbon dots for highly efficient oxygen photosensitization and as photo-oxidative nanozymes. ACS Appl Mater Interfaces 2018; 10(47): 40808-14.
[http://dx.doi.org/10.1021/acsami.8b15318] [PMID: 30387982]

[21]   Klaine SJ, Alvarez PJJ, Batley GE, *et al.* Nanomaterials in the environment: behavior, fate, bioavailability, and effects. Environ Toxicol Chem 2008; 27(9): 1825-51.
[http://dx.doi.org/10.1897/08-090.1] [PMID: 19086204]

[22]   Lee D, Cohen RE, Rubner MF. Antibacterial properties of Ag nanoparticle loaded multilayers and formation of magnetically directed antibacterial microparticles. Langmuir 2005; 21(21): 9651-9.
[http://dx.doi.org/10.1021/la0513306] [PMID: 16207049]

[23]   Voura EB, Jaiswal JK, Mattoussi H, Simon SM. Tracking metastatic tumor cell extravasation with quantum dot nanocrystals and fluorescence emission-scanning microscopy. Nat Med 2004; 10(9): 993-8.
[http://dx.doi.org/10.1038/nm1096] [PMID: 15334072]

[24]   Wei C, Lin WY, Zainal Z, *et al.* Bactericidal activity of TiO2 photocatalyst in aqueous media: toward a solar-assisted water disinfection system. Environ Sci Technol 1994; 28(5): 934-8.
[http://dx.doi.org/10.1021/es00054a027] [PMID: 22191837]

[25]   Hajkova P, Patenka PS, Horsky J, Horska I, Kolouch A. Antiviral and antibacterial effect of photocatalytic TiO2 films. Tissue Engineering 2007; 13(4): 908.

[26]   Ipe BI, Lehnig M, Niemeyer CM. On the generation of free radical species from quantum dots. Small 2005; 1(7): 706-9.
[http://dx.doi.org/10.1002/smll.200500105] [PMID: 17193510]

[27]   Moradi Alvand Z, Rajabi HR, Mirzaei A, Masoumiasl A, Sadatfaraji H. Rapid and green synthesis of cadmium telluride quantum dots with low toxicity based on a plant-mediated approach after microwave and ultrasonic assisted extraction: Synthesis, characterization, biological potentials and comparison study. Mater Sci Eng C 2019; 98: 535-44.
[http://dx.doi.org/10.1016/j.msec.2019.01.010] [PMID: 30813055]

[28]   Huang K, Dai R, Deng W, Lin L, Zhang A, Yuan X. Aqueous synthesis of CdTe quantum dots by hydride generation for visual detection of silver on quantum dot immobilized paper. Anal Methods 2017; 9(36): 5339-47.
[http://dx.doi.org/10.1039/C7AY01705G]

[29]   Williams DN, Saar JS, Bleicher V, Rau S, Lienkamp K, Rosenzweig Z. Poly(oxanorbornene)-Coated CdTe Quantum Dots as Antibacterial Agents. ACS Appl Bio Mater 2020; 3(2): 1097-104.
[http://dx.doi.org/10.1021/acsabm.9b01045] [PMID: 33215080]

[30] Akbari M, Rahimi-Nasrabadi M, pourmasud S, *et al.* CdTe quantum dots prepared using herbal species and microorganisms and their anti-cancer, drug delivery and antibacterial applications; a review. Ceram Int 2020; 46(8): 9979-89.
[http://dx.doi.org/10.1016/j.ceramint.2020.01.051]

[31] Kumari A, Khare SK, Kundu J. Adverse effect of CdTe quantum dots on the cell membrane of Bacillus subtilis : Insight from microscopy. Nano-Structures & Nano-Objects 2017; 12: 19-26.
[http://dx.doi.org/10.1016/j.nanoso.2017.08.003]

[32] Ananth DA, Rameshkumar A, Jeyadevi R, *et al.* Antibacterial potential of rutin conjugated with thioglycolic acid capped cadmium telluride quantum dots (TGA-CdTe QDs). Spectrochim Acta A Mol Biomol Spectrosc 2015; 138: 684-92.
[http://dx.doi.org/10.1016/j.saa.2014.11.082] [PMID: 25544184]

[33] Li X, Lu Z, Li Q. Multilayered films incorporating CdTe quantum dots with tunable optical properties for antibacterial application. Thin Solid Films 2013; 548: 336-42.
[http://dx.doi.org/10.1016/j.tsf.2013.09.088]

[34] Courtney CM, Goodman SM, McDaniel JA, Madinger NE, Chatterjee A, Nagpal P. Photoexcited quantum dots for killing multidrug-resistant bacteria. Nat Mater 2016; 15(5): 529-34.
[http://dx.doi.org/10.1038/nmat4542] [PMID: 26779882]

[35] Lai L, Li SJ, Feng J, *et al.* Effects of surface charges on the bactericide activity of CdTe/ZnS quantum dots: A cell membrane disruption perspective. Langmuir 2017; 33(9): 2378-86.
[http://dx.doi.org/10.1021/acs.langmuir.7b00173] [PMID: 28178781]

[36] Deng Q, Zhang W, Lan T, *et al.* Anatase TiO2 quantum dots with a narrow band gap of 2.85 eV based on surface hydroxyl groups exhibiting significant photodegradation property. Eur J Inorg Chem 2018; 2018(13): 1506-10.
[http://dx.doi.org/10.1002/ejic.201800097]

[37] Ahmed F, Awada C, Ansari SA, Aljaafari A, Alshoaibi A. Photocatalytic inactivation of *Escherichia coli* under UV light irradiation using large surface area anatase TiO_2 quantum dots. R Soc Open Sci 2019; 6(12): 191444.
[http://dx.doi.org/10.1098/rsos.191444] [PMID: 31903213]

[38] Fakhri A, Azad M, Fatolahi L, Tahami S. Microwave-assisted photocatalysis of neurotoxin compounds using metal oxides quantum dots/nanosheets composites: Photocorrosion inhibition, reusability and antibacterial activity studies. J Photochem Photobiol B 2018; 178: 108-14.
[http://dx.doi.org/10.1016/j.jphotobiol.2017.10.038] [PMID: 29131989]

[39] Fu G, Vary PS, Lin CT. Anatase TiO_2 nanocomposites for antimicrobial coatings. J Phys Chem B 2005; 109(18): 8889-98.
[http://dx.doi.org/10.1021/jp0502196] [PMID: 16852057]

[40] Sharma S, Umar A, Mehta SK, Ibhadon AO, Kansal SK. Solar light driven photocatalytic degradation of levofloxacin using TiO_2 /carbon-dot nanocomposites. New J Chem 2018; 42(9): 7445-56.
[http://dx.doi.org/10.1039/C7NJ05118B]

[41] Geng H, Jiang N, Li C, Zhu X, Qiao Y, Cai Q. Efficient photocatalytic inactivation of E. coli by Mn-CdS/ZnCuInSe/CuInS $_2$ quantum dots-sensitized TiO_2 nanowires. Nanotechnology 2020; 31(39): 395602.
[http://dx.doi.org/10.1088/1361-6528/ab8d6c] [PMID: 32340006]

[42] Kim YT, Han JH, Hong BH, Kwon YU. Electrochemical synthesis of CdSe quantum-dot arrays on a graphene basal plane using mesoporous silica thin-film templates. Adv Mater 2010; 22(4): 515-8.
[http://dx.doi.org/10.1002/adma.200902736] [PMID: 20217745]

[43] Kalasad MN, Rabinal MK, Mulimani BG. Ambient synthesis and characterization of high-quality CdSe quantum dots by an aqueous route. Langmuir 2009; 25(21): 12729-35.
[http://dx.doi.org/10.1021/la901798y] [PMID: 19711933]

[44] Murcia MJ, Shaw DL, Woodruff H, Naumann CA, Young BA, Long EC. Facile sonochemical synthesis of highly luminescent ZnS– shelled CdSe quantum dots. Chem Mater 2006; 18(9): 2219-25.
[http://dx.doi.org/10.1021/cm0505547]

[45] Choi SH, Song H, Park IK, *et al.* Synthesis of size-controlled CdSe quantum dots and characterization of CdSe–conjugated polymer blends for hybrid solar cells. J Photochem Photobiol Chem 2006; 179(1-2): 135-41.
[http://dx.doi.org/10.1016/j.jphotochem.2005.08.004]

[46] Priester JH, Stoimenov PK, Mielke RE, *et al.* Effects of soluble cadmium salts *versus* CdSe quantum dots on the growth of planktonic Pseudomonas aeruginosa. Environ Sci Technol 2009; 43(7): 2589-94.
[http://dx.doi.org/10.1021/es802806n] [PMID: 19452921]

[47] Garcia IM, Leitune VCB, Visioli F, Samuel SMW, Collares FM. Influence of zinc oxide quantum dots in the antibacterial activity and cytotoxicity of an experimental adhesive resin. J Dent 2018; 73: 57-60.
[http://dx.doi.org/10.1016/j.jdent.2018.04.003] [PMID: 29653139]

[48] Singh AK, Pal P, Gupta V, Yadav TP, Gupta V, Singh SP. Green synthesis, characterization and antimicrobial activity of zinc oxide quantum dots using Eclipta alba. Mater Chem Phys 2018; 203: 40-8.
[http://dx.doi.org/10.1016/j.matchemphys.2017.09.049]

[49] Fakhroueian Z, Harsini FM, Chalabian F, Katouzian F, Shafiekhani A, Esmaeilzadeh P. Influence of modified ZnO quantum dots and nanostructures as new antibacterials. Adv Nanopart 2013; 2(3): 247-58.
[http://dx.doi.org/10.4236/anp.2013.23035]

[50] Premanathan M, Karthikeyan K, Jeyasubramanian K, Manivannan G. Selective toxicity of ZnO nanoparticles toward Gram-positive bacteria and cancer cells by apoptosis through lipid peroxidation. Nanomedicine 2011; 7(2): 184-92.
[http://dx.doi.org/10.1016/j.nano.2010.10.001] [PMID: 21034861]

[51] Meshram JV, Koli VB, Kumbhar SG, Borde LC, Phadatare MR, Pawar SH. Structural, spectroscopic and anti-microbial inspection of PEG capped ZnO nanoparticles for biomedical applications. Mater Res Express 2018; 5(4): 045016.
[http://dx.doi.org/10.1088/2053-1591/aab917]

[52] Sun Z, Xie H, Tang S, *et al.* Ultrasmall black phosphorus quantum dots: synthesis and use as photothermal agents. Angew Chem Int Ed 2015; 54(39): 11526-30.
[http://dx.doi.org/10.1002/anie.201506154] [PMID: 26296530]

[53] Zhang X, Xie H, Liu Z, *et al.* Black phosphorus quantum dots. Angew Chem Int Ed 2015; 54(12): 3653-7.
[http://dx.doi.org/10.1002/anie.201409400] [PMID: 25649505]

[54] Kong N, Ji X, Wang J, *et al.* ROS-mediated selective killing effect of black phosphorus: mechanistic understanding and its guidance for safe biomedical applications. Nano Lett 2020; 20(5): 3943-55.
[http://dx.doi.org/10.1021/acs.nanolett.0c01098] [PMID: 32243175]

[55] Wang H, Yang X, Shao W, *et al.* Ultrathin black phosphorus nanosheets for efficient singlet oxygen generation. J Am Chem Soc 2015; 137(35): 11376-82.
[http://dx.doi.org/10.1021/jacs.5b06025] [PMID: 26284535]

[56] Lee HU, Park SY, Lee SC, *et al.* Black phosphorus (BP) nanodots for potential biomedical applications. Small 2016; 12(2): 214-9.
[http://dx.doi.org/10.1002/smll.201502756] [PMID: 26584654]

[57] Pal A, Sk MP, Chattopadhyay A. Recent advances in crystalline carbon dots for superior application potential. Materials Advances 2020; 1(4): 525-53.
[http://dx.doi.org/10.1039/D0MA00108B]

[58] Arshad F, Pal A, Rahman MA, Ali M, Khan JA, Sk MP. Insights on the solvatochromic effects in N-

doped yellow-orange emissive carbon dots. New J Chem 2018; 42(24): 19837-43.
[http://dx.doi.org/10.1039/C8NJ03698E]

[59] Arshad F, Sk MP. Aggregation-induced red shift in N,S-doped chiral carbon dot emissions for moisture sensing. New J Chem 2019; 43(33): 13240-8.
[http://dx.doi.org/10.1039/C9NJ03009C]

[60] Zheng XT, Ananthanarayanan A, Luo KQ, Chen P. Glowing graphene quantum dots and carbon dots: properties, syntheses, and biological applications. Small 2015; 11(14): 1620-36.
[http://dx.doi.org/10.1002/smll.201402648] [PMID: 25521301]

[61] Bing W, Sun H, Yan Z, Ren J, Qu X. Programmed bacteria death induced by carbon dots with different surface charge. Small 2016; 12(34): 4713-8.
[http://dx.doi.org/10.1002/smll.201600294] [PMID: 27027246]

[62] Namdari P, Negahdari B, Eatemadi A. Synthesis, properties and biomedical applications of carbon-based quantum dots: An updated review. Biomed Pharmacother 2017; 87: 209-22.
[http://dx.doi.org/10.1016/j.biopha.2016.12.108] [PMID: 28061404]

[63] Yang J, Zhang X, Ma YH, et al. Carbon dot-based platform for simultaneous bacterial distinguishment and antibacterial applications. ACS Appl Mater Interfaces 2016; 8(47): 32170-81.
[http://dx.doi.org/10.1021/acsami.6b10398] [PMID: 27786440]

[64] Jian HJ, Wu RS, Lin TY, et al. Super-cationic carbon quantum dots synthesized from spermidine as an eye drop formulation for topical treatment of bacterial keratitis. ACS Nano 2017; 11(7): 6703-16.
[http://dx.doi.org/10.1021/acsnano.7b01023] [PMID: 28677399]

[65] Verma A, Shivalkar S, Sk MP, Samanta SK, Sahoo AK. Nanocomposite of Ag nanoparticles and catalytic fluorescent carbon dots for synergistic bactericidal activity through enhanced reactive oxygen species generation. Nanotechnology 2020; 31(40): 405704.
[http://dx.doi.org/10.1088/1361-6528/ab996f] [PMID: 32498056]

[66] Stanković NK, Bodik M, Šiffalovič P, et al. Antibacterial and antibiofouling properties of light triggered fluorescent hydrophobic carbon quantum dots Langmuir–Blodgett thin films. ACS Sustain Chem& Eng 2018; 6(3): 4154-63.
[http://dx.doi.org/10.1021/acssuschemeng.7b04566]

[67] Liu J, Lu S, Tang Q, et al. One-step hydrothermal synthesis of photoluminescent carbon nanodots with selective antibacterial activity against Porphyromonas gingivalis. Nanoscale 2017; 9(21): 7135-42.
[http://dx.doi.org/10.1039/C7NR02128C] [PMID: 28513713]

[68] Thakur M, Pandey S, Mewada A, et al. Antibiotic conjugated fluorescent carbon dots as a theranostic agent for controlled drug release, bioimaging, and enhanced antimicrobial activity. J Drug Deliv 2014; 2014: 1-9.
[http://dx.doi.org/10.1155/2014/282193] [PMID: 24744921]

[69] Huang X, Zhang F, Zhu L, et al. Effect of injection routes on the biodistribution, clearance, and tumor uptake of carbon dots. ACS Nano 2013; 7(7): 5684-93.
[http://dx.doi.org/10.1021/nn401911k] [PMID: 23731122]

[70] Kováčová M, Marković ZM, Humpolíček P, et al. Carbon quantum dots modified polyurethane nanocomposite as effective photocatalytic and antibacterial agents. ACS Biomater Sci Eng 2018; 4(12): 3983-93.
[http://dx.doi.org/10.1021/acsbiomaterials.8b00582] [PMID: 33418799]

[71] Gagic M, Kociova S, Smerkova K, et al. One-pot synthesis of natural amine-modified biocompatible carbon quantum dots with antibacterial activity. J Colloid Interface Sci 2020; 580: 30-48.
[http://dx.doi.org/10.1016/j.jcis.2020.06.125] [PMID: 32679365]

CHAPTER 10

Photodynamic Therapy: A Viable Alternative Strategy to Control Microbial Invasions

Moushree Pal Roy[1,*]

[1] *Department of Microbiology, Ananda Chandra College, Jalpaiguri, West Bengal, India*

Abstract: Antimicrobial photodynamic therapy (aPDT) is a new-age therapeutic technique that by principle, focuses on the eradication of target cells by highly cytotoxic reactive oxygen species (ROS) generated through the activation of a chemical photosensitizer (PS) molecule with visible light of appropriate wavelength. The cytotoxic species can arise *via* two main mechanisms known as Type I and Type II photoreactions: the former leads to the generation of ROS and the latter to the formation of the singlet oxygen. These highly reactive oxidants can bring about instantaneous oxidation of a great array of biological molecules, causing havoc to the target cell. This technique provides significant advantages over conventional antimicrobial therapies in practice which are now facing the burning threat of growing complete resistance against them. To combat this world-wide health concern, new treatment strategies are the need of the time while ensuring no further rise of resistance against those alternative therapies, and aPDT appears to be highly promising in this aspect by fulfilling all the demands at the same time. It appears not only equally effective at killing both antibiotic-sensitive and multi-resistant bacterial strains, but also highly selective, non-invasive and rapid in action than other antimicrobial agents, and there have been no reports of resistance till date. The success of this phototherapy relies on several factors, including the target cell type, reaction conditions, and the type, molecular structure and cytolocalization of the PS; because its potency depends on the distribution, high reactivity and short lifetime of ROS as well as the PS itself in electronically excited states.

Keywords: Alternative therapy, Antibacterial, Antibiotic-resistance, Antibiotic-sensitive, Antifungal, Antimicrobial, Antiviral, Illumination, Gram-negative, Gram-positive, Oxidative stress, Photodynamic therapy, Phototherapy, Photoreactions, Photosensitization, Photosensitizer, Reactive oxygen species.

INTRODUCTION

Antimicrobial is a general term that refers to a group of agents used to destroy or inhibit microorganisms, mainly pathogenic ones. Thereby, these antimicrobial

---
* **Corresponding author Moushree Pal Roy:** Department of Microbiology, Ananda Chandra College, Jalpaiguri, West Bengal, India, E-mail: moushree.palroy@gmail.com

occupy the central position in therapeutic measures against microbial invasions; one of the major groups of antimicrobials is specifically antibiotics. Although different kinds of antimicrobial agents have been in use for centuries, the discovery of penicillin in 1928 by Alexander Fleming revolutionized the medical field. Since then, treatment for microbial infections has been mainly channeled through antibiotics as the most effective chemotherapeutic option, but the rapid emergence of resistant strains of bacteria has seriously limited the efficacy of many commonly used antibiotics to treat bacterial infections, leading to a desperate search for alternative therapy to combat health hazards [1].

Antibiotic resistance happens to be nothing new rather, it is basically an evolutionary mechanism for bacteria to fight against antibiotics, as all the natural antibiotics are substances secreted by a group of microorganisms to inhibit the other for gaining survival advantage. Hence, it is an inherent tendency of microorganisms, especially bacteria, to develop appropriate resistance against antibiotics in a short time just to be in the game. Many strains show a natural phenotype of low or no susceptibility to particular antibiotics, known as intrinsic resistance, whereas many susceptible ones acquire the resistance through mutations and horizontal gene transfer mechanisms mainly due to extensive and inappropriate use of antibiotics in the healthcare system [2]. Some of the antibiotic-resistant pathogens that have already emerged as major threats to public health are methicillin-resistant *Staphylococcus aureus* (MRSA), vancomycin-resistant *Enterococcus faecalis* (VRE), multidrug-resistant mycobacteria, Gram-negative pathogens and fungi [3]. Not only are resistance mechanisms emerging fast and spreading globally, but it also threatens our ability to treat common infections, resulting in prolonged illness, disability, and death. The gravitas of the situation has forced the World Health Organization (WHO) to declare in their report in 2014 [4] about the global surveillance of antibiotic resistance that "antibiotic resistance is no longer a prediction for the future; it is happening right now, across the world, and is putting at risk the ability to treat common infections in the community and hospitals. Without urgent, coordinated action, the world is heading towards a post-antibiotic era, in which common infections and minor injuries, which have been treatable for decades, can once again kill". Besides, antimicrobials are also essential in medical procedures such as surgery, organ transplantation, cancer chemotherapy, and diabetes management to control the probability of opportunistic infections. Therefore, antimicrobial resistance puts these procedures at very high risk, simultaneously increasing the overall cost of health care with lengthier hospital stays and more intensive care required.

The urgent need to get over the already-aggravated situation has drawn substantial research interest from all over the world into finding alternative methods of controlling microbial growth that would be clean enough to negate the threat of resistance against antibiotics having a lower probability of developing resistances themselves. A new-age approach to achieve this goal is antimicrobial photodynamic therapy (aPDT) which uses a non-toxic and light-sensitive dye, the so-called photosensitizer (PS), in combination with harmless visible light of the correct wavelength to excite the PS and oxygen that can selectively control bacterial infections [5]. Light has long been an effective agent for decreasing the microbial population and has been used in treating several medical conditions. Given the rise in antibiotic resistance and unknown consequences of long-term use of chemotherapeutic agents, photomedicine is gaining revived interest and being considered more convenient, safe, and efficacious. Antimicrobial photodynamic therapy (aPDT) is one such alternative therapy that fits quite promisingly for all the criteria to treat microbial infections.

The antimicrobial effectivity of PDT comes from the free radicals released from the photosensitizing agents upon stimulation with light. This therapy is defined as an oxygen-dependent photochemical reaction that occurs upon light-mediated activation of a photosensitizing compound leading to the generation of cytotoxic reactive oxygen species, predominantly singlet oxygen and can safely be used as a non-invasive therapeutic modality for the treatment of various infections by bacteria, fungi, and viruses [6]. Photosensitizes are commonly aromatic molecules with a central chromophore with variable auxochrome groups that are mainly responsible for further electron delocalization of the molecule, thereby changing its absorption spectrum [7]. Though originally an age-old practice, this therapy has primarily come around in recent years mostly as a cancer therapy, finally reaching out as a potential antimicrobial therapy due to exhibiting features favorable for the treatment of microbial infections, such as a broad spectrum of action, efficient inactivation of multi-antibiotic-resistant strains, low mutagenic potential, and the lack of selection of photo-resistant microbial cells [7].

BRIEF HISTORY

The inception of using electromagnetic radiation, especially ultraviolet radiation or visible light, for therapeutic purposes dates back to ancient times, practiced efficiently in ancient Greece, Egypt and India. But with time, it majorly lost into oblivion until it resurfaced in the Westerly world at the onset of the 20th century by a Danish physicist, Niels Finsen, who successfully demonstrated PDT by utilizing heat-filtered light from a carbon-arc lamp (The Finsen Lamp) treating a tubercular condition of the skin, called Lupus Vulgaris and consequently he won the Nobel Prize in Physiology or Medicine in 1903 [8]. In 1913, Meyer-Betz injected himself with hematoporphyrin (Hp, a photosensitizer), which led to general skin sensitivity when exposed to sunlight, and this is still a major issue with many photosensitizers [8].

The concept of cell death induced by the interaction of light and chemicals was first reported by Osar Raab, a medical student working with Professor Herman Von Tappeiner in Munich. While studying the effects of acridine on paramecia cultures, he noticed that cells of *Paramecium* spp. stained with acridine orange were destroyed upon exposure to bright light. So acridine orange can be considered as the first agent used for photodynamic action targeted against microorganisms. As a result of subsequent research, Von Tappeiner showed that oxygen was also essential for this activity, which he named as "Photodynamic action" [9]. Even the first application of photodynamic approach to treat cancer was reported by Tappenier and Jesionek in 1903 using topical eosin and white light on skin tumors [10]. Much later, Thomas Dougherty and Co-workers at Roswell Park cancer institute, Buffalo, New York, clinically tested PDT. In 1978, they published striking results in which they treated 113 cutaneous or subcutaneous malignant tumors and observed a total or partial resolution of 111 tumors. The active photosensitizer used in this clinical PDT trial was called Hematoporphyrin Derivative [10].

With these newly found uses, PDT was being explored in the 1970s for the selective treatment of cancerous cells and has successfully established itself as a prime tool for the treatment of various tumors and malignancies and also majorly for age-related macular degeneration, ophthalmologic disorders and in dermatology [3]. Further, PDT was approved by the Food and Drug Administration (FDA) in 1999 to treat pre-cancerous skin lesions of the face or scalp and has gradually emerged as a new non-invasive therapeutic option mainly for localized infections as having particular promise for controlling pathogens resistant to conventional antibiotic therapies [6].

BASIC THERAPEUTIC STRATEGY BEHIND APDT

Photodynamic therapy (PDT) relies on the interplay of visible light and a photosensitizer which, under photo-activation, generate short-lived cytotoxic species in the site [11]. aPDT has been successfully applied *in vivo* and *ex vivo* tissue or in biological materials (*e.g.*, blood sterilization), in animal models for treating localized infections (*e.g.*, surface wounds, burns, oral lesions, abscesses, middle ear infections) and also clinically studied for several dermatological infections (*e.g.*, viral papillomatosis lesions, acne, leishmaniosis and mycobacterial infections) [5]. After light stimulation, the PS gets converted from singlet to triplet state by an intersystem crossing (ISC) process, which, in turn, reacts with surrounding molecules to produce radical species and non-radical reactive species like hydrogen peroxide, or transfers its energy to molecular oxygen to produce singlet oxygen. All these oxygen species are capable of killing target cells by causing oxidative stress to different parts of the cell [12].

Reactive Oxygen Species (ROS) and Oxidative Stress

Reactive oxygen species (ROS) are powerful green oxidants. As the reduction of O_2 is extremely complicated, disproportionation and protonation of intermediate species are unavoidable, and usually, O_2 can be activated by adding energy towards the transformation of itself into ROS. Four oxidation states of O_2 are known: $(O_2)^n$, where n is 0 (dioxygen, O_2); +1 (dioxygen cation, O_2^+); −1 (superoxide ion, O_2^-); and −2 (peroxide dianion, O_2^{-2}) [13]. On the other hand, oxidative stress is a biological phenomenon caused by the production and accumulation of reactive oxygen species (ROS) in cells and/or tissues that fail to detoxify these highly reactive products, thereby suffering from damage [14]. Superoxide radicals ($O_2 \bullet -$), hydrogen peroxide (H_2O_2), hydroxyl radicals ($\bullet OH$), and singlet oxygen ($^1O_2$) are common ROS which is reactive intermediates with a highly positive redox potential (Table **1**) and are generated by-products of aerobic respiration and metabolism. When maintained at low concentrations, these reactive species play roles in some physiological processes such as in phagocytosis and in the synthesis of specific cellular structures; but if accumulated in excess, they may cause extremely aggressive oxidative damage to major cellular structures, especially membranes, lipids, proteins, lipoproteins and DNA [14]. Although antioxidant defenses, *e.g.*, enzymes superoxide dismutase (SOD) and catalase, largely counteract these ROS to mitigate damage, higher concentrations of some ROS may overwhelm these defenses (*e.g.*, H_2O_2, $O_2 \bullet -$), whereas there are no adequate defenses against even small amounts of some of them (*e.g.*, $\bullet OH$, $^1O_2$) [7]. The resulting damage may rapidly prove fatal to the organism, especially to bacteria, which are unicellular in nature. This is why our immune system effectively uses these microbicidal oxidants for the phagocytic killing of pathogens through a process known as a respiratory burst or oxidative burst, and this makes microbial infection followed by immune cell activation a prime trigger for endogenous free radical production. In fact, it plays a major role in the killing of a wide array of cells, including microorganisms such as bacteria, fungi, viruses and also malignant and non-malignant nucleated cells, thus protecting our body against them.

Table 1. Standard reduction potentials for biologically relevant molecules and reactive species.

Half Reaction	Electrode Potential
$HO^{\bullet} + e- + H^+ \rightarrow H_2O$	+2.31 V
$O_3 + 2e- + 2H^+ \rightarrow H_2O + O_2$	+2.075 V
$O^-_3 + e- + 2H^+ \rightarrow H_2O + O_2$	+1.9 V
$Co(III) + e- \rightarrow Co(II)$	+1.82 V
$CO_3^- + e- \rightarrow CO_3^{2-}$	+1.8 V

(Table 1) cont.....

Half Reaction	Electrode Potential
$H_2O_2 + 2e- + 2H^+ \rightarrow 2H_2O$	+1.76 V
$RO^\cdot + e- + H^+ \rightarrow ROH$ (alkoxyl)	+1.6 V
$O^{\cdot-} + e- + 2H^+ \rightarrow H_2O$	+1.46 V
$N_3^\cdot + e- \rightarrow N_3^-$	+1.3 V
$^1\Sigma_g O_2 + e- \rightarrow O_2^{\cdot-}$	+1.27 V
$N_2O_4 + e- \rightarrow NO^\cdot + NO_3^-$	+1.2 V
$HOCl + H^+ + 2e- \rightarrow H_2O + Cl^-$	+1.08 V
$Fe(III)(1,10\text{-phen})_3 + e- \rightarrow Fe(II)(1,10\text{-phen})_3$	+1.06 V
$HO_2^\cdot + e- + H^+ \rightarrow H_2O_2$	+1.06 V
$O_3 + e^- \rightarrow O_3^{\cdot-}$	+1.03 V
$ROO^\cdot + e^- + H^+ \rightarrow ROOH$ (alkylperoxyl)	+1.0 V
$NO_2^\cdot + e^- \rightarrow NO_2^-$	+0.99 V
$O_3(g) + e^- \rightarrow O_3^{\cdot-}$	+0.91 V
$^1\Delta_g O_2 + e- \rightarrow O_2^{\cdot-}$	+0.81 V
$N_2O_3^- + e^- \rightarrow NO^\cdot + NO_2^-$	+0.8 V
$^1\Delta_g O_2(g) + e- \rightarrow O_2^{\cdot-}$	+0.64 V
$SO_3^\cdot e^- \rightarrow So_3^{2-}$	+0.63 V
$H_2O_2 + e^- + H^+ \rightarrow H_2O + HO^\cdot$	+0.32 V
$O_2 + e\cdot^- + 2H^+ \rightarrow H_2O_2$	+0.36 V
$ONOO^\cdot + e^- \rightarrow ONOO^-$	+0.2 V
$Cu(II) + e^- \rightarrow Cu(I)$	+0.16 V
$2H^+ + 2e^- \rightarrow H_2$	0.00 V
$O_2 + e^- \rightarrow O_2^{\cdot}$	-0.18 V
$O_2(g) + e^- \rightarrow O_2^{\cdot}$	-0.33 V
$NAD^+ + e^- + H^+ \rightarrow NADH^-$	-1.58 V
$H_2O + e^- \rightarrow e_{aq}^-$	-2.87 V

[In aqueous solution (pH = 7), unless otherwise stated (g), at 25 °C and 1 M concentration.] SOURCE: Vatansever *et al.* (2013) [7].

Mechanism of Photosensitization During aPDT

The PDT technique utilizes three elements, *i.e.* , a photosensitizer molecule (PS), visible light and oxygen, to yield potent microbicidal oxidizing species that can readily bring about death to the target cells. This lethal process of cell killing is called photosensitization which is being popularized for treating microbial infections.

Light Sources

Photodynamic therapy requires a source of visible light to excite the photosensitizer molecules by exposing them to lower energy electromagnetic radiation of a specific wavelength. The majority of the photosensitizers (PSs) get activated by red light between 630 and 700 nm, corresponding to a light penetration depth of 0.5-1.5 cm, which also limits the depth of necrosis. The total light dose, dosage rates, and depth of destruction vary with the target tissue treated and photosensitizer used [15]. Mainly, three kinds of light sources are in use in aPDT:

Incoherent Light Sources or Gas Discharge Lamps (GDL): The first light source used in aPDT was the conventional bulbs of different types, including tungsten filament, quartz halogen, xenon arc, metal halide and phosphor-coated sodium lamps. Though they are often preferred for treatment of larger areas, these bulbs did not yield good results owing to the characteristic of polychromatic, strong thermal component, and incoherency. They transmit excessive heat to the illuminated area, which can lead to tissue damage. Therefore, other light sources have been tried to cause the molecular excitement [16].

Light Amplification by Stimulated Emission of Radiation (LASER): Laser is an exceptional source of radiation that is capable of producing extremely fine spectral bands, intense, coherent electromagnetic fields extending from the near-infrared to ultraviolet. The emitted laser light is characterized by electromagnetic waves of the same wavelength and direction that differs from conventional light with different wavelengths in all directions. For effective microbial killing, a laser with low power (≤100mV) or low intensity should be used. The first low-intensity laser used was a gas mixture of helium and neon (HeNe laser) emitted in the red spectrum (632.8 nm). Currently, a great majority of lasers use a crystal of semiconductor diode of gallium arsenide (GaAs) produced in the laboratory, which can be doped with various other elements, depending on the desired intensity or the intended purpose, *e.g.* gallium-aluminum-arsenide diode lasers (630–690, 830, or 906 nm), and argon laser (488–514 nm). They are easy to handle, portable, and cost-effective [17].

Additionally, liable PDT set up demands: (a) the right drivers, which are often expensive, even if reused by changing the semiconductor sources; (b) a proper set of lenses/mirrors to control and focus the optical beam;(c) the corresponding mechanical mounts, keeping the optical elements in a stable and defined position;(d) an optical spectrum analyzer, for the stability of the wavelength emitted by the laser source over time (as it may drift in case of non-accurate thermal control); (e) a power meter for ensuring the stability of the laser operating conditions even after months of usage; and (f) a properly calibrated camera to achieve spatial intensity distribution and to analyze the images obtained to make sure that the beam uniformity is maintained on the whole surface [5].

Light Emitting Diode (LED): LEDs are another alternative light source for aPDT and are light weight, economical and flexible. This device has an active semiconductor medium that emits visible light when energized; this light is formed when a certain voltage is applied between layers of semiconductor dopants [17]. LEDs are not monochromatic like lasers, can still produce high-intensity broad-spectral bands. In the LED, the light wave can be produced in various lengths and with adequate power, although the laser has a greater light penetration capacity into the tissue [18]. In addition to its relatively low cost and great versatility, LEDs can be arranged in various ways and in large quantities for irradiation of large areas [19].

The light sources used in PDT may vary depending on the location and form of the lesion, but are typically fiber-optic catheters terminated with cylindrical diffusers or lenses for flat-field applications. Uniform illumination is a prerequisite for accuracy in the calculation of the delivered dose. Devices for simultaneous monitoring of both light delivery and PS's fluorescence are highly desirable for the advancement of PDT into routine medical practice [20].

Photosensitizers and the Photophysical Processes

Here, the target microbial cells are first pre-impregnated with a photosensitizer dye (PS) and subsequently exposed to a specific wavelength of light [17]. The absorption of the light by the PS leads to a transition from its initial ground state (PS_0) to an energetically excited state ($^1PS^*$) that can relax to the more long-lived triplet state ($^3PS^*$) and interact with molecular oxygen enabling the PS to regain its ground state [5].

In the singlet state, a PS has the most stable electronic configuration, with its electrons spinning paired in low-energy orbitals. Upon application of a particular wavelength of electromagnetic radiation that matches their transition energy, the electron in the highest occupied molecular orbital (HOMO) of the PS jumps to the lowest unoccupied molecular orbital (LUMO), causing the PS to reach the unstable and short-lived excited singlet state [3]. This may rapidly lead to several processes such as fluorescence and internal conversion to heat, but the most critical of these to PDT is the reversal of the excited electron's spin, known as intersystem crossing, to the triplet state of the PS [21]. This triplet state is less energetic than the excited singlet state, but has a considerably longer lifetime, as the excited electron, now with a spin parallel to its former paired electron, may not immediately fall back down (as it would then have identical quantum numbers to that of its paired electron, thus violating the Pauli Exclusion Principle) [3]. Accordingly, the excited electron in the PS triplet state may first obtain correct spin orientation (a relatively slow process) and then fall to ground levels (phosphorescence), or the PS may interact with molecules abundant in its immediate environment in the process. Due to the Selection Rules which specify that triplet-triplet interactions are spin-allowed while triplet-singlet interactions are spin-forbidden, the PS triplet can react readily with molecular oxygen that is a triplet in its ground state [3]. This much longer lifetime (many μs as compared to a few ns) means the triplet PS can survive long enough to carry out chemical reactions, which would not have been possible with the excited singlet PS and hence dyes without a significant triplet yield may be highly absorbent or fluorescent but are not good PS [21].

An optimal photosensitizer must possess optimum photo-physical, chemical, and biological characteristics, including the following ones [20]:

a) Pure and chemically defined.

b) High quantum yield for singlet oxygen production.

c) Strong absorption with a high extinction coefficient at a longer wavelength (preferably between 630-700nm).

d) Optimum photochemical reactivity with minimal dark toxicity.

e) Should impart toxicity only when illuminated.

f) Desirable retention by the target tissue.

g) Should be readily excreted out from the body.

h) Can be synthesized from easily available precursors and be capable of formulation

i) Stable and readily dissolved in the tissue fluids.

Most of the PSs used for clinical purposes belong to the following basic structures [5, 20], and some of these structures are depicted in Fig. (**1**).

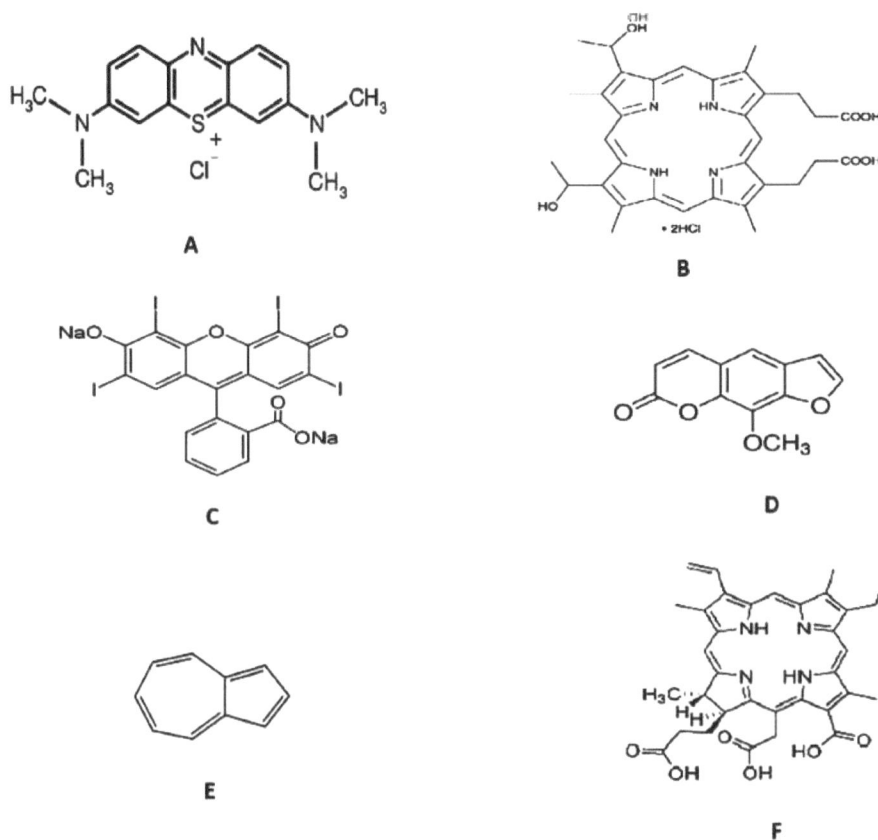

Fig. (1). Chemical structures of various photosensitizers. A. Methylene blue; B. Hematoporphyrin HCl; C. Erythrosine; D. Xanthotoxin; E. Azulene; F. Chlorin.

1. Tricyclic dyes with different meso-atoms *e.g.*, Acridine orange, proflavine, riboflavin, methylene blue, toluidine blue O, fluorescein, and erythrosin.

2. Tetrapyrroles. *e.g.*, Porphyrins and derivatives (Hematoporphyrin HCl, photofrin, benzoporphyrin derivatives), chlorophyll, phylloerythrin, and phthalocyanines.

3. Furocoumarins. *e.g.*, Psoralen and its methoxy-derivatives, xanthotoxin, and bergaptene.

4. Xanthenes. *e.g.*: Erythrosine

5. Monoterpene. *e.g.*: Azulene.

6. Chlorins. *e.g.*, Chlorine e6, stannous (IV) chlorine e6, chlorine e6-2.5 N-

methyl-d-glucamine (BLC1010), polylysine and polyethyleneimine conjugates of chlorine e6.

Photofrin and hematophyrin derivatives are called as first generation sensitizers; while 5-aminolevulinic acid (ALA), benzoporphyrin derivative, texaphyrin, and temoporfin (mTHPC) are second-generation PSs that have greater ability to generate singlet oxygen [22]. Topical ALA has been used to treat pre-cancer conditions, and basal and squamous cell carcinoma of the skin. Third-generation PSs include biological conjugates (*e.g.*, Antibody conjugate, liposome conjugate) [22]. In antimicrobial studies, two common PSs used are toluidine blue O and methylene blue. Both have similar chemical and physicochemical characteristics and are popularly used stains in microbiological and histological preparations. Toluidine blue O solution is blue-violet in color and it stains granules within mast cells and proteoglycans/glycosaminoglycans within connective tissues [6]. On the other hand, methylene blue is a redox indicator that is blue in an oxidizing environment and becomes colorless upon reduction. Methylene blue, when irradiated, has effectively killed the influenza virus, *Helicobacter pylori*, and *C. albicans*. Both stains have been proved to be very useful for the inactivation of both Gram-positive and Gram-negative periodontopathic bacteria. Tetracyclines, used as antibiotics in periodontal diseases, are also effective photosensitizers producing singlet oxygen [6].

Dark Activity

As discussed above, all the PS get excited by exposure to an appropriate amount of light of the particular wavelength and thus, various ROS species are formed that are responsible for cellular destruction in PDT. However, many researchers have reported an unusual aspect of the use of photosensitizer molecules, known as the dark activity or dark toxicity of PSs. This feature can be defined as the toxic effect of the PSs that is imparted in the absence of illumination. This indicates that in addition to their phototoxicity, many PSs can impose some other mechanisms to cause target cell damage where neither $^1O_2$ nor ROS species (which are formed only in the presence of light) are involved [23]. Dark activity basically depends on the concentration and type of the PS used, manifesting itself in different ways for various PSs, with illumination inevitably extending the harm. There are reports of dark toxicity of Rose Bengal (RB) dye against Gram-positive *Enterococcus faecalis, Bacillus* sp., *Listeria monocytogenes, Staphylococcus aureus* and Gram-negative *Pseudomonas aeruginosa;* of phloxine B against Gram-positive *Bacillus* spp. and *Listeria monocytogenes;* of high concentrations (>500 μg/mL) of acriflavin neutral against *E. coli,* whereas dark toxicity of malachite green against the same microorganisms has been found to be very low [24]. These observations ultimately intrigued a continuing search for some alternative methods for the excitation of PSs in the dark in order to overcome the problem with visible light, though cheapest, easiest and most effective in its way to activate PSs, being limited in its application owing to its restricted penetration into tissues. Ultrasonic waves, chemo- or bioluminescent light and non-ionizing radiofrequency electromagnetic waves are being considered as possible candidates to attract research attention as an alternative illumination to activate PSs against microorganisms [24].

Photochemical Reactions and Generation of ROS

PDT's oxidizing characteristics are centered on molecular oxygen (O_2). The ground state of oxygen is a triplet state, whereby the two outermost orbitals are unpaired but spin parallel. The stability of the PS in the singlet excited state ($^1PS^*$) actually dictates the intersystem crossing to the triplet and long-lived excited state ($^3PS^*$), which may involve many physical pathways [25]. Only this triplet state has the ability to undergo photochemical processes and interact with triplet state molecules such as molecular oxygen. When the PS is in the long-lived triplet state, it may interact with O_2 in two distinctly different ways: Type I and type II photoreactions. Type I and type II photoreactions are both oxygen-dependent and involve the formation of highly unstable reactive species such as radical cations or neutral radicals from the substrates and/or singlet oxygen ($^1O_2{}^1\Delta_g$) by direct energy transfer to molecular oxygen [26].

Type I Photoreactions: This type of reaction involve the transfer of electrons from the PS in an excited triplet state to molecular oxygen, in the presence of a suitable reducing agent, to produce reactive oxygen species such as superoxide anion(O_2^-), hydroxyl radical ($^.OH$), hydrogen peroxide (H_2O_2) and hydroxide ions (OH^-) (Fig. **2**) [25]. Type I reactions account for an electron and/or proton transfer, where the PS interacts directly with the cellular substrate (*i.e.* , lipids, proteins, nucleic acids, *etc* .) [5]. In this type of reactions, the PS in the excited triplet state extracts an electron from a reducing molecule (*e.g.* NADPH, guanine in nucleic acids, tryptophan and tyrosine in proteins), leading to the formation of a pair of radical anion ($^.PS^-$) and radical cations ($^.biomolecule^+$). In oxic conditions, this PS radical anion donates its extra electron to O_2, producing a superoxide anion radical ($^.O_2^-$) and restoring the PS [27].

Type II Photoreactions: In type II reactions, energy from the PS in the excited triplet state is directly transferred to molecular oxygen ($^3O_2$) to form more reactive singlet oxygen ($^1O_2$) in the excited state. Energy transfer to $^3O_2$ can occur only if both PS and oxygen (triplet ground state) are in the same triplet state [25]. This energy promotes one of the two unpaired electrons of $^3O_2$ to a high-energy orbital, also inverting its spin, thereby converting it into $^1O_2$ (Fig. **2**) [27]. This non-radical species is highly reactive toward electron-rich substrates such as aromatic rings, amines, and thioesters [28].

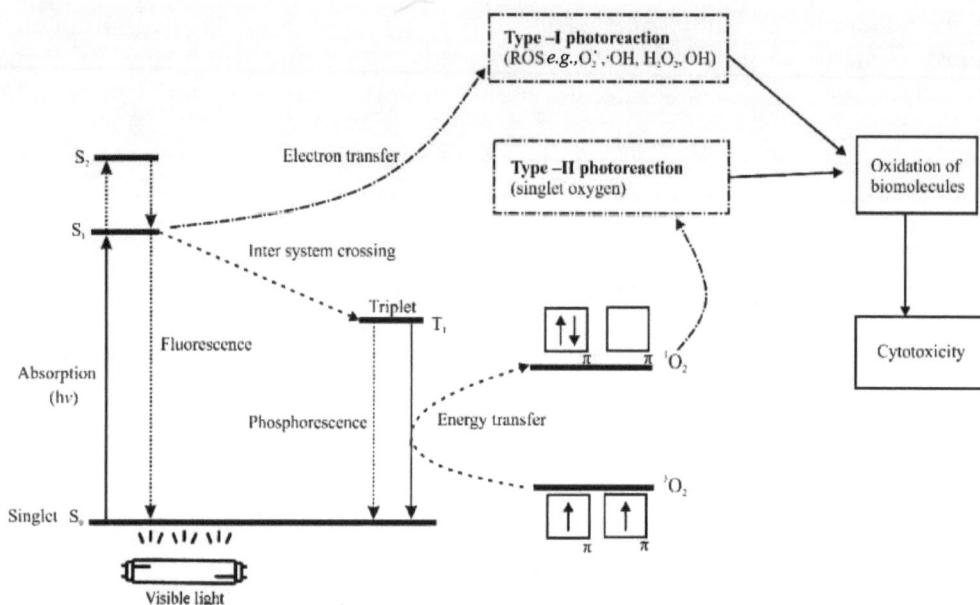

Fig. (2). Jablonski diagram showing the main events leading to type I and type II photochemical reactions, which eventually may result in oxidative cell damage. S_0, ground state of the photosensitizer (PS); S_1, first excited singlet state of PS; S_2, second excited singlet state of PS; T_1, first excited triplet state of PS; $^3O_2$, triplet oxygen; $^1O_2$, singlet oxygen. MODIFIED FROM: Zhang *et al.* 2018, Ormond and Freeman, 2013 [90, 93]

So, both types of photoreactions cause the formation of ROS, which leads to cellular damage observed during PDT. As type II reactions have been observed to occur more frequently, it is often believed that $^1O_2$ generated from a type II reaction is the major reason for the biological effect of PDT. This is because of the facts that not only energy transfer to O_2 occurs at a higher rate ($k \approx 1–3 \times 10^9$ $M^{-1} s^{-1}$) than electron transfer (*e.g.*, to give O_2^-, estimated as $k \leq 1 \times 10^7 M^{-1} s^{-1}$) but also singlet oxygen is more reactive than O_2^-; moreover, aerobic cells possess enzyme systems to counter superoxide, but antioxidant enzymes that eliminate singlet oxygen have not yet evolved, presumably because of its short lifetime [27]. However, recent studies indicated that radical species from the type I mechanism may lead to an amplified PDT response, particularly under low oxygen conditions; as for the type II pathway PDT effect relies greatly on oxygen availability. Although direct comparisons between the contributions of type I and type II mechanisms to PDT efficacy are difficult due to the complexity of ROS formation on environmental factors, it is clear that these two competing mechanisms can occur simultaneously [29]. In general, the photoreactions and their effects on the cells depend on several factors, including the kind of PS, the subcellular localization, the substrate, and molecular oxygen concentration within the target cells. Despite the fact that detailed mechanisms of these processes are not yet fully understood, it has been established that the photoreactions, regardless of the type, produce ROS that cause instant oxidation of biomolecules like lipids, proteins (especially amino acids), and nucleic acids (mostly guanine) in the cell. Studies also indicated that actively growing cells selectively uptake PS, and cell death can be restricted to regions where the light of the appropriate wavelength is applied, thereby allowing PDT to be applied selectively [5]. As microorganisms show the fastest generation time and growth rate, it was suggested that PDT could be effective against microbial cells. Moreover, Huang *et al.* [30] reported that Gram-negative bacteria are more susceptible to $\cdot OH$ than $^1O_2$ and so type I reaction is preferred against Gram-negative species.

Recent studies have proposed two new strategies of photoreactions, called as type III and IV, where the cytotoxicity occurs even in the absence of oxygen within the cellular structures [31].

Type III Photoreactions: The PSs used in this process are antioxidant carrier sensitizers (ACS) and have been shown to decrease the radical concentrations in the target cells and generate singlet oxygen [32]. Antioxidants, reducing agents able to scavenge oxidative species, generally have been shown to reduce the photodamaging activity of PDT. In contrast, few of them (*e.g.*, ascorbic acid, butyl-4-hydroxyanisole or α-tocopherol), when added with controlled conditions like amount, timing and presence of transition metals, have been reported to enhance the efficacy of PDT, especially against bacterial activity [33]. Therefore, these results may lay a promising path towards the development of a highly effective new therapeutic modality exploiting ACS molecules.

Type IV Photoreactions: In this type of reaction, the PS cannot bind to the molecular target, and after irradiation photoisomerization (*i.e.* , structural change between isomers caused by photoexcitation) may occur. This process causes intramolecular remodeling that facilitates PS binding to the cellular target [32].

Induction of the Oxidative Stress

Superoxide anion ($O_2^{\cdot-}$) radical is the most important ROS formed by the electron transfer reactions during type I photoreactions in which an electron is transferred to molecular oxygen.

$$O_2 + e^- \rightarrow O_2^{\cdot-}$$

Although superoxide is able to act both as a monovalent oxidant and a reductant, it does not show much reactivity in biological systems to cause much oxidative damage directly to nucleic acids, lipids or carbohydrates [34]. When acting as an oxidant, the superoxide radical can readily oxidize small molecules such as sulfite, tetrahydroflavins, leukoflavins, catecholamines, the enediolate tautomers of sugars, and can react with biologically relevant radicals, releasing potentially toxic cell-damaging products including organic hydrogen peroxide (H_2O_2) and others; for example, it reacts with another superoxide radical in a dismutation reaction in the presence of superoxide dismutase (SOD), in which one radical is oxidized to oxygen, and other is reduced to hydrogen peroxide [35].

$$O_2 + O_2^{\cdot-} + 2H_2O \rightarrow H_2O_2 + O_2$$

H_2O_2 is not a free radical but it can cause damage to the cell even at relatively low concentration (10 μM). It can readily diffuse through the biological membranes and can cause DNA damage by producing hydroxyl radical (OH^-) in the presence of transition metal ions despite having no direct effect on DNA. At higher concentrations, various enzymes, such as glyceraldehhyde-3-phosphate dehydrogenase, are inhibited [35].

Superoxide can also react with nitric oxide (NO^-) to produce peroxynitrite ($OONO^-$), with guanine neutral radical [G(–H)·] to form oxidatively modified guanine bases and with phenols (abundant in proteins as in amino acid tyrosine) to generate phenoxyl radicals [27]. Peroxynitrite, a causative agent of nitroxidative damage, can react with CO_2 and bicarbonates to yield nitrosoperoxy carbonate, a precursor of the carbonate radical anion (CO_3^-), which in turn can oxidise tyrosine and tryptophan as a one-electron oxidant [36]. It has been reported that hydrogen bonding to an amine group increases the electrophilicity of O_2^- and the reduction potential of the O_2/O_2^- couple, which accounts for a dramatic increase in O_2^- reactivity [37]. Not only free radicals but superoxide can also induce oxidative changes to protein molecules, such as [4Fe-4S] clusters of proteins, and as these proteins are indispensable for cellular respiration, destruction of their [4Fe-4S] clusters have detrimental repercussion. Moreover, as a reductant, superoxide can donate one electron to the metal ions (such as ferric iron or Fe^{3+} in the cluster) in the Fenton reaction (iron, copper, or other metals can also catalyze this reaction), so that the reduced metal (ferrous iron or Fe^{2+}) then catalyzes the breaking of the oxygen–oxygen bond of hydrogen peroxide to produce powerful oxidants, hydroxyl radical ($HO^{\cdot}$) and hydroxide ion (HO^-) [34].

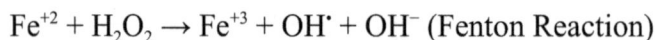

$$Fe^{+2} + H_2O_2 \rightarrow Fe^{+3} + OH^{\cdot} + OH^- \text{ (Fenton Reaction)}$$

The Fe^{2+} released from the [4Fe-4S] clusters would bind to anionic molecules, including proteins, nucleic acids, lipids and other cell membrane components and would be kept in a reduced Fe^{2+} state by cellular reductants. H_2O_2, which is relatively stable and can diffuse across membranes, would reach Fe^{2+} bound to biomolecules, and HO• would be generated at the spot [38].

The hydroxyl radical (HO•), the three-electron reduction product of molecular oxygen, is the most reactive species of oxygen due to its high reduction potential, has a very short half-life (10^{-9} s), reacts with organic molecules immensely fast almost at diffusion limit ($>10^9$ M^{-1} s^{-1}) and are primarily responsible for the cytotoxic effect in an aerobic organism [39]. This highly reactive radical can perform a number of oxidation reactions with an organic substrate, often by forming hydroxylated adducts or abstracting an electron, and the resulting oxidized molecule being a radical itself can amplify a chain reaction of oxidative damage inside a cell [34]. For example, HO• can react with the superoxide to form singlet oxygen, or it can react with ground-state oxygen to produce a peroxyl radical (ROO•) [40]. Hydroxyl radicals have very high rate constants with a majority of biomolecules, including nucleic acids (especially guanosine), lipids, sugars, polypeptides and proteins (especially thiamine). It initiates the primary stage of lipid hydroperoxidation generating a lipid radical (•Lipid) which triggers the chain reaction of lipid peroxidation on their acyl chains [40]. It can also add readily to unsaturated compounds, especially aromatic compounds, where it adds to a double bond, resulting in hydroxycyclohexadienyl radical. The resulting radical goes on a chain reaction, like a reaction with oxygen, to give peroxyl radical, or decomposition by water elimination to phenoxyl type radicals [39].

Singlet Oxygen ($^1O_2$), formed during the type II reaction, is a highly excited, meta-stable state of O_2 and is more reactive than $O_2^{•-}$ [35]. Molecular oxygen has two singlet (S_1) states ($^1\Delta_g$ and $^1\Sigma_g$) that differ by the arrangement of the π-antibonding orbitals and by their energy differences of 95 kJ/mol versus 158 kJ/mol with the ground state, respectively (Fig. **3**) [41].

The presence of electrons with opposite spins, which removes the spin restriction typical for the triplet (T_1) oxygen, makes singlet oxygen highly reactive, as it acts quickly on unsaturated carbon-carbon bonds, neutral nucleophiles such as sulfides and amines, and with anions, initially producing peroxides followed by other radicals in chain reactions that can launch a variety of destructive changes in the cell [27].

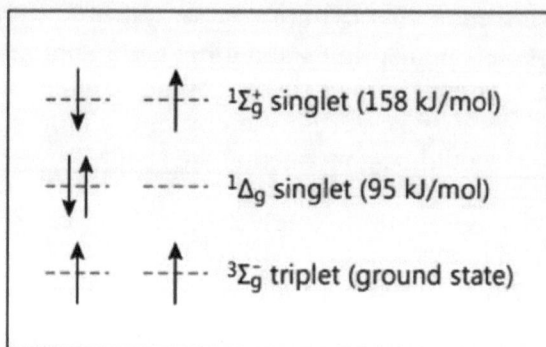

Fig. (3). Schematic representation of the electron configuration and energetic levels of ground and excited molecular oxygen states. SOURCE: Benov, 2015 [27].

The presence of electrons with opposite spins, which removes the spin restriction typical for the triplet (T_1) oxygen, makes singlet oxygen highly reactive, as it acts quickly on unsaturated carbon-carbon bonds, neutral nucleophiles such as sulfides and amines, and with anions, initially producing peroxides followed by other radicals in chain reactions that can launch a variety of destructive changes in the cell [27].

Recent studies suggest that high PDT efficacy results from a synergistic action of species generated by type I and type II reactions ($O_2^{\cdot-}$, $HO\cdot$ and $^1O_2$) [42]. The fact that the lifespan of PDT products is extremely short, allows only specific parts of the cells that are within a 20 nm radius being affected, and this makes localization of the PS instrumental in delivering the damage to particular cellular structures and, as well as in the success of the PDT treatment [43]. Proteins and largely amino acid residues (mainly cysteine, methionine, tyrosine, histidine, and tryptophan) in them are regarded as the main target for photooxidations. The reaction mechanisms are complex giving rise to a variety of end products, *e.g.*, cysteine and methionine get oxidized to sulfoxides, histidine yields a thermally unstable endoperoxide, tryptophan becomes N-formylkynurenine while tyrosine can undergo phenolic oxidative coupling [34]. On the other hand, unsaturated fatty acids, majorly present in membranes, are also targeted by $^1O_2$ and other ROS-inducing lipid peroxidation *via* separate mechanisms to initiate free-radical chain reactions, and this can lead to a bunch of deleterious outcomes such as direct damage of lipids as well as membrane integrity and functions, secondary modifications of proteins and polynucleotides, alterations in metabolism and cell signaling pathways, resulting ultimately in cell death [44]. DNA damage by photooxidation at the level of the nucleic bases or at the glycoside linkages or cross-linking to protein is one of the prime causes of cell death in PDT [45].

PDT AGAINST MICROORGANISMS

Studies have shown that aPDT is a powerful tool for controlling and killing pathogens. It can be used successfully against a wide array of harmful microorganisms, including viruses, bacteria, and fungi. Hence, PDT acts as an alternative antibacterial, antiviral and antifungal treatment, especially for drug-resistant strains, as it is not yet likely that microbial strains would develop resistance against the cytotoxic action of a variety of ROS molecules arising *via* cascades that have been weaponized by our immune system to resist microbial attacks from time immemorial.

aPDT for Bacterial Infections

According to the WHO, "Antibiotic resistance is one of the biggest threats to global health, food security, and development today, and it leads to higher medical costs, prolonged hospital stays, and increased mortality" [4]. For example, Gram-positive and Gram-negative 'superbugs' such as *Enterococcus faecium, Staphylococcus aureus, Klebsiella pneumoniae, Acinetobacter baumannii, Pseudomonas aeruginosa,* and *Enterobacter* species (the so-called 'ESKAPE' strains) are showing resistance towards almost all antibiotics [46]. This rapid and uncontrolled emergence of complete drug-resistant (CDR) and multiple drug-resistant (MDR) bacterial strains is instrumental in the accelerated search for alternative antibacterial techniques and bactericides. Studies showing encouraging results of photodynamic antibacterial therapy (PDAT) have placed the focus mainly on control of the resistant strains, including the 'superbugs' and also biofilm-associated infections. This method is gaining acceptance for multiple reasons like its exceptional non-invasiveness, broad antibacterial spectrum, and above all, less probability to grow drug resistance among bacteria owing to the fact that it does not involve direct interaction between PS and bacteria [47].

In fact, PDAT has several advantages over antibiotics [46]:

1. It is considered highly specific due to (a) the controlled administration of PSs to be taken up only by the target cells, (b) the ineffectiveness of PSs in the absence of light activation, (c) the site-directed irradiation of the infected area.

2. As a result of targeted application, damages to the cells other than the infected ones can be avoided.

3. Development of resistance against the PSs is less likely, even after repeated treatment with PDAT, due to several reasons: (a) the drug-light interval is too short for bacteria to develop resistance, (b) it is difficult for bacteria to 'sense' that the oxidative stress emanates from the otherwise non-toxic PS, so any metabolic

adaptations are directed elsewhere (*e.g.*, antioxidant defense machinery),(c) the cells becomes too damaged to confer cross-generation adaptivity, (d) unlike traditional antibiotics, the PDAT targets various cellular structures and different metabolic pathways at a time by producing an array of ROS [46].

For these reasons, the application of aPDT against bacterial infections has gained considerable research interest. Many studies have showed that the affectivity of PS depends on preferential binding and uptake by the target cells, and that again is dependent on the bacterial species owing to their distinctive cell surface biochemistry, especially among Gram-positive and Gram-negative species (Table **2**). It has been established that PDT gives better results against Gram-positive strains, which are much more susceptible to anionic and neutral PS because of the thick, porous peptidoglycan cell wall, than Gram-negative bacteria, which are less prone to intake exogenous compounds due to the outermost lipopolysaccharide layer acting as an extra permeability barrier [46]. The primary targets of PDAT are commonly various proteins, lipids and nucleic acids present in bacterial cells. The bactericidal action of PDAT has been reported in different bacterial species (both Gram-positive and Gram-negative strains) like *Staphylococcus aureus, Streptococcus pneumoniae, Streptococcus mutans, Streptococcus sanguis, Pseudomonas aeruginosa, Mycobacterium tuberculosis, Haemophilus influenzae, Helicobacter pylori, Enterococcus faecalis, Escherichia coli, Porphyromonas gingivalis, Prevotella intermedia, Aggregatibacter actinomycetemcomitans, Tannerella forsythia, Fusobacterium nucleatum, Porphyromonas gingivalis, P. intermedia, Propionibacterium acnes,* etc. [17, 46].

Table 2. The extra- and intracellular targets of some common photosensitizers.

Class	Name	Extracellular target	Intracellular target	Bacteria
Phenothiazinium	Methylene blue (MB)	Cell wall surface and membrane protein	Chromosomal DNA	*E. faecalis*
	Rose Bengal (RB)	Cytoplasmic membrane	DNA*	*E. coli*
	Toluidine blue O (TBO)	Lipopolysaccharides and outer membrane	ND	*P. aeruginosa*

(Table 2) cont.....

Class	Name	Extracellular target	Intracellular target	Bacteria
Porphyrin	5,10,15,20-tetrakis (1 - methylpyridinium-4- yl)porphyrin tetra-iodide (Tetra-Py -Me)	Lipopolysaccharides and outer membrane lipids	DNA*	*E. coli, Aeromonas salmonicida, Aeromonas hydrophila, Rhodopirellula* sp., *S. aureus, Trueperaradiovictrix, Deinococcus geothermalis, Deinococcus radiodurans*
	5,10,15,20-tetra (4-N,N,- -trimethylammo- niumphenyl) porphyrin	Cell wall and cytoplasmic membrane	Plasmid DNA	*E. coli*
	5,10,15,20-tetrakis(N-methyl-4-pyridyl): 21 H,23H-porphine (Tetra-Py -Me)	Outer membrane	ND	*E. coli*
	Hematoporphyrin monomethyl ether (HMME)	Cytoplasmic membrane	ND	*S. aureus*
Phthalocyanine	Zinc(II) phthalocyanine (ZnPc)	Outer membrane and cytoplasmic membrane	ND	*E. coli*
Fullerene	N-methylpyrrolidiniumCm fullerene iodide salt	Cytoplasmic membrane	ND	*S. aureus*

*DNA as target of APDT still requires further investigation. In most studies, it is not distinguished whether the DNA damage comprises chromosomal DNA or plasmid DNA; ND = not detected/not discussed. SOURCE: Liu *et al.*, 2015 [46].

Despite the variations in antibacterial effectivity in different studies, which may have occurred because of variability in the protocols followed, the efficient killing of target pathogens proved the proficiency of the technique. Still the practical application of any clinical process demands certain basic standards to be met and maintained to achieve the best possible result every time. Ideal phototherapy should exhibit high phototoxicity, minimum dark toxicity and invasiveness, high quantum yield of ROS, preferential association with target cells, suitable pharmacokinetics, and binding to and accumulation in the pathogenic species [48]. These criteria are majorly manipulated by adjusting two factors, *i.e.* , the concentration of the PS, which dictates the cytotoxic potential and the photoemission parameters [49].

aPDT for Viral Infections

Although most of the research attention has been focused on antibacterial or antifungal PDT, growing resistance to antiviral drugs has led to extensive studies on the photodynamic inactivation (PDI) of viruses as an alternative tool for treating viral infections. The susceptibility of viruses to PDT was reported in the 1930s, and the evolution of different parameters of the techniques, such as new potent PSs, upgraded illumination technologies, targeted delivery systems, *etc* ., have placed this modality in the front row for the inactivation of viruses just within the last 30 years [50]. The application of PDI has mainly been focused on the treatment of localized viral lesions, but systemic effects of such treatments triggering immune responses have also been reported recently [51]. PDI was first successfully reported in the early 1970s for treatment of herpes virus infection [52], followed by a great variety of reports about both DNA and RNA viruses and also both enveloped and non-enveloped viruses being effectively inactivated by this method *in vitro*, such as human papillomatosis virus (HPV), Herpes simplex virus (HSV), human immunodeficiency viruses (HIV), hepatitis B viruses (HBV), human T-lymphotropic viruses (HTLV) type I and II, cytomegalovirus (CMV), bovine viral diarrhea virus (BVDV), West Nile virus, Chikungunya virus, Influenza A virus (IAV), SARS Corona virus, Dengue virus (DENV), Crimena-Congo hemorrhagic fever virus, Parvo virus B19, blue tongue virus, human Adeno viruses, Zika virus, and many others [53]. Another aspect of PDI application is photodynamic disinfection of blood products which has acquired mentionable advancement, especially against HIV, hepatitis viruses, CMV, human parvovirus B19 and HTLV-I and II that pose major threats of viral contamination in blood and blood products; *in vitro* HIV inactivation by PDI has been highly promising [50]. Owing to the simplest structural organization of viruses, three molecular structures are targeted in antiviral PDI: nucleic acids (DNA or RNA), virus proteins and, if present, viral lipids. Moreover, studies also indicated that PDI may interfere with certain stages of the viral life cycle, as viruses at different life cycle stages show different levels of susceptibilities to PDI [54]. The success story of PDI does not stop at an elaborate list of long-known viruses but also continues with the emergent and resurgent varieties, *e.g.*, the Zika virus and, more recently, SARS-CoV-2. SARS-CoV-2, responsible for the ongoing Coronavirus disease, COVID-19 pandemic has shaken the whole world as well as the scientific community miserably. With no reliable treatment or vaccine available and researchers still working hard to find a proper way to fight back, symptomatic treatments are the only refuge. Few research findings show hope in this situation, one of them predicted capabilities of photobiomodulation (PBM) and aPDT, with the best outcome hopefully obtained from a combination of both methods. As the fight against COVID-19 is now majorly focused on virus removal, tissue oxygenation, and reduction or inhibition of cytokine storm caused by severe inflammation, a combined effect of PBM and aPDT may possibly achieve these goals with minimal interference with pharmaceutical methods [55].

However, most of these studies are actually *in vitro* investigations that would require effective applicational standardization and proper clinical trial. Presently, antiviral PDT attends to three major practical uses: (a) treatment of local HSV infections and (b) treatment of HPV infections (viral warts) with authorized PSs, such as δ-Aminolevulinic acid (ALA), Haematoporphyrin derivative (HPD) or methylene blue (MB), (c) blood product decontamination by reduction/removal of viruses (and bacteria) from blood products, mainly with the help of three commercially available PSs, *i.e.*, MB, riboflavin, and the psoralen derivative amotosalen [51].

aPDT for Fungal Infections

Almost 10–20% of the world population may be affected by recurring and chronic cases of mycoses, and fungal pathogens are often responsible for serious morbidity, malaise, social isolation, structural deformities, lowering body's defense system and even significant mortality, especially in systemic infections [56]. Conventional antifungal treatments are quite effective, but because of escalating resistance to antifungal drugs, treatment failure often due to poor delivery options and/or poor drug absorption inside the body, drug interactions and treatment-related toxicity, there is an increasing need for alternative therapy [57]. In this context, the observed effects of PDT on pathogenic yeasts (*e.g. Candida albicans, Kluyveromyces marxianus*), dermatophytes (*e.g. Trichophyton rubrum, T. mentagrophytes, T. tonsurans, Microsporum cookei, M. gypseum, and Epidermophyton floccosum*), nondermatophyte fungi (*e.g. Malassezia*) and other major fungal pathogens (*Cryptococcus* spp., *Coccidioides* sp.) have fueled the idea of placing it as a potent alternative therapy as it is cost-effective, highly selective and less likely to develop drug-resistant strains [58, 59]. The extensive research on antifungal PDT has been testing the range of its application on various fungi with a variety of PSs and irradiation protocols. Not only have there been no reports so far on the development of resistance to antifungal PDT, but also the treatment has not yet been associated with any mutagenic effects or genotoxicity *in vitro* [60]. PDT includes the systemic or topical administration of a PS, alongside the selective illumination of a target lesion with the light of the appropriate wavelength, in order to cause localized oxidative photodamage and subsequent cell death [56]. Many studies have presented PDT is highly effective in the destruction of fungi *in vitro* and that *in vitro* efficacy has also been confirmed *in vivo* using various animal models; moreover, human studies and clinical trials are also going on in many protozoa (Table **3**). [59]. However, although photodynamic approaches have been well established experimentally, especially for the treatment of certain cutaneous mycoses, no clinical treatment based on PDT has been licensed so far, mainly because there is limited information about its mechanism of action for specific pathogens as well as the risks to healthy tissues [59].

Table 3. Clinical trials using aPDT as treatment.

Disease	Fungus Species	aPDT	Additional Treatment	Final Outcome	References
Pityriasis versicolor	*Malassezia* species	ALA-PDT	-	Complete clearance	[61]
Onychomycosis	*T. rubrum*	ALA-PDT	Treatment with 40% urea ointment for12h prior to aPDT	Clinical and microbiological cures	[62]
Onychomycosis	-	ALA and red light	-	Significant improvement after treatments	[63]
Onychomycosis	*T. rubrum*	MB-light emission diode (MBLED)/PDT	-	clinical response significantly better in mild-to-moderate (100%) cases compared to severe onychomycosis (63.6%)	[64]
Malassezia infection	*M. folliculitis*	MAL (methyl 5-ALA)-PDT		Control of the tissue fungal burden	[65]
Onychomycosis	*F. oxysporum* and *A. terreus*	MAL 16% and LED	Treatment with 40% urea ointment for12h prior to aPDT	Clinical and microbiological cures	[66]
Denture stomatitis (DS)	*Candida* species	Photogem and LED	-	Mycological cures	[67]
Cutaneous Granuloma	*Candida albicans*	ALA)-PDT	Treatment with itraconazole for 1 month followed by two sessions of ALA-PDT	After 2 months of follow-up, the patient was completely cured	[68]
Sporotrichosis	*S. schenckii* complex	MB-PDT	Low dose itraconazole	Complete microbiological and clinical response	[69]
Chromoblastomycosis	*F. pedrosoi and C. carrionii*	MB and LED light		Control of the tissue fungal burdens	[70]

(Table 3) cont.....

Disease	Fungus Species	aPDT	Additional Treatment	Final Outcome	References
Interdigital mycosis	*Candida* or *Trichophyton species*	ALA-PDT	20% ALA preparation in Eucerin cream under an occlusive dressing	Clinical and microbiological recovery in 6 out of 9 patients after one (four cases) or four (two cases) treatments	[71]

[MODIFIED FROM: Baltazar *et al.*, 2015 [59].

aPDT against Parasites

Another promising field of application for aPDT is the treatment of parasites like protozoa. Light and PS-mediated killing have been proved fruitful for many human pathogenic protozoa (Table **4**); for example, *Plasmodium falciparum*, an etiological agent of deadliest form of malaria, reportedly killed by aPDT with an N-(4-butanol) pheophorbide derivative and silicon phthalocyanines such as Pc4; *Trypanosoma cruzi, the* causative agent of Chagas disease, killed by aPDT using PSs like hematoporphyrin and phthalocyanine [17]. In addition, human helminth eggs in sewage can be inactivated by mesosubstituted cationic porphyrin and light [72]. But substantial reports have been accumulated on the effective use of PDT for the treatment of leishmaniasis; use of four *meso*-tetra-cationic porphyrins reportedly inactivated *Leishmania major* promastigotes and macrophages infected with *Leishmania;* similar inactivation reports are there of water-soluble tetra-cationic zinc(II) *meso*-tetrakis (*N*-ethylpyridinium-2-yl) porphyrin photosensitizer against *Leishmania braziliensis* Al(III), and of Zn(II) phthalocyanines against *L. amazonensis* and other *Leishmania* promastigotes under red-light irradiation, and of sapphyrin and heterosapphyrin derivatives against *L. tarentolae* or *L. panamensis* amastigotes and promastigotes [73].

Table 4. Studies using aPDT against various parasites.

Disease	Species	aPDT	Final Outcome	References
Cutaneous leishmaniasis	*Leishmania major*	10% ALA preparation+ red light of 633 nm	A greater than 90% cure rate without evidence of recurrence	[74]

(Table 4) cont.....

Disease	Species	aPDT	Final Outcome	References
Cutaneous leishmaniasis	*L. major*	ALA-PDT	25-fold decrease in parasite load in infected mice after one PDT treatment	[75]
Cutaneous lesions of leishmaniasis	*L. amazonensis*	Methylene blue + non-coherent light source (low-cost PDT protocol)	Effective *in vitro* and *in vivo* studies upon local application of the photosensitiser	[76]
Cutaneous leishmaniasis	*Leishmania* sp.	Tetracationic porphyrins	inactivated infected macrophages and also the promastigote form	[76]
Cutaneous leishmaniasis	*Leishmania* sp.	(3,7-Bis(N,N-dibutylamino) phenothiaziniumbromide (PPA904) + 665-nm non-coherent light source or a 635-nm diode laser	high parasiticidal effect *in vivo* after single exposure in mice models	[77]
Cutaneous leishmaniasis	*Leishmania* sp.	δ-aminolevulinic acid-derived protoporphyrin IX + 665-nm non-coherent light source or a 635-nm diode laser	significant reduction of the parasite loads but also vigorous tissue destruction	[77]
Facial cutaneous leishmaniasis	*Leishmania tropica* (strain resistant to various therapeutic regimes)	-	Excellent results achieved	[78]
American trypanosomiasis or Chagas' disease	*Trypanosoma cruzi*	Silicon phthalocyanine Pc4 and thiazole orange-PDT	Inactivation of *T. cruzi* in donated blood *via* blood disinfection protocols	[79]

(Table 4) cont.....

Disease	Species	aPDT	Final Outcome	References
Chagas' disease	*T. cruzi*	Amotosalen (4'-aminomethyl-4,5'-8-trimethylpsoralen)-PDT	Complete inactivation of the infective form of *T. cruzi* both in fresh-frozen plasma and in platelet concentrates	[80]
Human African trypanosomiasis	*T. brucei*	MB-PDT	In citrated guinea pig blood parasites were immobilized causing no subsequent disease in mice	[81]
Malaria	*Plasmodium falciparum*	Cationic silicon phthalocyanine-PDT	Pathogen inactivation in blood products	[80]
Trichomonosis	*Tritrichomonas foetus*	-	Programmed cell death	[82]
Demodicosis	*Demodex folliculorum*	MAL-PDT	Achieved good results in 10 out of 17 patients, and fair results in another 4 patients	[83]

aPDT for Biofilms

Presently, another serious task of medical microbiology is to develop effective strategies to treat infections caused by microbial biofilm, as they account for up to 80% of all bacterial and fungal infections in humans [84]. Biofilm describes a sessile, complex, microbial community, in which cells are firmly attached to the substratum and embedded in an extracellular polymeric matrix, showing phenotypic and genotypic alterations for better adaptation towards environmental fluctuations and also resistance to antimicrobial strategies [85]. These are medically important because biofilm-associated pathogens are particularly resistant to antibiotic treatment, and thus novel antibiofilm approaches need to be developed. aPDT is coming up as a feasible alternative to combat clinically relevant biofilms such as dental biofilms, ventilator-associated pneumonia, chronic wound infections, oral candidiasis, and chronic rhinosinusitis [86]. Many studies have already proved aPDT as an effective tool for the inactivation of biofilm infections *in vitro* and *in vivo*. Oral infections such as periodontal, endodontic or mucosal infections and periimplantitis represent a superior field for the application of aPDT due to their localized nature [85].

CONCLUSION

aPDT has now become a prominent area of increasing research interest, not only in the medical field and cancer therapy, but various non-clinical fields are also being explored. The rapid progress in the aPDT research is instrumental in the widening of its horizon, for example, in the purification of contaminated water, bacterial control in food, especially for inactivation of microorganisms associated with food-borne illnesses, production of anti-infective medical devices and surfaces and pathogen inactivation in fish farming and many more [17]. Despite its huge prospect as a potent antimicrobial tool, aPDT still has some limitations to its wider deployment, including aggregation, hydrophobicity, and sub-optimal penetration capabilities of the photosensitizers, all of which decrease the production of ROS and lead to reduced therapeutic performance [87, 88]. Incorporating PSs into liposomes, micelles, or nanoparticles is leading the way to reduce the PS self-aggregation and enhance the targeted delivery of the PS, thereby leading to the potential expansion of aPDT [59]. Among these, nanoparticles have gained considerable research interest owing to their small size, functionalisable structure, and large contact surface, allowing a high degree of internalization by cellular membranes and tissue barriers [88, 89]. In addition, there have been few reports of adverse effects of PS application; for example, although overall tolerability of ALA-PDT has been good, in some cases, it showed side effects like burning sensation during irradiation, erythema, pain, edema and blistering while treating superficial mycoses [56]. Investigation correlated this problem with the skin penetrating difficulty of ALA and hence improving delivery systems and enhancement of skin penetration (using ultrasound, laser, microneedles and iontophoresis, chemical penetration enhancers like oleic acid and dimethyl sulfoxide, or various vehicles) showed better outcome. Liposomes, microscopic vesicles with membrane-like phospholipid bilayers surrounding an aqueous medium, are one of the best drug delivery systems for low molecular-weight PSs, such as ALA [56]. More preclinical and clinical studies are required in line to determine optimal delivery methods, potency, irradiation source, and accumulation in and removal from post-treatment [90, 91]. There has also been significant research on the rational design of PS molecules having desired pharmaceutical properties as well as getting the better of the drawbacks of traditional PS, such as poor chemical purity, long half-life, excessive accumulation into the skin, and low attenuation coefficients [92, 93].

Several innovative PSs, such as non-self-quenching PSs, conjugated polymer–based PSs, and nano-PSs, as well as advanced multifunctional aPDT systems such as *in situ* light-activated aPDT, stimuli-responsive aPDT, oxygen self-enriching enhanced aPDT, and aPDT-based multimodal therapy are being developed to overcome the inherent defects of aPDT (*e.g.*, no systemic therapeutic use, the limited penetration depth of light and hypoxic environment of infectious sites) [47, 94]. With the help of modern technology and rapid research expansion, aPDT is undoubtedly coming out as a formidable weapon in the fight against microbial invasions.

CONSENT OF PUBLICATION

Not applicable.

CONFLICT OF INTEREST

The author declares no conflict of interest, financial or otherwise.

ACKNOWLEDGEMENT

Declared none.

REFERENCES

[1]   Cock ie , Cheesman MJ, Ilanko A, Blonk B. Developing new antimicrobial therapies: Are synergistic combinations of plant extracts/compounds with conventional antibiotics the solution? Pharmacogn Rev 2017; 11(22): 57-72.
        [http://dx.doi.org/10.4103/phrev.phrev_21_17] [PMID: 28989242]

[2]   Maisch T. Resistance in antimicrobial photodynamic inactivation of bacteria. Photochem Photobiol Sci 2015; 14(8): 1518-26.
        [http://dx.doi.org/10.1039/C5PP00037H] [PMID: 26098395]

[3]   St Denis TG, Dai T, Izikson L, *et al.* All you need is light: Antimicrobial photoinactivation as an evolving and emerging discovery strategy against infectious disease. Virulence 2011; 2(6): 509-20.
        [http://dx.doi.org/10.4161/viru.2.6.17889] [PMID: 21971183]

[4]   WHO. Antimicrobial resistance: global report on surveillance. WHO Library Cataloguing-i--Publication Data 2014; pp. 1-232.

[5]   Bloise N, Minzioni P, Imbriani M, Visai L. Can nanotechnology shine a new light on antimicrobial photodynamic therapies? In: Tanaka Y, Ed. Photomedicine - Advances in Clinical Practice. London: IntechOpen Limited 2017.
        [http://dx.doi.org/10.5772/65974]

[6]   Koshi E, Mohan A, Rajesh S, Philip K. Antimicrobial photodynamic therapy: An overview. J Indian Soc Periodontol 2011; 15(4): 323-7.
        [http://dx.doi.org/10.4103/0972-124X.92563] [PMID: 22368354]

[7]   Vatansever F, de Melo WCMA, Avci P, *et al.* Antimicrobial strategies centered around reactive oxygen species – bactericidal antibiotics, photodynamic therapy, and beyond. FEMS Microbiol Rev 2013; 37(6): 955-89.
        [http://dx.doi.org/10.1111/1574-6976.12026] [PMID: 23802986]

[8]   Daniell MD, Hill JS. A history of photodynamic therapy. ANZ J Surg 1991; 61(5): 340-8.
[http://dx.doi.org/10.1111/j.1445-2197.1991.tb00230.x] [PMID: 2025186]

[9]   Von Tappenier H. Uber die WirkungfluoreszierenderStoffe auf InfusoriennachVersuchen von O. Raab. Munch Med Wochenschr 1900; 47: 5.

[10]   Dougherty TJ. Photodynamic therapy (PDT) of malignant tumors. Crit Rev Oncol Hematol 1984; 2(2): 83-116.
[http://dx.doi.org/10.1016/S1040-8428(84)80016-5] [PMID: 6397270]

[11]   Mahmoudi H, Bahador A, Pourhajibagher M, Alikhani MY. Antimicrobial photodynamic therapy: An effective alternative approach to control bacterial infections. J Lasers Med Sci 2018; 9(3): 154-60.
[http://dx.doi.org/10.15171/jlms.2018.29] [PMID: 30809325]

[12]   Darabpour E, Kashef N, Mashayekhan S. Chitosan nanoparticles enhance the efficiency of methylene blue-mediated antimicrobial photodynamic inactivation of bacterial biofilms: An *in vitro* study. Photodiagn Photodyn Ther 2016; 14: 211-7.
[http://dx.doi.org/10.1016/j.pdpdt.2016.04.009] [PMID: 27118084]

[13]   Hayyan M, Hashim MA, AlNashef IM. Superoxide Ion: Generation and Chemical Implications. Chem Rev 2016; 116(5): 3029-85.
[http://dx.doi.org/10.1021/acs.chemrev.5b00407] [PMID: 26875845]

[14]   Pizzino G, Irrera N, Cucinotta M, *et al.* Oxidative Stress: Harms and Benefits for Human Health. Oxid Med Cell Longev 2017; 2017: 1-13.
[http://dx.doi.org/10.1155/2017/8416763] [PMID: 28819546]

[15]   Kumar V, Sinha J, Verma N, Nayan K, Saimbi CS, Tripathi A. Scope of photodynamic therapy in periodontics. Indian J Dent Res 2015; 26(4): 439-42.
[http://dx.doi.org/10.4103/0970-9290.167636] [PMID: 26481895]

[16]   Yildirim C, Karaarslan ES, Ozsevik S, Zer Y, Sari T, Usumez A. Antimicrobial efficiency of photodynamic therapy with different irradiation durations. Eur J Dent 2013; 7(4): 469-73.
[http://dx.doi.org/10.4103/1305-7456.120677] [PMID: 24932123]

[17]   Rosa LP, da Silva FC. Antimicrobial Photodynamic Therapy: A New Therapeutic Option to Combat Infections. J Med Microbiol Diagn 2014; 3: 4.

[18]   Pagin MT, de Oliveira FA, Oliveira RC, *et al.* Laser and light-emitting diode effects on pre-osteoblast growth and differentiation. Lasers Med Sci 2014; 29(1): 55-9.
[http://dx.doi.org/10.1007/s10103-012-1238-5] [PMID: 23179312]

[19]   Asai T, Suzuki H, Kitayama M, *et al.* The long-term effects of red light-emitting diode irradiation on the proliferation and differentiation of osteoblast-like MC3T3-E1 cells. Kobe J Med Sci 2014; 60(1): E12-8.
[PMID: 25011637]

[20]   Khurana D, Sharnamma B, Rathore PK, Tyagi P. Photodynamic therapy- a ray towards periodontics. IQSR J Dent Med Sci 2014; 13(3): 64-71.

[21]   Lakshmanan S, Gupta GK, Avci P, *et al.* Physical energy for drug delivery; poration, concentration and activation. Adv Drug Deliv Rev 2014; 71: 98-114.
[http://dx.doi.org/10.1016/j.addr.2013.05.010] [PMID: 23751778]

[22]   Dave D, Desai U, Despande N. Photodynamic Therapy: A View through Light. Journal of Orofacial Research 2012; 2(2): 82-6.
[http://dx.doi.org/10.5005/jp-journals-10026-1019]

[23]   Al-Omari S. Toward a molecular understanding of the photosensitizer–copper interaction for tumor destruction. Biophys Rev 2013; 5(4): 305-11.
[http://dx.doi.org/10.1007/s12551-013-0112-4] [PMID: 28510111]

[24]   Nakonechny F, Nisnevitch M. Aspects of Photodynamic Inactivation of Bacteria, Microorganisms.

IntechOpen 2019.
[http://dx.doi.org/10.5772/intechopen.89523]

[25] Abrahamse H, Mfouo TI. Photodynamic therapy, a potential therapy for improve cancer management. 2018.
[http://dx.doi.org/10.5772/intechopen.74697]

[26] Baptista MS, Cadet J, Di Mascio P, *et al.* Type I and Type II Photosensitized Oxidation Reactions: Guidelines and Mechanistic Pathways. Photochem Photobiol 2017; 93(4): 912-9.
[http://dx.doi.org/10.1111/php.12716] [PMID: 28084040]

[27] Benov L. Photodynamic therapy: current status and future directions. Med Princ Pract 2015; 24 (Suppl. 1): 14-28.
[http://dx.doi.org/10.1159/000362416] [PMID: 24820409]

[28] Nava HR, Allamaneni SS, Dougherty TJ, *et al.* Photodynamic therapy (PDT) using HPPH for the treatment of precancerous lesions associated with barrett's esophagus. Lasers Surg Med 2011; 43(7): 705-12.
[http://dx.doi.org/10.1002/lsm.21112] [PMID: 22057498]

[29] Ding H, Yu H, Dong Y, *et al.* Photoactivation switch from type II to type I reactions by electron-rich micelles for improved photodynamic therapy of cancer cells under hypoxia. J Control Release 2011; 156(3): 276-80.
[http://dx.doi.org/10.1016/j.jconrel.2011.08.019] [PMID: 21888934]

[30] Huang L, Xuan Y, Koide Y, Zhiyentayev T, Tanaka M, Hamblin MR. Type I and Type II mechanisms of antimicrobial photodynamic therapy: An *in vitro* study on gram-negative and gram-positive bacteria. Lasers Surg Med 2012; 44(6): 490-9.
[http://dx.doi.org/10.1002/lsm.22045] [PMID: 22760848]

[31] Scherer KM, Bisby RH, Botchway SW, Parker AW. New Approaches to Photodynamic Therapy from Types I, II and III to Type IV Using One or More Photons. Anticancer Agents Med Chem 2017; 17(2): 171-89.
[http://dx.doi.org/10.2174/1871520616666160513131723] [PMID: 27173966]

[32] de Oliveira AB, Ferrisse TM, Marques RS, de Annunzio SR, Brighenti FL, Fontana CR. Effect of Photodynamic Therapy on Microorganisms Responsible for Dental Caries: A Systematic Review and Meta-Analysis. Int J Mol Sci 2019; 20(14): 3585.
[http://dx.doi.org/10.3390/ijms20143585] [PMID: 31340425]

[33] Li L, Cho H, Yoon KH, Kang HC, Huh KM. Antioxidant-photosensitizer dual-loaded polymeric micelles with controllable production of reactive oxygen species. Int J Pharm 2014; 471(1-2): 339-48.
[http://dx.doi.org/10.1016/j.ijpharm.2014.05.064] [PMID: 24939615]

[34] Castano AP, Demidova TN, Hamblin MR. Mechanisms in photodynamic therapy: part one—photosensitizers, photochemistry and cellular localization. Photodiagn Photodyn Ther 2004; 1(4): 279-93.
[http://dx.doi.org/10.1016/S1572-1000(05)00007-4] [PMID: 25048432]

[35] Phaniendra A, Jestadi DB, Periyasamy L. Free radicals: properties, sources, targets, and their implication in various diseases. Indian J Clin Biochem 2015; 30(1): 11-26.
[http://dx.doi.org/10.1007/s12291-014-0446-0] [PMID: 25646037]

[36] Carballal S, Bartesaghi S, Radi R. Kinetic and mechanistic considerations to assess the biological fate of peroxynitrite. Biochim Biophys Acta, Gen Subj 2014; 1840(2): 768-80.
[http://dx.doi.org/10.1016/j.bbagen.2013.07.005] [PMID: 23872352]

[37] Nagy P, Kettle AJ, Winterbourn CC. Superoxide-mediated formation of tyrosine hydroperoxides and methionine sulfoxide in peptides through radical addition and intramolecular oxygen transfer. J Biol Chem 2009; 284(22): 14723-33.
[http://dx.doi.org/10.1074/jbc.M809396200] [PMID: 19297319]

[38] Imlay JA. The molecular mechanisms and physiological consequences of oxidative stress: lessons from a model bacterium. Nat Rev Microbiol 2013; 11(7): 443-54.
[http://dx.doi.org/10.1038/nrmicro3032] [PMID: 23712352]

[39] Maurya PK. Animal biotechnology as a tool to understand and fight aging. In: Verma AS, Singh A, Eds. Animal Biotechnology. Academic Press 2014; pp. 177-91.

[40] Toyokuni S, Ikehara Y, Kikkawa F, Hori M. Regulation of cell membrane transport by plasma. In: Toyokuni S, Ikehara Y, Kikkawa F, Hori M, Eds. Plasma Medical Science. Academic Press 2019; pp. 173-247.

[41] Fridovich I. Oxygen: how do we stand it? Med Princ Pract 2013; 22(2): 131-7.
[http://dx.doi.org/10.1159/000339212] [PMID: 22759590]

[42] Dąbrowski JM, Arnaut LG, Pereira MM, *et al.* Combined effects of singlet oxygen and hydroxyl radical in photodynamic therapy with photostable bacteriochlorins: Evidence from intracellular fluorescence and increased photodynamic efficacy *in vitro*. Free Radic Biol Med 2012; 52(7): 1188-200.
[http://dx.doi.org/10.1016/j.freeradbiomed.2011.12.027] [PMID: 22285766]

[43] Allison RR, Moghissi K. Photodynamic therapy mechanisms. Clin Endosc 2013; 46(1): 24-9.
[http://dx.doi.org/10.5946/ce.2013.46.1.24] [PMID: 23422955]

[44] Negre-Salvayre A, Auge N, Ayala V, *et al.* Pathological aspects of lipid peroxidation. Free Radic Res 2010; 44(10): 1125-71.
[http://dx.doi.org/10.3109/10715762.2010.498478] [PMID: 20836660]

[45] Cadet J, Ravanat JL, TavernaPorro M, Menoni H, Angelov D. Oxidatively generated complex DNA damage: Tandem and clustered lesions. Cancer Lett 2012; 327(1-2): 5-15.
[http://dx.doi.org/10.1016/j.canlet.2012.04.005] [PMID: 22542631]

[46] Liu JN, Bu WB, Shi JL. Silica coated upconversion nanoparticles: A versatile platform for the development of efficient theranostics. Acc Chem Res 2015; 48(7): 1797-805.
[http://dx.doi.org/10.1021/acs.accounts.5b00078] [PMID: 26057000]

[47] Jia Q, Song Q, Li P, Huang W. Rejuvenated Photodynamic Therapy for Bacterial Infections. Adv Healthc Mater 2019; 8(14): 1900608.
[http://dx.doi.org/10.1002/adhm.201900608] [PMID: 31240867]

[48] Tim M. Strategies to optimize photosensitizers for photodynamic inactivation of bacteria. J Photochem Photobiol B 2015; 150: 2-10.
[http://dx.doi.org/10.1016/j.jphotobiol.2015.05.010] [PMID: 26048255]

[49] Galo ÍDC, Carvalho JA, Santos JLMC, Braoios A, Prado RP. The ineffectiveness of antimicrobial photodynamic therapy in the absence of preincubation of the microorganisms in the photosensitizer. Fisioter Mov 2020; 33: e003304.
[http://dx.doi.org/10.1590/1980-5918.033.ao04]

[50] Costa ACBP, Rasteiro VMC, Pereira CA, Rossoni RD, Junqueira JC, Jorge AOC. The effects of rose bengal- and erythrosine-mediated photodynamic therapy on *Candida albicans*. Mycoses 2012; 55(1): 56-63.
[http://dx.doi.org/10.1111/j.1439-0507.2011.02042.x] [PMID: 21668520]

[51] Wiehe A, O'Brien JM, Senge MO. Trends and targets in antiviral phototherapy. Photochem Photobiol Sci 2019; 18(11): 2565-612.
[http://dx.doi.org/10.1039/C9PP00211A] [PMID: 31397467]

[52] Felber TD, Smith EB, Knox JM, Wallis C, Melnick JL. Photodynamic inactivation of herpes simplex: Report of a clinical trial. JAMA 1973; 223(3): 289.
[http://dx.doi.org/10.1001/jama.1973.03220030027005]

[53] Schneider K, Wronka-Edwards L, Leggett-Embrey M, *et al.* Psoralen Inactivation of Viruses: A

Process for the Safe Manipulation of Viral Antigen and Nucleic Acid. Viruses 2015; 7(11): 5875-88.
[http://dx.doi.org/10.3390/v7112912] [PMID: 26569291]

[54]   Sobotta L, Skupin-Mrugalska P, Mielcarek J, Goslinski T, Balzarini J. Photosensitizers mediated photodynamic inactivation against virus particles. Mini Rev Med Chem 2015; 15(6): 503-21.
[http://dx.doi.org/10.2174/1389557515666150415151505] [PMID: 25877599]

[55]   Fekrazad R. Photobiomodulation and Antiviral Photodynamic Therapy as a Possible Novel Approach in COVID-19 Management. Photobiomodul Photomed Laser Surg 2020; 38(5): 255-7.
[http://dx.doi.org/10.1089/photob.2020.4868] [PMID: 32326830]

[56]   Liang YI, Lu LM, Chen Y, Lin YOUKUN. Photodynamic therapy as an antifungal treatment. Exp Ther Med 2016; 12(1): 23-7.
[http://dx.doi.org/10.3892/etm.2016.3336] [PMID: 27347012]

[57]   Shen JJ, Jemec GBE, Arendrup MC, Saunte DML. Photodynamic therapy treatment of superficial fungal infections: A systematic review. Photodiagn Photodyn Ther 2020; 31: 101774.
[http://dx.doi.org/10.1016/j.pdpdt.2020.101774] [PMID: 32339671]

[58]   Ragàs X, Agut M, Nonell S. Singlet oxygen in *Escherichia coli*: New insights for antimicrobial photodynamic therapy. Free Radic Biol Med 2010; 49(5): 770-6.
[http://dx.doi.org/10.1016/j.freeradbiomed.2010.05.027] [PMID: 20638940]

[59]   Baltazar LM, Ray A, Santos DA, Cisalpino PS, Friedman AJ, Nosanchuk JD. Antimicrobial photodynamic therapy: An effective alternative approach to control fungal infections. Front Microbiol 2015; 6: 202.
[http://dx.doi.org/10.3389/fmicb.2015.00202] [PMID: 25821448]

[60]   Prates RA, Kato IT, Ribeiro MS, Tegos GP, Hamblin MR. Influence of multidrug efflux systems on methylene blue-mediated photodynamic inactivation of Candida albicans. J Antimicrob Chemother 2011; 66(7): 1525-32.
[http://dx.doi.org/10.1093/jac/dkr160] [PMID: 21525022]

[61]   Kim YJ, Kim YC. Successful treatment of pityriasis versicolor with 5-aminolevulinic acid photodynamic therapy. Arch Dermatol 2007; 143(9): 1209.
[http://dx.doi.org/10.1001/archderm.143.9.1218] [PMID: 17875898]

[62]   Piraccini BM, Rech G, Tosti A. Photodynamic therapy of onychomycosis caused by *Trichophyton rubrum*. J Am Acad Dermatol 2008; 59(5) (Suppl.): S75-6.
[http://dx.doi.org/10.1016/j.jaad.2008.06.015] [PMID: 19119130]

[63]   Sotiriou E, Koussidou-Eremonti T, Chaidemenos G, Apalla Z, Ioannides D. Photodynamic therapy for distal and lateral subungual toenail onychomycosis caused by *Trichophyton rubrum*: Preliminary results of a single-centre open trial. Acta Derm Venereol 2010; 90(2): 216-7.
[http://dx.doi.org/10.2340/00015555-0811] [PMID: 20169321]

[64]   Souza LWF, Souza SVT, Botelho ACC. Distal and lateral toenail onychomycosis caused by *Trichophyton rubrum*: treatment with photodynamic therapy based on methylene blue dye. An Bras Dermatol 2014; 89(1): 184-6.
[http://dx.doi.org/10.1590/abd1806-4841.20142197] [PMID: 24626676]

[65]   Lee JW, Kim BJ, Kim MN. Photodynamic therapy: New treatment for recalcitrant *Malassezia* folliculitis. Lasers Surg Med 2010; 42(2): 192-6.
[http://dx.doi.org/10.1002/lsm.20857] [PMID: 20166153]

[66]   Gilaberte Y, Aspiroz C, Martes MP, Alcalde V. Espinel-IngroffA, Rezusta A. Treatment of refractory fingernail onychomycosis caused by nondermatophytemolds with methylaminolevulinate photodynamic therapy. J Am Acad Dermatol 2011; 65: 669-71.
[http://dx.doi.org/10.1016/j.jaad.2010.06.008] [PMID: 21839332]

[67]   Mima EG, Vergani CE, Machado AL, *et al.* Comparison of Photodynamic Therapy *versus* conventional antifungal therapy for the treatment of denture stomatitis: A randomized clinical trial.

Clin Microbiol Infect 2012; 18(10): E380-8.
[http://dx.doi.org/10.1111/j.1469-0691.2012.03933.x] [PMID: 22731617]

[68] Cai Q, Yang L, Chen J, Yang H, Gao Z, Wang X. Successful Sequential Treatment with Itraconazole and ALA-PDT for Cutaneous Granuloma by *Candida albicans*: A Case Report and Literature Review. Mycopathologia 2018; 183(5): 829-34.
[http://dx.doi.org/10.1007/s11046-018-0267-4] [PMID: 29767317]

[69] Gilaberte Y, Aspiroz C, Alejandre MC, *et al.* Cutaneous sporotrichosis treated with photodynamic therapy: An *in vitro* and *in vivo* study. Photomed Laser Surg 2014; 32(1): 54-7.
[http://dx.doi.org/10.1089/pho.2013.3590] [PMID: 24328608]

[70] Lyon JP, de Maria Pedroso e Silva Azevedo C, Moreira LM, de Lima CJ, de Resende MA. Photodynamic antifungal therapy against chromoblastomycosis. Mycopathologia 2011; 172(4): 293-7.
[http://dx.doi.org/10.1007/s11046-011-9434-6] [PMID: 21643843]

[71] Calzavara-Pinton PG, Venturini M, Capezzera R, Sala R, Zane C. Photodynamic therapy of interdigital mycoses of the feet with topical application of 5-aminolevulinic acid. Photodermatol Photoimmunol Photomed 2004; 20(3): 144-7.
[http://dx.doi.org/10.1111/j.1600-0781.2004.00095.x] [PMID: 15144392]

[72] Wainwright M, Baptista MS. The application of photosensitisers to tropical pathogens in the blood supply. Photodiagn Photodyn Ther 2011; 8(3): 240-8.
[http://dx.doi.org/10.1016/j.pdpdt.2011.04.001] [PMID: 21864797]

[73] Deda DK, Iglesias BA, Alves E, Araki K, Garcia CRS. Porphyrin Derivative Nanoformulations for Therapy and Antiparasitic Agents. Molecules 2020; 25(9): 2080.
[http://dx.doi.org/10.3390/molecules25092080] [PMID: 32365664]

[74] Asilian A, Davami M. Comparison between the efficacy of photodynamic therapy and topical paromomycin in the treatment of Old World cutaneous leishmaniasis: A placebo-controlled, randomized clinical trial. Clin Exp Dermatol 2006; 31(5): 634-7.
[http://dx.doi.org/10.1111/j.1365-2230.2006.02182.x] [PMID: 16780497]

[75] Kosaka S, Akilov OE, O'Riordan K, Hasan T. A mechanistic study of delta-aminolevulinic acid-based photodynamic therapy for cutaneous leishmaniasis. J Invest Dermatol 2007; 127(6): 1546-9.
[http://dx.doi.org/10.1038/sj.jid.5700719] [PMID: 17218937]

[76] Bristow CA, Hudson R, Paget TA, Boyle RW. Potential of cationic porphyrins for photodynamic treatment of cutaneous Leishmaniasis. Photodiagn Photodyn Ther 2006; 3(3): 162-7.
[http://dx.doi.org/10.1016/j.pdpdt.2006.04.004] [PMID: 25049150]

[77] Dai T, Huang YY, Hamblin MR. Photodynamic therapy for localized infections—State of the art. Photodiagn Photodyn Ther 2009; 6(3-4): 170-88.
[http://dx.doi.org/10.1016/j.pdpdt.2009.10.008] [PMID: 19932449]

[78] Sohl S, Kauer F, Paasch U, Simon JC. Photodynamic treatment of cutaneous leishmaniasis. J Dtsch Dermatol Ges 2007; 5(2): 128-30.
[http://dx.doi.org/10.1111/j.1610-0387.2007.06177.x] [PMID: 17274779]

[79] Wagner SJ, Skripchenko A, Salata J, O'Sullivan AM, Cardo LJ. Inactivation of *Leishmania donovani infantum* and *Trypanosoma cruzi* in red cell suspensions with thiazole orange. Transfusion 2008; 48(7): 1363-7.
[http://dx.doi.org/10.1111/j.1537-2995.2008.01712.x] [PMID: 18422841]

[80] Baptista MS, Wainwright M. Photodynamic antimicrobial chemotherapy (PACT) for the treatment of malaria, leishmaniasis and trypanosomiasis. Braz J Med Biol Res 2011; 44(1): 1-10.
[http://dx.doi.org/10.1590/S0100-879X2010007500141] [PMID: 21152709]

[81] Do Campo R, Moreno SNJ, Cruz FS. Enhancement of the cytotoxicity of crystal violet against *Trypanosomacruzi* in the blood by ascorbate. MolecBiochemParasitol 1998; 27: 241-7.

[82] Soares da Silva N, Ferreira AM, Galvão CW, Etto RM, Soares CP. Cell death after photodynamic

therapy treatment in unicellular protozoan parasite tritrichomonas foetus. In: Inada NM, Buzzá H, Cristina Blanco K, Dias LD, Eds. London: IntechOpen Limited 2017.
[http://dx.doi.org/10.5772/intechopen.94140]

[83]   Bryld LE, Jemec GBE. Photodynamic therapy in a series of rosacea patients. J Eur Acad Dermatol Venereol 2007; 0(0): 070605092649003-.
[http://dx.doi.org/10.1111/j.1468-3083.2007.02219.x] [PMID: 17894705]

[84]   Høiby N. A short history of microbial biofilms and biofilm infections. Acta Pathol Microbiol Scand Suppl 2017; 125(4): 272-5.
[http://dx.doi.org/10.1111/apm.12686] [PMID: 28407426]

[85]   Cieplik F, Tabenski L, Buchalla W, Maisch T. Antimicrobial photodynamic therapy for inactivation of biofilms formed by oral key pathogens. Front Microbiol 2014; 5: 405.
[http://dx.doi.org/10.3389/fmicb.2014.00405] [PMID: 25161649]

[86]   Hu Z, Li J, Liu H, Liu L, Jiang L, Zeng K. Treatment of latent or subclinical Genital HPV Infection with 5-aminolevulinic acid-based photodynamic therapy. Photodiagn Photodyn Ther 2018; 23: 362-4.
[http://dx.doi.org/10.1016/j.pdpdt.2018.07.014] [PMID: 30048762]

[87]   Hamblin MR. Photodynamic therapy and photobiomodulation: Can all diseases be treated with light? In: Guenther BD, Steel DG, Eds. Elsevier 2018; pp. 100-35.
[http://dx.doi.org/10.1016/B978-0-12-803581-8.09688-0]

[88]   Bekmukhametova A, Ruprai H, Hook JM, Mawad D, Houang J, Lauto A. Photodynamic therapy with nanoparticles to combat microbial infection and resistance. Nanoscale 2020; 12(41): 21034-59.
[http://dx.doi.org/10.1039/D0NR04540C] [PMID: 33078823]

[89]   Perni S, Prokopovich P, Pratten J, Parkin IP, Wilson M. Nanoparticles: their potential use in antibacterial photodynamic therapy. Photochem Photobiol Sci 2011; 10(5): 712-20.
[http://dx.doi.org/10.1039/c0pp00360c] [PMID: 21380441]

[90]   Ormond A, Freeman H. Dye Sensitizers for Photodynamic Therapy. Materials (Basel) 2013; 6(3): 817-40.
[http://dx.doi.org/10.3390/ma6030817] [PMID: 28809342]

[91]   Das S, Tiwari M, Mondal D, Sahoo BR, Tiwari DK. Growing tool-kit of photosensitizers for clinical and non-clinical applications. J Mater Chem B Mater Biol Med 2020; 8(48): 10897-940.
[http://dx.doi.org/10.1039/D0TB02085K] [PMID: 33165483]

[92]   Garland MJ, Cassidy CM, Woolfson D, Donnelly RF. Designing photosensitizers for photodynamic therapy: strategies, challenges and promising developments. Future Med Chem 2009; 1(4): 667-91.
[http://dx.doi.org/10.4155/fmc.09.55] [PMID: 21426032]

[93]   Zhang J, Jiang C, Figueiró Longo JP, Azevedo RB, Zhang H, Muehlmann LA. An updated overview on the development of new photosensitizers for anticancer photodynamic therapy. Acta Pharm Sin B 2018; 8(2): 137-46.
[http://dx.doi.org/10.1016/j.apsb.2017.09.003] [PMID: 29719775]

[94]   Hu T, Wang Z, Shen W, Liang R, Yan D, Wei M. Recent advances in innovative strategies for enhanced cancer photodynamic therapy. Theranostics 2021; 11(7): 3278-300.
[http://dx.doi.org/10.7150/thno.54227] [PMID: 33537087]

SUBJECT INDEX

A

Acid(s) 5, 12, 61, 105, 106, 61, 130, 131, 147, 148, 228
  acetic 130
  ascorbic 148, 228
  benzoic 106
  bile 12
  chloroauric 147
  cholic 61
  conjugated linoleic 5
  propanoic 105
Action, photodynamic 218
Antibody 34, 35, 40
  phage display (APD) 40
  production technology 34, 35
Anti-cancer 110, 115, 116
  activity 110, 115, 116
  agents 115
Anti-fungal 101, 105, 116, 119
  activities 101, 105, 119
  agents 105, 116
Antigen-presenting cells (APCs) 5
Antimicrobial 57, 58, 59, 60, 64, 65, 66, 67, 68, 69, 70, 82, 93, 94, 147, 149, 154, 157, 158
  activity 58, 59, 60, 65, 66, 67, 68, 69, 93, 94, 147, 149, 154, 157, 158
  properties 57, 59, 64, 68, 70, 82, 154
Antioxidant 228, 233
  carrier sensitizers (ACS) 228
  defense machinery 233
Aspergillus 105, 106, 107, 108, 146, 147
  flavus 105, 106, 107, 146, 147
  niger 105, 107, 108
Atherosclerosis 12
Athlete's foot 105
ATP synthase 156

B

Bacillus 60, 85, 94, 112, 114, 117, 118, 146, 158, 159, 198
  megaterium 198
  subtilis 60, 85, 94, 112, 114, 117, 118, 146, 158, 159
Bacteria 54, 82, 83
  antimicrobial-resistant 54
  biofilm-forming 82, 83
Bacterial growth 84, 147, 197, 199
  inhibited 199
Bacterial infections 76, 77, 82, 190, 191, 192, 202, 203, 216, 217, 232, 233
  resistant 190, 191
Bactericidal activity 59, 93, 147, 156, 159, 173, 191, 193, 198, 199
Biofilms, microbial 240
Bovine viral diarrhea virus (BVDV) 235

C

Cancer cells, melanoma 112, 117
Carbon quantum dots (CQDs) 190, 191, 192, 196, 204, 205, 206, 207, 208, 209
Chagas disease 238
Chemical vapor deposition (CVD) 175
Chemotherapy 115, 196, 206
  photodynamic anti-microbial 206
  photothermal antibacterial 196
Chikungunya virus 235
Chromoblastomycosis 237
CMV infections 45
Complementarity-determining region (CDR) 34, 37, 232
Coronavirus disease 235
Corticotropin-releasing hormone (CRH) 19
COVID-19 pandemic 235
Crohn's disease 56
Crystal field theory (CFT) 102
Curcuma longa 150
Cytokines, proinflammatory 7, 20
Cytomegalovirus 235